DRAWING THE LINE

BORDER REGIONS SERIES

Series Editor: Doris Wastl-Walter

In recent years, borders have taken on an immense significance. Throughout the world they have shifted, been constructed and dismantled, and become physical barriers between socio-political ideologies. They may separate societies with very different cultures, histories, national identities or economic power, or divide people of the same ethnic or cultural identity.

As manifestations of some of the world's key political, economic, societal and cultural issues, borders and border regions have received much academic attention over the past decade. This valuable series publishes high quality research monographs and edited comparative volumes that deal with all aspects of border regions, both empirically and theoretically. It will appeal to scholars interested in border regions and geopolitical issues across the whole range of social sciences.

Drawing the Line

Nature, Hybridity and Politics in Transboundary Spaces

JULIET FALL
University of British Columbia, Canada

ASHGATE

Published by
Ashgate Publishing Limited
Gower House
Croft Road
Aldershot
Hants GU11 3HR
England

Ashgate Publishing Company
Suite 420
101 Cherry Street
Burlington, VT 05401-4405
USA

Ashgate website: http://www.ashgate.com

British Library Cataloguing in Publication Data
Fall, Juliet
Drawing the line : nature, hybridity and politics in transboundary spaces. - (Border regions series)
1.Transfrontier conservation areas - Management
2.Transfrontier conservation areas - Political aspects
3.Boundaries 4.Nature - Effect of human beings on
I. Title
333.9'516

Library of Congress Cataloging-in-Publication Data
Fall, Juliet, 1974-
Drawing the line : nature, hybridity, and politics in transboundary spaces / by Juliet Fall.
p. cm. -- (Border regions series)
Includes index.
ISBN 0-7546-4331-X
1. Transfrontier conservation areas. 2. Boundaries--Philosophy. I. Title. II. Series.

S944.5.P78F26 2005
333.72--dc22

2004028577

ISBN 0 7546 4331 X

Printed and bound in Great Britain by Antony Rowe Ltd, Chippenham, Wiltshire.

Contents

List of Figures

List of Tables

"Edges, Borders, Boundaries, Brinks and Limits have appeared like a team of trolls on their separate horizons. Short creatures with long shadows, patrolling the Blurry End"
(Arundati Roy, 1997, 'The God of Small Things', Flamingo, London, p. 3)

Acknowledgements

A number of people helped me along this exploration of boundaries and each deserves sincere thanks. A book is always more of a collective enterprise than the single author suggests, building on the writings, ideas, friendship and support of many people.

First of all, I should like to extend my sincere thanks to the countless people who helped me on my way during my travels during the fieldwork in France, Germany, Italy, Poland, Slovakia, Romania and Ukraine. Without their kindness and welcome, this book would never have been written. They opened up their offices and homes to a stranger with a notebook, offering food, shelter and thoughtful answers to my endless questions. Special mention to Marian who first showed me a blue slug in Slovakia, a creature that became almost as mythical to me as the Chimera in Chapter 10: endemic to the Eastern Carpathians and quietly charismatic in the transboundary spaces I so enjoyed exploring.

In another setting, I extend my thanks to Antoine Bailly who not only offered me a job for five years at the University of Geneva, he also wisely granted me the freedom to explore ideas while offering practical support and advice; Martin Price, when I was at Oxford and then in Geneva, provided careful feedback while bridging the boundaries of academia and the wider world, providing an inspiring example in the process and opening many doors; Joe Painter's kindness, help and insight came at just the right time and was instrumental in keeping me on track during my year in the Department of Geography in Durham, providing a sounding block, pertinent references and support to ease and assist the writing process; Bernard Debarbieux and Jean Ruegg brought different backgrounds and insights into the final stages of the PhD, offering constructive comments and advice in the final stages of writing. Doris Wastl-Walter deserves special thanks as she was instrumental, through support and mentoring, in getting this book into print. Her unfailing support was unique and inspirational. I am also grateful to Barbara Morehouse who provided excellent constructive feedback on transforming the thesis into a book.

Special thanks also go to Anne Fournand, Alexandre Gillet, Irène Hirt, Katharina Isele, Régina Leiggener, Matthieu Petite and Gilles Rudaz who made my life at the University of Geneva such a pleasant succession of delightful tea breaks; many thanks also to Grégoire Métral and Valérie November who started out as mentors and became precious friends; to Andrea Kofler, my boundary-soul mate in Bern; to Wolfgang Zierhofer in Basle, my friend and colleague across the Röstigraben (the Swiss grilled potato boundary); to Sophie Rey and Cristina Del Biaggio for their excellent assistance with the thankless task of interview transcribing.

The practical support and mentoring of the 'Carrière Académique' programme for promoting women in academia at the University of Geneva, run jointly by Geneviève Billeter and Eliane Barth, was also crucial in encouraging me to continue in this field. Two wonderful women doing valuable work.

During the year in the Department of Geography at the University of Durham, Stuart Elden, Emma Mawdsley and Martin Pratt deserve special thanks for making me feel welcome, providing ideas, fresh air and friendship when I was writing like a hermit. In the Department of Geography at the University of British Columbia, Derek Gregory, Trevor Barnes, Merje Kuus and Juanita Sundberg provided feedback for the last haul of the book, ideas and support in easing my move across the Atlantic.

Almost last and so far from least, endless thanks to my mother, Angela, a loving source of support, a wise and graceful role model; my father, Merrick, who transmitted his love of travel, language and discovery and organised many intrepid tours to nurture my curiosity about the world, crossing boundaries and giving them meaning; my brother, Caspar, who trod the paths of academia, showing it could be done while calmly inspiring his little sister; and Thomas, who makes walking, cycling and rowing across boundaries a companionable delight and whose love and acceptance continues to give me strength at each stage of the journey.

The repeated generous financial assistance of the Fonds National Suisse de la Recherche Scientifique, in the shape of two research fellowships at the University of Durham and the University of British Columbia (and the University of California, Los Angeles) is gratefully acknowledged. The Société Académique de Genève generously provided financial support for some of the fieldwork. In bringing this book into being, both Valerie Rose and Rosalind Ebdon at Ashgate provided fantastic help, gentle reminders and efficient coaching. I am most grateful to them both for their help and support, as well as for their patience.

Every effort has been made to trace the copyright owners of material reproduced in this book. However, if advised of any inadvertent omissions, the publisher and I would be pleased to incorporate proper acknowledgements in future editions.

To the East Carpathian blue slug, crossing boundaries of good taste.
I met her in the undergrowth while researching this book:
charismatic but unsung, embodying marginal spaces.

Chapter 1

Introduction

This book was symbolically born on two mountain summits, as I stood gazing down to the valleys below: a perplexed geographer recovering from a walk. On the first, in 1997, towering above the city of Mostar in Bosnia-Herzegovina, I looked down on a divided city. I walked among the abandoned strongholds, picking up discarded tins of food in the dry scrub. Soldiers had fired from the mountain, aiming at the famous bridge, enforcing divisions along the front line. The city was split in two, divided by bitter enmity. Tanks enforced the peace. As I surveyed the city, a United Nations official explained how people coped and organised their daily life in a town cut down the middle. On the second peak, in the Carpathians in 1998, I looked over undulating green forests along the border between Slovakia and Ukraine. A large strip of trees had been cleared, searing the green for miles on end. A little further away, a rusting barbed-wire fence was still regularly patrolled. Politics were inscribed on the landscape, entrenching difference. A national park guard stood next to me, showing how bison climbed over the fence in winter, when snow lay on the ground.

Both mountain panoramas involved boundaries and imaginary lines inscribed in space by human beings. In both places, these abstract boundaries had very real effects in the daily lives of people. For a geographer, both scenes suggested possible research topics, leading in different directions. Climbing down from the second mountain, I wondered what the bison thought when confronted with a fence. Did she calmly ponder, waiting for it to snow? Was her strategy substantially different from that of people in Bosnia waiting for peace to erase divisions? What did it fundamentally mean to suggest that 'nature knows no boundaries', although humans spend inordinate amounts of time constructing them? Was this question, suggesting a division between 'nature' and 'culture', essentially misguided? Musing about my role, I was reminded of a quote by Paasi who clearly laid out the task assigned to the discipline:

> *Geography had to produce and reproduce territoriality, to create and establish boundaries on the horizontal continuum of absolute space, nature and culture that prevailed on the globe, and to establish the stories of creating a distinction between us and the Other*
>
> *(Paasi 1996 : 21).*

In this sentence, Paasi suggested that the function of geography as a discipline and practice was to establish boundaries, creating distinctions between individuals, enshrining and structuring how we relate and behave in space. Despite a great admiration for Paasi's work, I felt that this sentence did not go far enough: 'space, nature and culture' could not be considered an unproblematic continuum. The bison was subtly breathing down my neck, suggesting her role had been forgotten in the catchy formula. Despite the wealth of writing about territory, territoriality and boundaries, there was a lack of engagement and conceptualisation of nature in the broadly political literature in geography. After climbing up and down many other mountains, this book was written as an indirect response to the bison.

Bison, goats and bandwagons

The objects examined in this book are protected areas – national parks, nature reserves, biosphere reserves and the like – established in locations that span international political boundaries. Local projects have existed around the world carrying out 'transboundary' projects within protected areas for many years, yet this only became a fashionable cause on an international level around the beginning of the 1990s. In an increasingly competitive world of organisations vying for limited funding, it has become a leading paradigm, portrayed as the current Big Thing in nature conservation. The World Parks Congress held in South Africa in September 2003 – the global meeting of the international protected area movement that takes place every decade – was fittingly titled *Benefits Beyond Boundaries*. Transboundary issues held centre stage, referred to in the opening keynote speech by Nelson Mandela. In the past five years, few international meetings on protected areas have taken place without at least some acknowledgement, working group or discussion on transboundary issues.

Transboundary protected areas have indeed been seized upon with vigour. Few people remain unconvinced when told that thanks to such areas, African elephants, European mountain goats or South American panthers can migrate 'as nature intended'. A wealth of articles has appeared on the subject, largely produced by international conservation organisations or non-specialist journals such as National Geographic. This popular enthusiasm has generated a self-quoting clique, with key references systematically appearing unchallenged within the literature, as new authors jump on the bandwagon quoting others in a closed circle, with little fieldwork informing the discussion. These have furthered the suggestion that transboundary protected areas are unproblematic territorial objects that exist because they can be listed. Different terms for such areas are used in the literature, including 'transfrontier' (Zbicz and Green, 1997), 'transborder' (Hamilton *et al.*, 1996), 'transboundary', 'peace parks', or the more neutral 'internationally adjoining protected area' (Zbicz 1999d : 2). I have chosen the term 'transboundary' as the most appropriate, influenced by boundary studies.

Instead of a celebration of these projects as a *fin-en-soi*, this book suggests a critical and political approach, using transboundary protected areas to discuss issues related to boundaries, identity and cooperation. Such a discussion is overdue, as the surge in enthusiasm for creating such areas has not been followed by clear analytical discussion. In particular, the difficulty of carrying out 'transboundary' projects in practice has been blamed on lack of funds or time, rather than on deeper social, cultural, political or conceptual factors. In order to move beyond such shortcomings and identify what these factors are, this book suggests deconstructing the complex socio-spatial processes by focussing on boundaries. This involves identifying what it means to 'draw a line' within different academic and professional traditions, as well as in different cultural contexts, transposing this to protected areas. The implications of such differences are discussed in the context of postmodernism in which stark dualisms – and therefore boundaries – are taken as conceptually transcended.

Protected areas are rarely discussed as territorial objects or spatial scenarios. This conceptual slant allows this book to question what initially appears in Paasi's quotation as an unproblematic division between space, nature and culture. Previous political studies of protected areas have mostly been carried out in the field of environmental geopolitics, something of an offshoot of critical geopolitics applied to the environment. In a study of protected area boundaries in Trinidad, for example, Sletto argues that 'boundary making in the name of conservation has become an increasingly complex social act, shaping and reflecting local and state practice and relations of power between local, national and international actors' (Sletto 2002 : 184). Although there have been tentative attempts to use approaches within environmental geopolitics to examine the construction of boundaries to protected areas, this has taken place in the absence of a comprehensive conceptual and practical link between writings on boundaries and reflections on nature. This book forges such a link, arguing that the complex transboundary character of the objects studied requires it. In particular, it explores the relevance of linking up social approaches to nature and critical geopolitical approaches to boundaries. It discusses how this has further implications for how space and territory are conceptualised within geography. More concretely, it helps to answer a series of questions where nature, territory and identity appear intertwined. Why did a German biologist tell me he couldn't work with his colleagues in France because they didn't understand what nature really was? What did it mean when a Polish forester accused his Slovak counterpart across the border of interfering with nature by feeding deer in the winter? Why did none of the transboundary sites have a common management plan when all the people I spoke to suggested this was crucial? Why did it matter that Italian ornithologists who had participated in the creation of a protected area stretching across two countries were jubilant when birds released in France came to live in Italy? Why was I intrigued when national park managers said they were more progressive because they cut down trees in a more natural way than in the neighbouring country? How had nature come to be associated with reason? How was nature being used to ground politics?

Structure of the book

> *One of the delights of a life as a geographer is that, as a profession or a passion, it provides a copper bottomed excuse to spend time in the oddest corners of the world*
> *(Haggett 1990 : 95).*

The 'corners of the world' is one of those wonderful expressions that harks back to former worldviews – similar to the idea that the sun rises and sets, true to a pre-Copernican ritual. If 'odd corners' are marginal spaces, spaces of exclusion or periphery, then borderlands certainly qualify. They qualify not because these spaces of Otherness are far away from a mythical central point, different from those surrounding the barely disguised omniscient observer. Rather, boundaries and strangeness are inextricably linked, as Self and Other intermingle in ritual separations, fusions and hybridisations. It is because boundaries provide such a copper bottomed excuse to perform, question and celebrate 'oddness' that they are worthy of research.

Boundaries and boundary regions are not in themselves original topics for research. Within geography, studies of boundaries and boundary effects have taken many forms, although the wealth of approaches can be divided into three main questions: Do boundaries matter? If boundaries matter, how can they be overcome? And finally, how are such boundaries constructed? (Van Houtum 2000 : 73). These three general questions lead to three very different lines of enquiry. This book broadly lies within the field of the third question, additionally discussing what boundaries mean or imply once they have been constructed.

The critical tradition that considers boundaries as more than simple lines traced along some predetermined pattern is the starting point. The action of 'drawing a line' is seen as a complex process intimately related to issues of power, identity and control. Boundaries need addressing critically in all their complexity if the spatial and social processes taking place within protected areas are to be understood. The problematic placing of boundaries, both symbolic and concrete, plays a part in governing the shifting understanding of what is Self (inside) and Other (outside) as it reflects on 'society' and 'nature' being territorialized as distinct ontological domains (Whatmore 2002 : 61). This involves addressing who the archetypal 'Other' defined by a boundary is, or who / what lies on the 'other side' of a line. What constitutes 'nature' is contingent on social practices and culturally-specific definitions. Thus the boundaries between the distinct domains of 'nature' and 'culture' are defined and negotiated differently in different contexts, constructed discursively by the different relations and links within heterogeneous social networks that include both human and non-human actors (see Haraway 1985; Braun & Wainwright 2001; Whatmore 2002). The following paragraphs present an outline of the argument, demonstrating how approaches to boundaries, territoriality and 'social nature' are bound together.

This book draws from fieldwork carried out in five 'transboundary protected areas' around Europe. An overview of the methodology adopted in the field is laid out briefly in a box at the end of the third chapter. Brief descriptions of the case studies can be found in the appendices (Appendix I), providing background information to avoid repeating factual details throughout the discussion.

Boundaries and territoriality

Establishing a protected area implies the reinvention and redefinition of both social and spatial practices by a variety of actors involved and affected by the process. Individual and collective spatial identities are negotiated between an increasingly complex set of local, national and international actors. In Chapter 2 (Drawing Lines, Constructing Spaces), I discuss the concept of boundary as a socio-spatial phenomenon, the redefinition of which has an impact on collective territorial identities. Drawing on the pool of Foucaldian notions of power and critical geopolitics (Raffestin 1986; Paasi 1996; Albert 1998; Ò Tuathail 1998; Newman 1999), boundary making is an act of power establishing a spatial entity. This process is made visible within a discernible discourse. The focus is on the role boundaries play in the construction of space, as identities are constructed by symbolic and material boundaries that define Self and Other. Shifting territorial discourses enshrine and construct spatial entities. I specifically explore the constructed nature of scientific knowledge and the differing uses made of both biophysical and societal arguments in defining boundaries. The establishment of a spatial entity implies the territorialisation of a given territory by those involved in the process by defining, and subsequently managing and appropriating the area, implying the use of differing discourses in the construction of space. I use Paasi's discussion of concepts such as place and region (Paasi 1991; 1996 & 1998) to explore how spatial concepts are used to construct space. This is linked to the process of *spatial* socialisation that implies the construction and attachment of significance to demarcations and boundaries by a variety of actors.

One approach to studying such dynamic processes has been to identify three steps, identifying territorialisation, de-territorialisation and re-territorialisation (Raffestin 1997: 100; Albert 1998: 61; Paasi 1998: 69; see also Deleuze & Guattari 1992), sometimes known as TDR or de/re-territorialisation (Whatmore 2002: 60). Dividing it into three stages seems to imply the existence of a point zero, a time in which the spatial slate is scrubbed clear before the process can restart. This is unsatisfactory as the process is evidently ongoing. The term '(re)territorialisation' has also been used, a typographical trick that avoids the problem of resetting the clock and implicitly focuses on the dynamic nature of the process.

Territorialisation, with or without a prefix, is always dynamic, shifting and changing in the face of modified power relations. For Raffestin, 'the spatio-temporal container within which power relations are born must be seen as a whole. Therefore, the boundary or frontier is not only a spatial but also a temporal phenomenon. The matrix ("quadrillage") is not solely territorial since it is also temporal: activities are regulated, organised, controlled both within space and time, in a place and at a specific time, or over a certain extent and over a stretch of time. This simultaneous construction of both space and time has been too often forgotten or perhaps has not been put forward enough. This has led to a formal treatment of boundaries which are treated either superficially or negligently, considering they constitute one of the bases of spatial practices'[1] (Raffestin 1980 : 152). With Raffestin, I suggest that if boundaries are reified power, and key elements in the construction of territories, then changes in these boundaries shake up the balance, bringing about (re)territorialisation. 'To destroy or erase previous boundaries is to disorganise territoriality and consequently to lay open to question the daily existence of populations'[2] (Raffestin 1980 : 156). Thus the creation or redefinition of boundaries are far from innocent and represent key challenges to a status quo. This focus on times of 'crisis' during which boundaries are contested, demonstrates the dynamic character of boundaries and boundary definition. Different forms of knowledge and different conceptions of boundaries begin to emerge, hinting at the struggles involved in constructing spatial entities.

Boundaries and hybrid nature

Protected areas are territorial projects uniquely bound up with the idea of 'nature'. It is not enough to study them only as lines on maps, or as jurisdictional designations, without engaging with their historical association with notions of 'wilderness' removed from 'civilisation'. Such notions refer to an essential boundary between 'nature' and 'culture' that must be addressed if protected areas are to be understood. Instead of accepting this distinction as fundamental, I follow in Castree's footsteps in suggesting that 'politics is always geographical. (. . .) Time and again, "nature" is identified with certain spaces – such as "rural" spaces (. . .), zones of "wilderness", and formally designated "nature reserves". (. . .) This equation of nature with certain delimited geographical territories where environmental politics can focus many of its energies is problematic: for these spaces are neither wholly natural nor merely zones where certain social actors impose their culturally specific ideas of what nature is supposed to be' (Castree & MacMillan 2001 : 220). Protected areas are therefore unique examples through which to (re)introduce explicitly the 'political' into discourses of 'nature', transcending what Gregory has called imaginative achievements or 'conceptual constructions, the product of imaginative cuts in the fabric of the world' (Gregory 2001: 87).

Rather than negating the physical materiality of the world, this constructivist position supposes that nature can only be known through culturally-specific systems of meaning and signification. This implies that human representations of nature are not simply 'mirrors of nature', but instead are 'cultural products freighted with numerous biases, assumptions and prejudices' (Castree & MacMillan 2001 : 209). In Chapter 3 (Divide and Rule: Defining the Boundaries of Protected Areas), I therefore analyse protected areas as spatial objects constructed around a concrete set of arguments and assumptions. These can be identified through changes in the way nature has been perceived and identified as having spatial dimensions that require management. This includes a discussion of 'biosphere reserves', a model for nature conservation and sustainable development that builds on the idea of protected areas, created by the United Nations Educational, Scientific and Cultural Organisation (UNESCO) in the early 1970s. The brief history of the model explores the underlying conceptual assumptions contained within the proposed spatial model of zonation. This is followed by a discussion of new trends, including the increased enthusiasm for 'transboundary' protected areas.

The discussion in subsequent chapters focuses primarily on protected area managers, individuals working within protected area administrations in various roles who are involved in 'drawing lines': people who define the boundaries of protected areas. By exploring the particular discourses put forward by these actors in different situations, I illustrate how each stage of boundary definition is imbued with power struggles and conflicts as actors call upon a variety of biophysical or societal arguments. In Chapter 4 (Science, Politics and Legitimacy in the Design of Protected Areas), I explore both the return of the idea of 'natural' boundaries and the support for the idea of a societal definition of boundaries linked to human spatial practices, building on ideas introduced earlier. I suggest that because the assimilation of biophysical and societal boundaries has attained the status of a sacred myth in protected area planning, it is promoted as the main objective of successful design. This is linked to what Massey has called 'the assumption of (. . .) an isomorphism between space and society' (Massey 2001: 10). The conceptually doubtful and ontologically impossible reconciliation of boundaries of different natures echoes the one-time popularity attained by the idea of 'natural boundaries', leading down politically reactionary paths. Initially, by drawing on the authority conferred by belief in the self-evident accomplishments of science, protected area managers seek to reinforce their own position as legitimate custodians of the land. Thus in order to cope with the uncertainties inherent in such spatially complex scenarios, managers seek to reinforce decision-making structures and administrations through appropriate zonation and thereby their own authority over the land.

Because boundaries are inherently resisted as forms of reified power, they are inevitably contested. Conflicts about them are wider contests about the control of space by different actors: a conflict about differing territorial discourses. Drawing heavily from the fieldwork, conflicts are examined in Chapter 5 (Contested Boundaries and Complex Spatial Scenarios) as struggles for legitimacy. I examine these as contrasting understandings of insider and outsider linked to conflicting spatial discourses and differing conceptions of nature. Reflecting a background largely in the natural sciences, such discourses reflect the modernist distinction between 'nature' and 'culture'. As such, nature is seen as conceptually removed from Self: a fundamental Other beyond an unbreachable boundary. Managers thus position themselves within distinct territorial entities (protected areas) that they see as conceptually removed from the surrounding humanised landscape, despite official policy seeking greater integration. Paradoxically, however, by strongly associating with these entities, they position themselves as belonging to them: both insider and outsider.

Building on this paradox, I discuss how increasingly complex spatial scenarios emerge in which competing interests over an area vie for legitimacy.

These identity discourses are constructed by managers based around notions of insider/outsider. In a review of the spatial and institutional complexity in protected areas, I suggest that the metaphor of 'New Medievalism' is pertinent. This term was coined to describe an overlapping of various authorities on the same territory (Bull 1977; Anderson 1996; Albert 1998). Here, I use it to frame the discussion. This metaphor shows the difficulty of ascribing one administration or institution to one space, illustrating the contested nature of boundaries in protected areas, as well as the emerging forms of resistance to them.

Chapter 6 (Constructing Transboundary Entities) is divided into two parts. Initially, I discuss what is implied by the creation of a 'transboundary' protected area. This spatial and institutional process carried out across several sovereign countries increases the balancing act between competing interests. The definition of the boundaries of these areas is likewise additionally problematic, implying a process of (re)territorialisation implying negotiation between an increasingly complex set of local, national and international actors. Until now, when analysing transboundary protected areas, researchers have drawn on theories of international cooperation from the fields of international relations and political science. This has led to a failure to identify the nature of (re)territorialisation that implies a common reinvention and redefinition of both social and spatial practices. In this chapter, it is not so much 'cooperation' that is discussed, but rather the specifically spatial aspects of the process of establishing a transboundary entity ('Cooperation' is examined more comprehensively in Chapter 7). The second part of the chapter is divided into three sections analysing three differing territorial discourses that hold different assumptions about boundaries, identity and cooperation, respectively international organisations, local protected areas and individual managers.

'Cooperation' as a dynamic process is distinct from the explicit construction of a shared spatial entity. Thus in order to avoid equating the two, cooperation is discussed in Chapter 7 (Cooperation: Understanding Acceptance and Resistance), isolating the non-spatial dimensions while recognising the interaction between 'working together' and 'constructing new spaces'. This leads to a focus on the joint processes of acceptance and resistance to the ideas and consequences of cooperation. Rather than relying on abstract ideas of cooperation, the case studies are understood as ongoing processes of identity construction in which individuals give meaning to discursive boundaries. Throughout the discussion, different definitions of cooperation are seen to inform the choices and actions of protected area managers. In order to understand the negotiated temporal and spatial dimensions, I explore assumptions regarding implicit and explicit power relations between actors. Issues of acceptance and resistance, integration and distinction, are discussed on a social and spatial level. Because of the inherent weaknesses of the existing literature, I suggest that more lateral analyses are needed. Rather than being an unproblematic process that leads to higher spatial integration, cooperation is identified as an unscripted and negotiated process that leads to unsuspected social and spatial results.

In order to illustrate some of the problematic aspects of constructing a transboundary entity, Chapter 8 (Mapping a Bounded Other) considers maps produced by protected area administrations. This is linked to the function of conventional cartography in transforming space into a legible and ordered territory, thereby institutionalising it. Maps are useful illustrations of the process of constructing an Other, providing a graphical illustration of one aspect of (re)territorialisation though the institutionalisation of space (Paasi 1996 : 7; Ò Tuathail 1996 : 5). Using a series of maps from the case study areas, the changes in the perception of the Other are identified, laying emphasis on the way boundaries are represented graphically. This process illustrates the rhetoric of integration and distinction, illustrating the politically charged nature of seemingly-objective maps. I argue that the resulting maps have, in certain cases, attained mythical status, appearing and promoted repeatedly as icons representing the success of cooperation. While such maps represent crucial steps in the process of (re)territorialisation, some of the crucial and more problematic transboundary issues are swept aside when they are elevated to the status of sacred icons.

Naturalising metaphors appear frequently in discourses on transboundary protected areas. The potent and seductive metaphors of 'boundless nature' are broadly taken to be unproblematic by protected area managers: a green version of the 'borderless world' myth. In Chapter 9 (The Myth of Boundless Nature), I specifically examine the myth of boundless nature as an example of a sedimented myth. Managers are in a paradoxical position: on one hand many adhere to a belief in the immutable and primeval 'boundlessness' of nature, while on the other they face stark difficulties when applying such ideas to practical management as people, not nature, need engaging with. In analysing certain elements of protected area managers' discourses that seek to naturalise spatial entities, 'politics' (re)appear in a domain initially considered to be unproblematically 'natural'. Nature is no longer considered in itself, assuming that facts such as apparent boundlessness, 'speak for themselves' (Castree 2001 : 5). Building on previous chapters discussing the definition of boundaries, transboundary zonation is examined. This is linked to institutional issues, discussing the irruption of culture and difference in the zonation process. I conclude by suggesting that some of the more radical approaches to 'social nature' offer the promise of transcending the practical problems posed by the stark confrontation of biophysical and societal conceptions of boundaries.

In Chapter 10 (Drawing Lines in Hybrid Spaces) I further consider how 'nature' is enabled and made intelligible through discourse. I demonstrate that the 'social nature' approach holds potential for moving beyond the 'cultural' sticking points that are seen as obstacles to the creation of transboundary protected areas. In contrast to the previous chapter, I explore a countervailing yet coexisting myth: that cultural differences are somehow expressed 'within nature'. The contention is that nature and wilderness are intrinsically discursive constructions spun between people and animals, plants and soils, documents and devices in heterogeneous social networks (Whatmore 2002 : 14). This performative conception of nature not only has relevance to examining the boundary between wild / civilized, or human / animal but also has wider implications for the study of spatialised difference.

This book suggests that the authority of protected area managers is repeatedly challenged by the specifically 'transboundary' nature of the interactions due to the intrinsically complex hybridity of boundaries. The opposition between the fundamentally binarist discourses of managers whose professional backgrounds upholds the modernist dualism of 'nature' and 'culture' contrasts with the effective hybridity of protected areas. Thus the discussion of transboundary situations illustrates the multifaceted nature of boundaries, indicating that transboundary protected areas are much more complex than simple encircled areas stretching across international boundaries. Rather than unproblematically defining one Self and one Other, these entities create multiple Selves and multiple Others in a overlapping and conflicting patchwork of multi-scalar identities. In Chapter 11 (Conclusion), it follows that the concepts of territory and (re)territorialisation need to be explicitly revisited in the light of this hybridisation. In response to this, the concept of chimeric territory is suggested as a conceptual tool to decipher the complexity: a mythical beast that combines different animals within one body. Like the Carpathian blue slug that poetically provided a colourful lure for exploring marginal spaces, the Chimera provides an excuse to explore ideas of space in a creative way.

Notes

1 Personal translation from: 'l'enveloppe spatio-temporelle dans laquelle prennent naissance les rapports de pouvoir est un tout. Ainsi donc, la limite ou la frontière ne ressortissent pas seulement à l'espace mais encore au temps. En effet, le quadrillage n'est pas exclusivement territorial, il est aussi temporel puisque les activités sont réglées, organisées, contrôlées s'expriment tout à la fois dans l'espace est dans le temps, en un lieu et en un moment donnés, sur une certaine étendue et pendant une certaine durée. Cette construction simultanée de l'espace et du temps a été trop oubliée ou peut-être pas assez mise en évidence et il en est résulté un traitement formel des limites; elles sont abordées légèrement voire négligemment alors qu'elles constituent une des bases des pratiques spatiales'.

2 Personal translation from: 'Détruire ou effacer les limites anciennes c'est désorganiser la territorialité et par conséquent mettre en question l'existence quotidienne des populations'.

PART I
DEFINED BOUNDARIES

Chapter 2

Drawing Lines, Constructing Spaces

Casting the line

There is nothing quite as satisfying as drawing a line, delimiting an area of space and giving it a name: making order out of chaos. A mere scribble becomes a map. Drawing a line, dividing two areas has not lost its force as symbolic and concrete Empires on different scales continue to be carved out in the world. The process is the same at any scale: when carving up the world or when designing flower beds. Lines and boundaries construct space.

In this chapter, I outline why boundaries must be understood as complex spatial and social phenomena that construct and structure space and to what extent their definition (where they are drawn) is linked up with the idea of creating 'ideal' spatial entities. A boundary is the generic term for a linear spatial discontinuity that structures a given portion of two-dimensional space by dividing it into two. More simply, a boundary 'demarcates two entities, or two parts of the same entity, which are then said to be in contact with each other' (Smith 2000 : 7). I examine the variety of arguments that have been used to define boundaries, noting the difference between biophysical and societal arguments which represent an ontological split between realist and anti-realist conceptions of space. The recurring positivist temptation of finding 'natural' or 'rational' political boundaries is explored. The tensions in the way boundaries are defined is then examined in the wider context of the construction of space and of the need to identify and define spatial entities. In the second part of the chapter, I look at the role boundaries play in defining spatial entities, dwelling on the different discourses – and spatial ideals – that underpin how space has been constructed. This sets the scene for understanding how certain spatial discourses have been instrumentalised within protected area design, both implicitly and explicitly.

The nature of boundaries

> *Boundary, n. In political geography, an imaginary line between two nations, separating the imaginary rights of one from the imaginary rights of the other*
>
> *(Bierce 1911).*

At different times and places, the term boundary has meant many things to different people. Ambrose Bierce, in his famous *Devil's Dictionary*, suggests a wonderfully cynical definition. In a curious twist, but without missing the joke, I take his definition seriously in exploring how imaginary lines actually come into being. While I discuss the concept from a geographical perspective, this is not the only field through which it has been studied. Approaches to boundaries can be roughly divided into two broad fields: those derived from a social science perspective, considering boundaries as predominantly social phenomenon linked to the anthropic construction of space and those derived from ecological and biogeographic traditions, defining boundaries as physical limits inscribed in the landscape at various scales. This is of course a huge simplification, but it remains a coherent and useful one, a least as a starting point.

Lines, boundaries and frontiers

Boundaries, borders, frontiers and limits all indicate both a line and the idea of separation into two units. The emergence of the concept has been traced back to the Roman *limes imperii* which separated the civilised world from the 'barbarians' living beyond the *fines* of the Empire. The *limes* was thus a fortified line, rather than a political boundary, while the *fines* was akin to the frontier, boundary zone, or marches, between both worlds (Bodénès 1990 : 10): temporary stopping places where the potentially unlimited expansion was halted (Kratochwil 1986 : 35; see also Febvre, quoted in George 1974 : 196). The evolution of the word was subsequently closely associated with military history and the birth of the notion of sovereignty and the modern nation-state in the 15th and 16th centuries (for more comprehensive historical reviews see Ratzel 1897, Ancel 1938, Prescott 1978 & 1987 or Foucher 1991).

In much of the literature the terms boundary and frontier are used interchangeably, despite recurring attempts to stress the differences, often conflictingly (Prescott 1987 : 1 and Anderson 1996 : 7–9 come to opposite conclusions). Paasi however notes that there is 'a thorough agreement that whereas a frontier is a zone of contact, an area, a boundary is a definite line of separation – and not merely a line demarcating legal systems but a line of contact between power structures which at least partly manifest themselves in territorial frames' (Paasi 1996 : 25; see Sletto 2002 for *rebordering*: changes in location or meaning. In Francophone geography, see Raffestin 1986 for *limites*; Brunet 1967; Hubert 1993; Gay 1995 for *discontinuités*). More innovative approaches include considerations of the role of boundaries in identity construction and the definition of self (Barth 1995; Said 1995; Paasi 1996; Velasco-Graciet 1998), viewing boundaries and boundary zones as spaces of interaction, transcending stark dualisms, both metaphorical and spatial.

The topological approach, considering boundaries as geometrical objects, can be traced with relative assurance to Friedrich Ratzel's *Politische Geographie*, first published in 1897.[1] Although much of this work is a theory of the State, the underlying search for a geographical model of political behaviour leads to an argument based on topological considerations of points, lines and surfaces. As Hussy writes '*Politische Geographie* suggests a particularly simple and usable theory. It is based on just three concepts that reflect the Euclidian triad of "point, surface, line" i.e. position, area and frontier'[2] (Hussy in 1988 : IV). This reference to material elements is linked to invisible elements linking actors, implying power games, tensions, complex relationships and ultimate motives for controlling space. Thus boundaries are peripheral organs of control, representing an organic dimension of the State. Smith, drawing from cognitive sciences and mathematical theory, develops this topological approach and suggests a division based on geometrical reasoning. Starting with the idea that boundaries are, in topological parlance 'Jordan curves', i.e. that 'the boundary of a geopolitical or administrative entity must be free of gaps and must nowhere intersect itself' (Smith 1995 : 475), he builds a simplified typology of spatial boundaries. This is based on an opposition between physical boundaries – which he calls *bona fide* boundaries and boundaries produced by human demarcation called *fiat* boundaries (Smith 1995 : 475).

The first category of boundaries covers those that are 'in the things themselves. They would exist (and did already exist) even in the absence of all delineating or conceptualizing activity' (Smith 1995 : 475), such as the inside and outside of an apple. *Bona fide (Latin: 'in good faith')* boundaries thus 'exist independently of all human cognitive acts – they are a matter of qualitative differentiations or discontinuities in the underlying reality' (Smith 1995 : 476), independently of any observer or point of view. *Bona fide* boundaries yield a notion of contact between two surfaces which originates in mathematical reasoning but which can easily be applied to geometric, and also geographical, reasoning. Influenced by a realist position, this conception of boundaries largely informs the natural sciences and remains problematic when removed from the purely abstract.[3] This is however different from the idea of 'natural boundaries', i.e. the use of biophysical elements as sources or justifications for the definition of political boundaries.

The second form of boundaries defined by Smith derives from human decision or *fiat (Latin: 'let it be')*, those 'which exist only in virtue of the different sorts of demarcations effected cognitively by human beings' (Smith 1995 : 476). The terminology is 'designed to draw attention to the sense in which the latter owe their existence to acts of human decision or fiat, to laws or political decrees, or to related human cognitive phenomena' (Smith 1995 : 476). For reasons relating to human cognition, 'our cognition of external reality involves the systematic imposition of boundaries of many sorts, including fiat boundaries that may be more or less ephemeral. One important motor for the drawing of ephemeral fiat boundaries is perception, which as we know from our experience of Seurat paintings has the function of articulating reality in terms of sharp boundaries even when such boundaries are not genuinely present in the autonomous (which is to say mind-independent) physical world' (Smith 2000 : 5).

This distinction in the ontological nature of boundaries is echoed in much of the writings on boundaries. As such, the distinction is interesting in that it draws attention to ontologically different conceptions of space, translated into epistemologically different orientations. Broadly speaking, the social sciences have had, as a central paradigm, the negation of *bona fide* boundaries and in contrast have tended to stress the relevance of *fiat* boundaries, reflecting the view that all boundaries are social constructions: the result of human conceptual articulations. Proponents of the natural sciences, such as biology or landscape ecology, on the other hand, have tended to emphasise *bona fide* boundaries and to develop theories, statistical methodologies and typologies which are based on and affirm the concrete existence of such boundaries in the biophysical world.

Boundaries are intrinsically connected to the objects they define, since '*bona fide* and fiat boundaries share a fundamental property: they are ontologically parasitic on (i.e. cannot exist in isolation from) their hosts, the entities they bound. This is a common feature that a comprehensive treatment of boundary phenomenon should emphasize' (Smith 2000 : 13). Smith notes that 'one important reason for conceiving fiat objects and fiat boundaries as created entities (rather than as entities picked out or discovered within the pre-existing totality of all relevant geometrically determined possibilities) turns on the fact that there are fiat boundaries which coincide (occupy an identical spatial location) throughout their total length' (Smith 1995 : 478). Thus the boundary of the city of Hamburg and that of the German land of the same name coincide exactly. But they are not identical and could in principle diverge. In other words, they do not belong exclusively to the entity they bound, but rather shift according to the nature of the object considered. This point is important since two spatial entities, such as a national park and a biosphere reserve, can coincide on the ground yet bound different coextensive entities. Two separate entities may have different management structures, creating a practical challenge for managers. Such complex and conflicting geometrical and institutional scenarios are discussed in detail later on in Chapter 5, when the term 'New Medievalism' (Albert 1998 : 56) is tentatively resurrected to describe situations of overlapping and conflicting spatial entities.

Drawing the line

Conceptually categorising boundaries is one thing, defining them – drawing lines – is another, implying in practice struggles for legitimacy and negotiation between a variety of actors, each mobilising different spatial discourses. Such arguments or discourses follow different trends reflecting different ontological positions. 'Natural' or 'rational' techniques for defining boundaries, for instance, implicitly and uncritically seek to base choices on supposedly uncontroversial pre-existing geographical objects, while negotiated processes follow a different overtly political approach, reflecting the scientific traditions of their proponents. While in practice boundary definition is much more fuzzy than this first stark biophysical / societal divide initially suggests, such a distinction is useful in identifying trends. Subsequent chapters offer more nuances, while suggesting that caricatured discourses continue to lurk.

Definition – delimitation – demarcation

Boundary definition involves choices about where to 'draw the line' in creating spatial entities. Much of the literature on boundary definition is concerned with international boundaries and the disputes and conflicts concerning them (Prescott 1978; Anderson 1996; Foucher 1991; Pratt 2000). In this geopolitical tradition, a boundary 'separates areas subject to different political control or sovereignty' (Prescott 1987 : 1), often described as a multi-stage process (*Preparation, decision and execution*: Lapradelle 1928 in Prescott 1987; *allocation, delimitation, demarcation and administration* Jones 1945 in Prescott 1987; *definition, delimitation and demarcation* de Blij 1980 : 84). Here, a boundary is defined as a geographical object defined by negotiation, not as something pre-existing and ready to be identified through rational processes. This is important since there is a strong tendency – or temptation – to justify choices of boundaries as absolute, when in reality the decision of where to draw the line is arbitrary and subjective: intensely political in the widest sense. Therefore definition and delimitation are considered largely synonymous, involving negotiation between actors and resulting in verbal descriptions or lines on maps, although the former term is preferred here; demarcation is the physical marking on the ground, e.g. with boundary posts, signs or fences. The other existing terms are not used.

The temptation of 'natural' boundaries

As stated above, *bona fide* boundaries must not be confused with the idea of 'natural' boundaries, a notion that has haunted geography since the first maps were drawn. 'Natural' boundaries have a long history and were found on maps such as those drawn by Caesar or Strabo. Strabo, in his 17-volume description of the entire known world, emphasised the size of rivers and mountains which showed the 'natural' limits of the Roman Empire. In this case, what is considered to be a *bona fide* boundary (a river, a mountain) becomes a *fiat* boundary (a state boundary).

> *It is difficult to think of an inhabited place, a desert space, as something that limits itself through the power of its supposed natural boundaries. This is true even when one is confronted with barriers that appear insurmountable: the sea, the desert, mountains, rivers, forests*[4]
>
> *(Zanini 1997 :18).*

The idea of a natural justification for political boundaries passed down from Antiquity, re-emerging in the 17th Century through the view that the position of each State, set out within its limits, was predetermined by Providence or Nature. Chapuzeau, for instance, developed the idea of the State as citadel, with rivers and mountains as its natural moats and fortifications: 'a river is a constantly-moving body of fresh water that is most often used as the natural division of kingdoms and provinces',[5] likewise, mountains 'serve as a thick and impassable wall between provinces, similar to rivers acting as moats'[6] (Chapuzeau 1667 quoted in Georges 1974 : 196). This concept also found favour with the Napoleonic ideal of 'natural', linear boundaries, conceptualised in the charmingly-named *potamologie:* the myth of river frontiers as divinely ordained, a notion much in vogue in 18th century France. Thus boundaries were something predestined that needed conquering and establishing, as a form of God-given gift. 'This predestination made people believe for a long time that the inherent artificiality of a boundary, of a frontier, could find its true origin and its ideal image in the physical barriers that nature had scattered on the earth'[7] (Zanini 1997 : 19).

The French Revolution popularised the idea of natural boundaries, offering a vision considered more egalitarian than the former 'unjust' historical and hereditary boundaries. This marked the return to a proto-historical, pre-royalist world-view, as found in such Revolutionary slogans as 'Freedom knows no boundaries'[8] (Bodénès 1990 : 19). Such a notion occurred in the famous speech made by Danton proclaiming that France's boundaries were divinely ordained: 'its limits are marked by Nature: we reach them on all four points of the horizon, on the Rhine, on the Ocean, in the Alps and the Pyrenees'[9] (quoted in Smith 1997 : 394–403). At the end of the 19th Century, Ratzel was also tempted by the idea of a 'natural' boundary, but avoided suggesting that boundaries should necessarily follow biophysical elements, preferring instead to review the relative merits of coasts, mountains, deserts and lakes as possible boundary lines. He argued that nature provided a series of suggestions that could subsequently be chosen as a basis for defining a political boundary. Thus: 'there are cases where nature has divided space along lines that can be transformed into boundaries'[10] (Ratzel 1897, translation 1988 : 350).

Likewise, images of boundaries as membranes of biological organisms also made use of naturalising arguments, transforming spatial entities into forms of super-organisms, defined unproblematically (Morin in Paasi 1996 : 24). Ratzel's conception of boundaries stemmed from his idea of 'natural domains': 'the state whose boundaries are defined in the most natural way is that which coincides with a natural domain, to the extent that the outside barrier does not only rest externally on the boundary, but also internally'[11] (Ratzel 1897, translation 1988 : 350). Similarly, in 1907, Lord Curzon of Keddleston had no trouble affirming that: 'I have already accepted the broad distinction between Natural and Artificial Frontiers, both as generally recognised, and as scientifically the most exact' (Curzon 1907), referring to boundaries which were dependent on, or independent of, physical features. Despite carefully qualifying his argument historically, Curzon was much attacked for this notion on the grounds that all boundaries were artificial, and that the implication of the expression 'natural' was that such boundaries were intrinsically more appropriate than boundaries not based on the physical landscape (Prescott 1978).

Both Ratzel and Curzon's approaches to boundaries reflected their experience of imperial expansion. Ratzel, in particular, dealt with the creation and historical evolution of States, introducing the idea of dynamic change and expansion: since boundaries were intrinsically dynamic, they could expand, just as land extends to fill in the sea or a forest grows to cover a field. Ratzel's theory of the *propagation space* of a given group linked him to theories of *Lebensraum*, much favoured subsequently by the Third Reich. This transformation 'was a spatialization of the imperialistic desires of the small community of militaristic males who felt Germany was castrated by the Versailles Treaty' (Ò Tuathail 1996b : 47), starkly and somewhat crudely illustrating how ideas of 'natural' boundaries can lead down politically indefensible paths.

Thus the idea of 'natural' boundaries rests on a deterministic view of the influence of topography on political organisation, feeding 'on the extreme value given to linear configurations, which can be both real or apparent obstacles'[12] (Foucher 1991 : 97). Gay noted similarly that 'certain natural features have proven more interesting to boundary makers than others. (...) This use of arbitrary natural features is problematic. Firstly, it does not guarantee the precision of the boundary. A watershed is only sharp when the slope is steep. As for rivers, these are surely the most problematic features. Their changes take place on the scale of a human life'[13] (Gay 1995 : 10). However, in both these quotes, it is not natural boundaries *per se* that are rejected, but their practical translation into political boundaries. In other words, it is a pragmatic rejection, not a conceptual one in which the theoretical and political underpinnings of natural boundaries are rejected. This is a crucial difference, with important political implications. Yet however dubious such ideas may seem when pointed out explicitly, this notion of a political plan being inscribed 'within nature' is still very much alive, appearing in the most unsuspected guises. As is discussed in the second part of this chapter, such ideas are worryingly (re)appearing in conservation literature in a new guise, linked to determining a scale for environmental planning around natural criteria.

The temptation of 'rational' boundaries

Another temptation is the spectre of 'rational' boundaries, defined according to a pre-determined, rational programme. Many attempts have been made to establish more 'rational' principles on which to base the drawing of boundaries. Anderson suggests various examples (Anderson 1996 : 110), in particular the creation of the *départements* in France, during the early years after the French Revolution. This involved proposing 80 units whose boundaries would be defined by population, territory and wealth, creating a grid of comparable-sized administrative entities throughout the country. The guiding principle was that 'the capital of the most important administrative unit should be within a day's journey from its outer limits, and the subdivisions, the cantons, should have convenient and, as far as possible, equal access to central government' (Anderson 1996 : 110). Foucher notes that this conception was intimately linked to the French perception of the nation-state, giving strong symbolic value to the geometrical creation of France as a hexagonal entity, in which 'boundaries are indicated and put forward. A map is not read, it is an icon that is piously revered'[14] (Foucher 1991 : 97).

L'Hexagone was thus both the shape of the nation, and the term used to describe it in everyday language, imbuing geometry with meaning. The core-criterion for the new units was size, population and wealth, with the complex shape of the final object having symbolic value. This example echoes other attempts made to define rational spatial units, laying emphasis on the area defined rather than on the boundaries themselves.

Other examples have included attempts to use *fiat* objects to define boundaries, unlike the *bona fide* elements such as rivers or mountain ridges which are used to justify 'natural' boundaries. In his 1907 lecture, Curzon discussed what he called 'artificial frontiers' 'by which are meant those boundary lines which, not being dependent upon natural features of the earth's surface for their selection, have been artificially or arbitrarily created by man' (Curzon 1907), but nevertheless following a rational logic. Examples include the division of Antarctica into segments radiating from the South Pole (Beck 1994 : 84) or geometric boundaries in Africa or North America following meridians and geometric lines (Prescott 1987 : 63; Smith 2000).

Spatial discourses: drawing lines – constructing spaces

The nightingales divinely sing;
And lovely notes, from shore to shore,
Across the sounds and channels pour;
Oh, then a longing like despair
Is to their farthest caverns sent;
For surely once, they feel we were
Part of a single continent
(Matthew Arnold)[15]

There is something satisfying in the idea of a single continent emerging from the waters at the creation of the Earth, as described in many founding myths or represented on maps depicting – less poetically – continental drift: no boundaries, no separations and no need for a book on drawing lines. Since that primeval day, however, much time and energy has been spent in dividing space both politically and academically. In academia, the desire to create divisions in order to understand the dynamics operating within a given space has produced a variety of concepts, some of which have had direct political translations.

Boundaries define spatial entities. The social and linguistic role of geography as an academic discipline has been to construct meaning and export it outside scientific institutions. The 'language of geography' consists of certain key words which are reproduced and modified in the practice of geographers: this essentially consists of a struggle to put forward legitimate definitions for these categories (Paasi 1996 : 22). Drawing largely from Paasi (1996), I explore the historically constructed nature of these created and recreated spatialized categories or conceptual totems. These coexisting discourses indicate the constructed nature of scientific knowledge and the differing uses that can be made of both biophysical and societal arguments in constructing space.

Divisions of space should not be reduced simply to a matter of scale, like Russian dolls, but rather taken as spatial entities of different natures, each corresponding to a particular approach, with its own internal logic. An analysis of the construction of spatial entities within geography must start by examining the predominant assumptions about boundaries inherent in the discipline. Spatial entities – place, region, territory, for a start – are certainly neither universal nor eternal[16] and instead represent divergent spatial discourses, constructed by various actors, contributing to the construction of space. The relative success of each of these within various national and disciplinary schools of geography is witness to this.[17] Paasi has convincingly argued that 'region' and 'place' are historically constructed, created and recreated categories which can be understood in connection with the construction of territories and boundaries. The discussion here specifically emphasises the concepts of territory, region and bioregion, since the discourses underpinning them are most often instrumentalised (implicitly or explicitly) in protected area design. This lays the foundation for an informed critique of the spatial assumptions behind the choices made in defining such areas.

Constructing space, defining territory

The boundaries of spatial entities are linked to power and identity: boundaries are the focus of power relations, inscribing social projects into space. A boundary thus expresses the limits of a territorial project, a territory, as a structuring element (Raffestin 1980 : 148; Fall 1997 : 5). Boundaries are more than political borders since a 'boundary does not only delimitate a territory strictu sensu but rather a spatio-temporal envelope, that is to say that it simultaneously organises an operational time and space, a place within which a relational system can exist'[18] (Raffestin 1974 : 27). These relations include relations to others as well as to an area, as the web of relations that 'groups, and in consequence those who belong to them, entertain with exteriority and alterity'[19] (Raffestin 1986 : 92). Spatial entities are therefore central to understanding the social construction of spatiality (*social* spatialization) (Paasi 1996).

Laying emphasis on the historical as well as the spatial dimension, Paasi notes that territorial units are historical products, 'not merely in their physical materiality, but also in their socio-cultural meanings' (Paasi 1996 : 3). The corresponding concept of *spatial* socialization is also important in this context. This is defined as the process through which individual actors and collectivities are socialized as members of specific territorially bounded spatial entities and through which they more or less actively internalise collective territorial identities and shared traditions (Paasi 1996 : 8). Spatial entities are taken to be specific discourses significant in the process of spatial socialization.[20] Paasi lays emphasis on the role of rhetoric in this process, understood as referring to the forms of persuasive argument put forward by various actors.

Societal approaches refer to the identity shared by a certain group, whether conceptualised as a sense of common territoriality, the shared feeling of belonging to a certain territory, or as a notion of regional identity or regional 'personality'. Such notions are evidently more complicated than a simple combination of unproblematic non-overlapping areas. Identities are intrinsically 'fragmented and may involve a variety of spatial scales' (Entrikin 1994 : 113) and therefore are much more complex than adjacent patches. Identities 'become further complicated through their linkages with cultural definitions of insider and outsider, core and frontier, and self and other. (...) This rich cultural dimension draws attention to the limits of treating place and region as purely spatial objects or as simply outcomes of social forces, and instead encourages their reinterpretation as part of the complex and densely textured moral geographies of modern societies' (Entrikin 1994 : 113). In other words, the societal dimensions of spatial divisions cannot be addressed in terms of discernible spatial units; they must rather be seen as dynamic and often shifting, requiring multiple levels of identification. The world is no more a global village than it is a patchwork; rather it is simultaneously both. Yet the idea of unproblematic entities continues to intrigue, as it did when the notion of 'region' first gained favour.

The seductive regional ideal

The regional approach, developed principally within the sub-discipline of regional geography, was one academic and political attempt to divide space into coherent units in order to analyse and describe them. This rested on the notion that such divisions existed inherently and physically, awaiting discovery and description. 'At a very early stage naturalists and geographers were aware of the existence of divisions linked with the influence of relief, soils and climate, which are easily decipherable from the way (...) formations (of vegetation) are divided' (Claval 1998 : 42). The division of space into coherent units derived from Vidal de la Blache, a founder of the classical French school of geography in the early 20th century. 'The Vidalian approach held the view that each fragment of the earth's space contained its own internal logic as far as the physical environment and human response were concerned' (Vidal de la Blache, 1910 in Thompson 1998, intro to Claval 1998: x). Massey astutely notes that 'region' is a deeply sedimented hegemonic concept that remains problematic because 'the essentialism through which regions are defined and the lack of recognition of their constitutive inter-relationality can be seen as part of a modernist *Zeitgeist* which had a much more general purchase on the imagination' (Massey 2001 : 10).

Vidal built his descriptions of different regions initially on geological structure and natural conditions and moved on to provide analysis of the lifestyles of different groups. More than simply listing disconnected regions, he suggested links between them. Vidal justified his approach of studying human societies in their individual environments by saying that 'man [sic] has been, in our lands, a faithful disciple of the soil. The study of this soil will then contribute to shedding light on the character, mores and tendencies of the inhabitants'[21] (Vidal de la Blache 1903 : 4). His methodology was simple, based on meticulous description of the various elements present, in the true spirit of the Enlightenment: 'in order to achieve precise results, this study must be reasoned: that is to say that it must link up the current aspect of the soil with its composition and geological past. Let us not fear to spoil thereby the impression that radiates from the lines of the horizon and the exterior aspect of things. On the contrary. A knowledge of their causes means that the order and harmony of things can be better enjoyed'[22] (Vidal de la Blache 1903 : 4). Vidal therefore pre-empted the criticism that his approach removed the beauty or mystery of a region, suggesting instead that technical understanding enhanced it.

Claval noted that such a tradition was derivative since 'he borrowed from Carl Ritter the desire to underline the links between regions, a taste for analysing position related to place, an interest in the role topographical configurations pay in these relations'[23] (Claval 1979 : XV). However, unlike Ratzel, who offered a conceptual framework in which to set his analysis of spatial relations, Vidal did not conceptualise his approach, leaving it implicit that his precise and astute descriptions offered sufficient explanation for the specificity of each region. While the region itself was initially defined by natural criteria, societal elements gave it its individual colour. Such a vision implied a division of space into units, laying the foundation for their interrelations within a nation: regional description became a political project of nation-building. These meticulous descriptions of regions and *pays* found a wide audience and political support in France in the early 20th century, building a coherent picture of the nation as a natural assemblage of individual units. 'The flowering of this approach was expressed in a whole suite of regional monographs, produced by contemporaries and disciples of Vidal, characterised by scholarly depth, elegance of style and an historical approach which elucidated the formation of cultural landscapes' (Thompson in Claval 1998 : x). This tradition was based on a fundamentally rural subject, however and proved inadequate for coping with the rapid evolution of society. The golden age of French regional geography passed and the attempt to establish the distinctive personality of regions was essentially dropped.

Wolves in sheep's clothing

Despite the critiques, the regional approach is far from dead. It has crept back within what is known as the bioregional approach, building on the concept of ecosystem. In 1992, the Convention on Biological Diversity (UNTS 3069, 1992), endorsed the 'ecosystem approach' as the primary framework for action under the Convention, clarifying this concept during the fourth meeting of the Conference of the Parties in February 2000 in Montreal. The 'ecosystem approach' is more than simply about ecosystems since that original concept is not explicitly set in a spatial context (see for instance Tansley 1935). The term ecosystem denotes 'a dynamic complex of plant, animal and micro-organism communities and their non-living environment interacting as a functional unit' (UNTS 3069 : Article 2), that is to say a 'partly bounded system, with most of the interactions inside it' (Glowka *et al.* 1996 : 20). These are not explicitly spatial definitions. However the text of the Convention transforms this into a spatialised concept in explaining that 'ecosystems can be small and ephemeral, for example, water-filled tree holes or rotting logs on a forest floor or large and long-lived like forests or lakes. Ecosystems commonly exist within ecosystems. Consequently, the user of the term has to define the level used in each case. Biologists are often concerned with smaller-scale ecosystems, but for conservation purposes larger units (such as particular forests, grasslands or coral reefs) are generally used' (Glowka 1996 : 21). The term 'habitat' is slightly more precise in spatial terms: 'the place or type of site where an organism or population naturally occurs' (UNTS 3069 : Article 2), although this is still not particularly precise.

Although the term ecosystem is not spatially defined and does not therefore 'necessarily correspond to the terms "biome" or "ecological zone"', two other fashionable terms, but rather 'can refer to any functioning unit at any scale' (UNESCO 2000c : 3), it is being increasingly adopted as a basis for spatial planning: like a political compromise or a guiding principle, vague enough to be versatile. The ecosystem approach is 'based on the application of appropriate scientific methodologies focused on levels of biological organization, which encompass the essential structure, process, functions and interactions among organisms and their environment' (UNESCO 2000c : 3). Quite how such 'appropriate' levels of biological organization are defined is not specifically explained, but they 'should be determined by the problem being addressed' (UNESCO 2000c : 3). Maltby, however, notes that 'it is applied within a geographic framework defined primarily by ecological boundaries' (Maltby 1999 : 17), a quasi-pleonastic definition. Fashionable offshoots include 'ecosystem management, bioregional planning, ecoregion-based conservation, the ecosystem approach, an ecosystem-based approach, a bioregional approach, integrated conservation and development projects, biosphere reserves, watershed management, landscape ecology, integrated coastal zone management' (McNeely 1999 : 21).

While the ecosystem approach is not defined very precisely, the concept of 'bioregion' is one notion that has been thought to offer a spatial framework within which such an approach can be set, with direct links back to the regional approach examined earlier. As with other new and politically-motivated management concepts, the definition remains rather vague. The World Resources Institute (WRI), a non-governmental organisation active in issues related to global resource management, has attempted to define the bioregion as 'a land and water territory whose limits are defined not by political boundaries, but by the geographical limits of human communities and ecological systems' (WRI 2000 : online). Similarly, after reviewing several sources using the term, Miller defines it as 'the geographic area that local communities and governments consider being the management unit' (Miller 1999b : 11). Aberley puts the societal concerns first: 'a combination of cultural, social, economic, and ecosystem considerations determine the scope and scale of a bioregion' (Aberley 1994, quoted in Miller 1999b : 11). It is therefore an entity defined using a mix of criteria combining both biophysical elements and the practices of the local human communities, something that initially seems like a workable compromise.

The size of a bioregion is seen to vary according to the principles defined by the ecosystem approach, with the added criterion of a 'sense of belonging' among human populations. A bioregion is therefore expected to require more than a simple management on the spatial level, but also to include a political vision, irrespective of existing political jurisdictions. The WRI criteria start with a straightforward list of biophysical attributes, noting that the area must be large enough to 'maintain the integrity of the region's biological communities, habitats, and ecosystems; support important ecological processes, such as nutrient and waste cycling, migration, and stream flow; meet the habitat requirements of keystone and indicator species' (WRI 2000 : online), i.e. it must encompass an area whose boundaries are defined by biophysical criteria. More surprisingly, however, it must also 'include the human communities involved in the management, use, and understanding of biological resources. It must be small enough for local residents to consider it home' (WRI 2000 : online). Therefore, it must also be defined along societal boundaries, as a function of the cultural composition and practices of existing groups. The vital implicit argument here is that such boundaries will necessarily coincide, so that each group will live within a coherent ecological entity. Thus a bioregion 'must have a unique cultural identity and be a place in which local residents have the primary right to determine their own development' (WRI 2000 : online). Brunckhorst and Rollings even state that such a spatial division, following what they define as ecological and social functions, leads to 'logical management zones' (Brunckhorst and Rollings 1999 : 62). This implies that such entities already exist and are waiting to be identified and administered as management units, rather than areas defined by *fiat*. In addition, lest it be thought that proponents of bioregions ignore existing political jurisdictions, an additional note states that 'in special cases, a bioregion might span the borders of two or more countries' (WRI 2000 : online). The boundaries of such an entity should therefore be defined according to 'soft perimeters characterized by its drainage, flora and fauna, climate, geology, human culture, and land use' (Brunckhorst and Rollings 1999 : 59).

The spatial organisation of elements within the bioregion follows a model divided into four main areas: core areas, buffer or transition zones, corridors and finally the matrix. This follows a pattern similar to that suggested in the early 1970s within the biosphere reserve model (Chapter 3). The implementation of the bioregion concept follows an ideal set of criteria for increasing the 'conservation, study, and sustainable use of biodiversity' (WRI 2000 : online). These broadly follow both spatial and institutional criteria, suggesting a political programme for management and study within a given area. Spatial criteria are additionally defined: 'Large, biotically viable regions (...) containing structure of cores, corridors and matrices. (...) Ideally such sites, which may already be designated as protected areas, are linked by corridors of natural or restored wild cover to permit migration and adaptation to global change. Both the core sites and corridors are nested within a matrix of mixed land uses and ownership patterns' (WRI 2000 : online). It is a fascinating mixture, bounded by both biophysical and societal discourses. On one hand, it advocates boundary definition using 'natural boundaries', similar in nature to Ratzel's biogeographic determinism or Curzon's natural boundaries; on the other, it suggests defining the boundaries according to a given group's sense of place, incorporating notions of territoriality. Like Vidal de la Blache's regional perspective, bioregions imply that 'each fragment of the earth's space contained its own internal logic as far as the physical environment and human response were concerned' (Thompson 1998 & Claval 1998 : x). Such ideas include the full involvement of 'stakeholders', as well as integration with other institutions and local organizations, not excluding the possible reorganisation of administrative boundaries to coincide with the new ones of the bioregion. This is rarely suggested openly, being implied instead under the heading of a need for new partnerships and modalities of collaboration.

One revealing suggestion is to use Geographic Information System (GIS) technology 'to help stakeholders envision their region and its distinctive features clearly' (WRI 2000 : online) i.e. in addition to the rivers or mountain ranges, as suggested by Ratzel or other proponents of natural boundaries, there is now an additional, quantified technological means to justify the definition of the boundaries of the bioregion. This should not, however, be mistaken for a change in the nature of the argument, which remains fundamentally deterministic. The use of such technology opens up the possibility of creating a map which will reinforce the unity of the area in the eyes of the local population. Maps can be tools for defending a political project, which subtly transform a representation of reality into reality itself (see Chapter 8). Thus GIS technology has a dual appeal: techniques can be used to help define the boundaries of the bioregion 'rationally' and subsequently produce maps which will present it as a coherent unit. Vidal de la Blache would have leapt for joy. At the same time, the bioregional idea defends an institutional and political vision. It is to be understood as a comprehensive level of management, which can support values and proposals that 'can be shared cooperatively between public and private entities, or fully community-based' (WRI 2000 : online). In other words, it is no less than a return to Vidal's idea of the region, as imagined in 1910.

The political assumptions behind bioregions

In addition to considerations of decentralisation and spatial scales appropriate to decision-making, the term 'bioregionalism' also extends to a potent political vision. It is a particular incarnation of a 'grassroots', 'bottom up' approach led by communities themselves, primarily in North America, but increasingly in Australia and Europe. Far from simply requiring ecologically-appropriate land management, it has a far-reaching agenda, representing an experiment in self-sufficiency, which rejects all forms of centralized authority. Its proponents refer to their portion of space as 'homeland,' defined as 'a geographic space that encompasses their water sources and other key ecological features, food production, forests and wilderness, villages and infrastructure' (WRI 2000 : online). This is undoubtedly consistent with the ecosystem approach and the bioregion, with its emphasis on decentralisation and devolution. Its implementation however is in some ways much less clear politically than it may seem.

McCloskey, a life-long advocate and political campaigner for the Sierra Club, argues that the political orientation of such ideas is in reality not always consistent with their apparent support of sustainability and environmental goals and that the interplay between natural science and social science may produce politically dubious creations. Reviewing a number of articles appearing in the Journal *Society and Natural Resources*, he suggests that 'in pushing doctrines that amount to devolution, localism, and voluntarism, too many social scientists are adopting an ideology that is remarkably similar to that advocated by the political right (. . .) their reflexive refrain is "let states and localities handle it"; "keep the federal government out"; "give us less regulation"' (McCloskey 2000 : 11), i.e. while advocating programmes more usually associated with left-leaning liberals, proponents of bioregionalism are in reality defending an ideology closer to the other end of the spectrum. The assumptions of bioregional management may therefore be transformed, subtly but unwittingly, into a political programme based on debatable foundations.

One of the strongest critiques of bioregionalism questions the assumptions that guide how bioregions are bounded. In bioregionalism, it is assumed that the earth can be divided into distinctive and discrete ecosystems, each system or region exhibiting a unique pattern of geographical characteristics and life forms, with corresponding human communities reflecting these 'natural' boundaries (Olsen 2001 : 73). This postulates that unique and distinctive social and ecological communities need defending, rooted in their specificity. Olsen argues that 'such a defence of particularism can all too easily become a narrow and politically dangerous idea opposed to one of modernity's greatest legacies – a commitment to the notion of a universal humanity which, by its very definition, is non-rooted' (Olsen 2001 : 74). Olsen argues that some of these ideas can have unintended and disturbing political manifestations as they 'migrate' to very different cultural and political contexts. While noting that there are striking differences in the intentions and visions of bioregionalists and proponents of right wing ecology, he argues that within right wing ecology the environmental crisis is seen as a crisis of 'up-rootedness', in which there is 'the breaking up of the natural bonds between a culture and its ecological community' (Olsen 2001 : 74). Specifically, this means that bioregionalism calls for 'respecting the distinctiveness of every culture and its unique ecosystem, protecting the purity and integrity of each culture and the landscape in which it is embedded' (Olsen 2001 : 74) and thus 'bioregionalism means that people who are born into a place have rights over all others' (Berg in Olsen 2001 : 74). Olsen therefore notes that while bioregionalism has the admirable desire to reclaim a sense of home in nature, a feeling of rootedness has a significant dark side as well, which 'may result in such things as violent separatism, apartheid, and ethnic cleansing' (Olsen 2001 : 82).

Old wine in a new bottle

Thus the ideological mesh comes full circle. Ratzel's biogeographic determinism, with its ideas of a natural portion of the earth for a given group, a *Vaterland* for the *Volk*, returns in more innocent guises, but with a sting in its tail. The bioregion, a natural child of the ecosystem approach, has given birth to a less presentable offspring. A consideration of boundaries based on ecological and biogeographic justifications, far from being a new, cutting-edge vision, must be recognised, surprisingly, as no more than old wine in a new bottle.

Drawing on the methodologies of the natural sciences transformed into management tools, the ecosystem approach and the concept of bioregion attempt to link the biophysical with the societal. However sincere, such attempts do not escape falling into the trap of biophysical determinism. While the initial desire for interdisciplinarity is no doubt sincere, such a process implies more than sprinkling social and political concepts onto a naturalistic mindset. There is nothing to be gained from superficiality, which rapidly leads back, in circles, to the fruitless debates in the 19th and early 20th century regarding determinism and natural boundaries.

The debate illustrates the difficulties of translating what is essentially a form of fact-based, scientific and applied knowledge of biophysical processes into a political project. This link between ecology and politics is specifically developed by Olsen in his critique of bioregionalism: 'One of the most prominent bioregionalists, Kirkpatrick Sale, in his book *Dwellers in the Land*, attempts to justify bioregionalism by conceptualising it as somehow in accordance with nature. Sale argues his point by referring to the way nature operates and indicates that bioregionalism is the most desirable social and political arrangement because it is the most natural. Thus, for example, Sale writes that bioregionalism is grounded in several "ecological laws" – among them "decentralization". The problem with this kind of argument, of course, is that justifying a political vision by reference to the "laws of nature" or the "lessons of ecology" is on decidedly shaky intellectual ground. (...) Here it is not merely that such formulations are philosophically suspect; they also lend themselves to all kinds of political abuse' (Olsen 2001 : 80–81).

Moving beyond dualisms

The return to naturalistic arguments within bioregional literature is in stark contrast to the 'reinvention' of nature and the heralded end of the founding modernist dualism between 'nature' on one hand and 'culture' on the other. This innovative literature has led to a different and much more interesting critical reappraisal of the relationship between humans and their environment. Refuting the essential modernist division between biophysical and societal approaches, and arguing for a fusion between the two – as prevalent within some of the bioregional writing – is dangerous. The ecological approach 'has progressively taken over the field but without always modifying its dominant biological discourse. (...) The extraordinary success of ecology outside its own domain leads through scientific vulgarisation and para- and extra-scientific approaches to analogous and reductionist positions. These are especially perverse when applied to the uncontrollable field of politics'[24] (Bertrand 1992 : 120).

Attempting to merge ontologically distinct types of boundaries, 'natural' and 'political' boundaries, is barking up the wrong tree. The challenge is instead to unmask taken-for-granted distinctions such as that separating the biophysical and the societal. Instead, 'nature', rather than being separate from 'society', is taken as always social and political: 'nature has never been simply "natural" – whether it's "wilderness", resources, "natural hazards", or even the human body. Rather, it is *intrinsically* social, in different ways, at different levels, and with a multitude of serious implications' (Castree 2001 : 5). The second contention is that because of this, 'the all-too-common habit of talking of nature "in itself", as a domain which is by definition non-social and unchanging, can lead not only to confusion but also the perpetuation of power and inequality in the wider world' (Castree 2001 : 5).

Nature must no longer be considered 'in itself', assuming that facts speak for themselves 'once geographers have adopted the "correct" perspective' (Castree 2001 : 5). Scientific discourse, writes Whatmore, 'is vulnerable to critical scrutiny only by getting up close and tracing its (un)making through the laborious assemblage of interpretative communities, ritual words and phrases, documentary precedents and professional protocols; performative achievements that are always partial, contestable and incomplete' (Whatmore 2002 : 61). Therefore, nature and wilderness are intrinsically discursive constructions and must be analysed as such: 'the notion of wilderness being fleshed out here is a relational achievement spun between people and animals, plants and soils, documents and devices in heterogeneous social networks which are performed in and through multiple places and fluid ecologies' (Whatmore 2002 : 14). Steering away from the dubious temptations of natural boundaries, these much more fertile approach will guide and inform the discussion throughout the following chapters.

From spatial entities to hybrid boundaries

This chapter explored the historically constructed nature of created and recreated spatialized categories. These coexisting discourses indicated the constructed nature of scientific knowledge and the differing uses that have been made of both biophysical and societal arguments in dividing space. Such divisions were shown to be more than differently scaled constructions, instead revealing ontologically different standpoints: different spatial ideals. In order to argue this, I discussed the different approaches to defining the boundaries of spatial entities.

'Territory' and 'region', both profoundly societal concepts with different shades of emphasis on biophysical features, laid emphasis on socio-cultural meanings of space. They were not self-evident, discernible units, but instead had dynamic and shifting boundaries built around multiple levels of identification. In contrast, 'bioregions' stemming from the ecosystem approach were based on the idea that material (though not always immediately discernible) boundaries defined them. These different forms of spatial entities were seen to belong to different ontologies, switching between anti-realist and realist positions, at times problematically attempting to combine both.

In the light of calls within international conservation movements for aligning political boundaries with 'natural' ones, I illustrated the difficulties and dangers of translating what is essentially a form of fact-based, scientific and applied knowledge of biophysical processes into a political project. This was shown to rapidly lead back, in circles, to the fruitless debates in the 19th and early 20th century regarding determinism and natural boundaries. Instead of getting trapped in such a debate, I referred to more recent attempts to do away with the boundary between (societal) 'culture' and (biophysical) 'nature' that suggested a far more radical and promising approach. These writings on 'social nature' suggest that nature, rather than being separate from culture, is intrinsically social and political in different ways, at different levels and with a multitude of serious implications. Boundaries are necessarily hybrid, transcending dualisms, reflecting the constructed nature of the entities they bound. Spatial entities can therefore only be intrinsically hybrid.

Notes

1 The original German uses the term 'Grenze', that could variously be translated into English as 'boundary, border or frontier'. This is translated here as 'boundary', reflecting a more generic terminology.
2 Personal translation from: 'Politische Geographie propose une théorie éminemment simple et utilisable. Trois concepts suffisent à l'articuler, trois, parce qu'ils mobilisent, (...) la triade euclidienne "point, surface, ligne": à savoir, position, étendue, frontière'.
3 Smith suggests that coast lines or rivers may be considered bona fide boundaries, but such a notion is far from straightforward, as these are more often zones of transition. The legal low-water tide, sometimes heralded as uncontroversial, remains a somewhat artificial construct (Pratt 2003, *pers.comm.*).
4 Personal translation from: 'E difficile pensare a un luogo abitato, a uno spazio desertico, come a qualcosa che si autolimiti grazie alla potenza di quelli che si suppone debbano essere i suoi confini naturali. Anche di fronte a quelle barriere che in apparenza sembrano insormontabili: il mare, il deserto, le montagne, i fiumi, le foreste'.
5 Personal translation from: 'un fleuve est une eau douce qui court sans cesse ... et qui sert le plus souvent de division naturelle aux royaumes et aux provinces'.
6 Personal translation from: 'servent d'un mur épais et impénétrable entre les provinces, comme les fleuves servent de fossé'.
7 Personal translation from: 'questa predestinazione ha fatto credere a lungo che l'artificiosità propria di un confine, di una frontiera, potesse trovare la sua vera origine e la sua imagine ideale nelle barriere fische che la natura ha disseminato sulla terra'.
8 Personal translation from: 'la liberté n'a pas de frontière'.

9 Personal translation from: 'ses limites sont marquées par la Nature; nous les atteindrons toutes des quatre points de l'horizon, du côté du Rhin, du côté de l'Océan, du côté des Alpes et des Pyrénées'.

10 Personal translation from: 'il est des cas où la nature a découpé linéairement des espaces, qui se laissent transformer en frontières'.

11 Personal translation from: 'le pays le plus naturellement ceint est celui qui coïncide avec un domaine naturel, de telle sorte que la clôture ne repose pas qu'extérieurement sur la frontière, mais également de l'intérieur, grâce à la cohésion de ce qu'elle referme'.

12 Personal translation from: 's'alimente d'une valorisation extrême de l'intérêt porté à des configurations linéaires, chaînes de montagne et fleuves, toujours grossies dans la représentation cartographique, alors qu'elles peuvent être que des obstacles apparents, ou fictifs'.

13 Personal translation from: 'certaines configurations naturelles ont, plus que d'autres, intéressé les traceurs de frontières ou les hommes politiques. (...) Cette utilisation des accidents naturels pose de nombreux problèmes. D'abord elle ne garantit pas la précision de la limite. Une ligne de partage des eaux n'est nette que lorsque la pente est forte. Quand aux fleuves, il s'agit sûrement du support le plus problématique. Leurs divagations s'effectuent à l'échelle d'une vie humaine'.

14 Personal translation from: 'les frontières sont montrées, données à voir. La carte ne se lit pas; c'est une icône, que l'on révère, pieusement'.

15 Matthew Arnold (1909) 'Empedocles on Etna, and other poems, by A': 1852. Reprinted in 'Poems of Matthew Arnold 1840–1867. 1909. O.U.P, Oxford. p.135.

16 Indeed, in the context of 'globalisation', debates have sometimes focused on the proclaimed end of the nation-state and the advent of a borderless world. This rather pointless discussion is not resurrected nor engaged with here (Ohmane 1995; for a critical discussion see Anderson *et al.* 1995 or Paasi 1998 : 70–71).

17 See for instance Debarbieux 1999 on differences in the use of the concept of territory in Francophone and Anglophone traditions.

18 Personal translation from: 'la frontière ne délimite pas seulement un territoire stricto sensu mais bien davantage une enveloppe spatio-temporelle, c'est à dire tout à la fois un aménagement du temps et de l'espace opératoire, lieu de la réalisation d'un système de relations'.

19 Personal translation from: 'que les groupes, et par conséquent les sujets qui y appartiennent, entretiennent avec l'extériorité et l'altérité'.

20 However, this is not to say that spatial entities are necessary to constructing spatiality, as Lussault (2000) has convincingly shown. ?

21 Personal translation from: 'l'homme a été, chez nous, le disciple longtemps fidèle du sol. L'étude de ce sol contribuera donc à nous éclairer sur le caractère, les mœurs et les tendances des habitants'.

22 Personal translation from: 'Pour aboutir à des résultats précis, cette étude doit être raisonnée: c'est-à-dire qu'elle doit mettre en rapport l'aspect que présente le sol actuel avec sa composition et son passé géologique. Ne craignons pas de nuire ainsi à l'impression qui s'exhale des lignes de l'horizon, de l'aspect extérieur des choses. Tout au contraire. L'intelligence des causes en fait mieux goûter l'ordonnance et l'harmonie'.

23 Personal translation from: 'il a emprunté à Carl Ritter le souci de souligner les rapports de région à région, le goût d'analyser la position relative des lieux, les configurations topographiques et l'intérêt pour leur rôle dans la vie de la relation'.

24 Personal translation from: 'prends le relais par l'intégration remontante et sans toujours modifier son discours biologique dominant. (...) L'extraordinaire succès

de l'écologie hors de son domaine propre conduit par banalisation scientifique et surtout para- et extra-scientifique à des comportements analogiques et réductionnistes qui sont d'autant plus pervers qu'ils tendent à s'exercer dans le champ incontrôlable du politique'.

Chapter 3

Divide and Rule: Defining the Boundaries of Protected Areas

Fencing in the wild

The previous chapter explored the historically constructed nature of created and recreated spatialized categories, indicating the constructed nature of scientific knowledge and the differing uses that have been made of both biophysical and societal arguments in dividing space. I illustrated the difficulties of translating what is essentially a form of fact-based, scientific and applied knowledge of biophysical processes into a political project. Building on this, I analyse protected areas – areas set aside for purposes linked to nature conservation – as spatial entities constructed discursively by an evolving set of arguments and assumptions. The establishment of protected areas is therefore analysed as a process of social spatialization (Paasi 1996) implying a variety of actors. Changes within the way such areas have been constructed echo changes in the way nature has been placed and constructed as containing spatial dimensions that require management. This can be identified by looking at the construction of the spatial at the level of the social imaginary and in the form of interventions in the landscape: drawing lines, deciding what is on the inside and on the outside.

In this approach, nature must not be considered 'in itself', assuming that 'facts' such as apparent boundlessness, speak for themselves once geographers have adopted the correct perspective (Castree 2001 : 5). Instead, scientific discourse – such as that surrounding the design of protected areas – 'is vulnerable to critical scrutiny only by getting up close and tracing its (un)making through the laborious assemblage of interpretative communities, ritual words and phrases, documentary precedents and professional protocols; performative achievements that are always partial, contestable and incomplete' (Whatmore 2002 : 61). The diversity of the assemblage, when explored, therefore yields material with which to explore the contradictions between boundlessness and boundaries. A quote from an article on protected areas, for example, draws on several very strong images related to boundaries:

> *Many places in the world where clusters of protected areas already exist are along international boundaries. Often this has been intentional, as central governments have sought to preserve military buffer zones and keep settlements away from their frontiers. In other instances, it has simply resulted from inaccessibility due to distance or lack of roads, such as in the Amazon region. But nature does not recognize political boundaries. In many cases, ecosystems have been severed by arbitrarily drawn political boundaries,*

> *while species continue to migrate across those borders as they always have, oblivious to customs regulations*
>
> *(Zbicz 1999 : 15).*

Boundless, passive nature brutally severed by political boundaries, almost violated in its holiness; arbitrary international boundaries that directly brutalise nature's integrity and the corresponding suggestion that 'natural' boundaries would be more coherent; and a violated, feminised landscape, in which animals remain oblivious to the human action on their home territory.

Zbicz draws heavily on the image of ideal nature as primitive, untouched, existing as a form of timeless Eden before being brutalised by human action. These images are not innocent or benign but rather contribute to perpetuating sets of ideas about nature that have serious consequences. In an analysis of frontier mythologies, Waitt and Head (2002) note the link between what they call frontier mythology and absence of human action. They note that 'primitivism is also an integral part of the frontier mythology, because, following the linear historical narrative of "civilisation", these are marginal locations placed outside the ambit of "human society". Instantly, marginal places become unpeopled landscapes or at best peopled by "uncivilised" humans' (Waitt & Head 2002 : 337), adding a subtext of race. Another of these subtexts relates to gender. Nesmith and Radcliffe have argued that although environmental thinking seems politically progressive by virtue of being 'green', 'it's shot through with highly patriarchal notions of the environment as something to be "protected" or something that is intrinsically "nurturing"' (Nesmith & Radcliffe 1997 in Castree 2001 : 11). These strands will not be examined here. Instead, the focus is on the meaning and consequence of stating that 'ecosystems are severed', translating a biophysical image into a political programme.

Naturalising metaphors are systematically used in protected area projects and are broadly taken to be unproblematic by those involved. In the literature on transboundary issues, the idea that international boundaries sever nature is particularly pervasive. Yet, paradoxically, there is never any suggestion that protected areas, themselves spatial entities based on defined boundaries, are in any way performing similar acts of violence on nature by deciding where to locate the wild. Nature, in these discourses, is inevitably seen to be boundless, even in cases where human action and existence in the landscape is alluded to.

The myth of boundless nature is an interesting example of a myth that has 'sedimented', that has evolved into autonomous components of the everyday stock of knowledge which is taken for granted by a society (Aho 1990 : 22 in Paasi 1996 : 13). Through sedimentation, word and myth come to have lives of their own, detached from the original act of mythmaking. This has important consequences on how protected area managers behave (see also Chapter 8). In the previous chapter, I noted that boundaries were always political. Likewise, nature is always social and political: 'nature has never been simply "natural" – whether it's "wilderness", resources, "natural hazards", or even the human body. Rather, it is *intrinsically* social, in different ways, at different levels, and with a multitude of serious implications' (Castree 2001 : 5). Because of this 'the all-too-common habit of talking of nature "in itself", as a domain which is by definition non-social and unchanging, can lead not only to confusion but also the perpetuation of power and inequality in the wider world' (Castree 2001 : 5).

Divide and rule: evolving paradigms

The boundaries to protected areas must therefore not be taken for granted or reified. Rather than creating unproblematic 'natural' spaces, the definition of boundaries in the landscape formally reifies the modernist duality of nature and culture. Underlying discourses define these areas according to a combination of biophysical and societal arguments. The objective here is to question this dualism, particularly examining its link to the discourse of nature as the Other. This is enmeshed with the struggle of people and organisations involved in designing such areas to put forward legitimate definitions for such spatial entities.

Protected areas can be analysed by examining their boundaries and the changes these have undergone since first designated. Such an analysis could of course be set within a thorough historical perspective, drawing on contemporary thinking about the relationship between societies and nature through time. Such an undertaking is however beyond the scope of this book. A brief historical analysis is suggested instead, emphasizing the changes linked to boundaries. In this chapter, I start by arguing that the changes within the 'protected area movement' since the end of the 19th Century can be better understood as changes in the conception of the boundaries to such areas. In the second part, I examine the emergence of the concept of transboundary planning for protected areas and end with a discussion of some of the critical voices questioning the idea that such areas are overwhelmingly beneficial. The purpose is not to be exhaustive, nor to review all that has been written on the topic, but rather to indicate strong moments and turning points.[1]

Sacred groves and landscapes

The idea of protecting an area from human impact has existed around the world in different forms for centuries, before European and North American people decided to legally define areas and designate them protected. All around the world people dependant on natural resources managed their local environments in various more or less sustainable ways. Gadgil writes that many small-scale societies exhibit 'a number of practices of restraint in the use of biological resources that promote conservation of biodiversity' (Gadgil 1996 : 349). He lists a number of practices in various populations around the world that indicate a respect for certain species or habitats. Such societies 'regulate habitat transformation by protecting samples of natural communities on sacred sites (e.g. sacred groves, sacred ponds)' (Gadgil 1996 : 349, see also Craven 1993 : 23). Thus people living within a given space shape the landscape through their daily activities by selecting specific zones for precise purposes.

An example will suffice to give some idea of how protected areas are linked to notions of communal land. One Melanesian community physically demarcates its land and can precisely point out the boundaries to it: 'These boundaries are marked by stones hidden by their ancestors, by totem trees planted several generations previously, and by the villagers' legends of the dispersal of their people following the emergence of the first couple from the rocks of the volcano peaks, an event misted over by distance in time and mythology' (Lees 1993 : 69). The whole of the land is deemed sacred, containing its people's sacred inheritance of resources such as timber, animals, plants and soil. Yet specific sites are additionally set aside within it for a particular purpose.

Traditional societies must not systematically be mistaken for an idyllic Garden of Eden, with 'primitive' humans living in symbiosis with nature. Cordell warns against the perils of such romanticism: 'Indigenous societies probably were and are neither significantly better nor worse than European societies at preserving their environments' (Cordell 1993 : 68). Writing about Australia, Cordell argues that 'the traditional tenure systems at issue here, which have come down through the ages, are not panaceas for environmental degradation; they are not formulas for maintaining communities in some ideal state of isolation and equilibrium with their lands' (Cordell 1993 : 68). Growth and movement of population are major factors in bringing about change in traditional practices, as are economic and social factors. Traditional philosophies on how to care for the land and create *de facto* protected areas might not be directly applicable today, although these are undoubtedly influencing contemporary protected area policy. Certainly, the idea that there exists an original state of grace in which nature and culture were undistinguished remains conceptually potent, if only as a myth.

Hunting preserves for the rich and royal

Historically, the idea of sacred groves protected by and for the benefit of local communities existed in parallel to other systems of land management in which certain benefits were reserved for specific elites. In Europe, rather than only having communal forests, large land owners or monarchs decided to reserve portions of their lands for recreation in the form of hunting or rivers for fishing. Harroy notes that 'at the most, hunting had, in certain cases, made game animals so scarce that certain monarchs or powerful aristocrats established their own personal hunting reserves which were strictly guarded against poaching. In doing so, these great land owners were in some cases unconsciously preparing the beginnings of subsequent natural reserves such as Fontainebleau, Rambouillet, the Royal Forests of Great Britain, or even the hunting grounds of the dukes of Savoy, now Gran Paradiso National Park' (Harroy 1974 : 25, see also Gadgil 1996 : 354).

Such protected lands were for those who could afford time for recreation, preserving privileges in specially designated lands. Such a system presupposed the possibility of enforcing legal protection of the area to prevent poaching, as well as a specific workforce employed to protect such privileges. Insiders and

Outsiders were defined by social class and belonging, not – as was subsequently the case – by their relative 'naturalness'. Ironically such an elitist system often directly preserved unique ecosystems, subsequently designated 'national' parks or other types of protected areas symbolically for all people. In a curious twist of history, some of these areas were subsequently used as hunting grounds for political elites under subsequent regimes in Europe.

The first national narks

The idea of wilderness was crucial to understanding the birth of the national park movement in the 19th Century, representing wild pristine nature, untouched by human hands, and of essence separate from human society. 'The presumption was that the wilderness was out there, somewhere in the Western heart of America, awaiting discovery, and that it would be the antidote for the poisons of industrial society' (Schama 1995 : 7). Such a notion implied the existence of its opposite, that is to say nature exploited, transformed by human action, and having thereby lost some of its original characteristics. This dichotomy implied considering whether there needed to be a boundary between pristine wilderness and modified, humanised stretches of land, or whether such a notion was unnecessary, or unhelpful. The American naturalist John Muir was a fervent proponent and defender of the idea of wilderness, rejecting the idea of an imposed boundary, preferring to see nature as an infinite, boundless entity. As such, the idea of protected areas contradicted his idealised vision of nature as ungraspable or unlimited and consequently boundless. Confining it spatially in a reserve was therefore morally wrong.

The creation in 1864 of a protected area in Yosemite 'as a sacred significance for the nation' (Schama 1995 : 7) however marked the birth of the idea of protected areas. These were established for 'the preservation of scenic beauty and the protection of natural wonders so that they could be enjoyed by people' (Hales 1989 : 139). In 1872, Yellowstone became the first official 'national park', followed by Yosemite in 1890. Boundless nature and pragmatic protection were thus combined. The creation of Yosemite as 'a democratic terrestrial paradise' (Schama 1995 : 7) enshrined the idea of the necessity of encircling nature by creating a legally established boundary. This was not only to protect it from outside depredations but rather to keep it untouched yet available for human contemplation.

The boundary defined an area of aesthetically pleasing landscape available for human enjoyment, setting aside land in the form of a 'vignette of primitive America' (Hales 1989 : 139). Park boundaries were therefore taken to be 'walls against which profane activities would founder, providing within sanctuary to the human spirit' (Hales 1989 : 140), delimiting an area for enjoyment and inspiration, designed for people, not nature. The means for doing so was 'to draw a boundary around the elements that were enjoyable or inspirational and preserve them unchanged' (Hales 1989 : 140). For Schama such bounded sites encompassed religious as well as aesthetic ideals: 'like all gardens, Yosemite presupposed barriers against the beastly. But its protectors reversed conventions by keeping the animals in and the humans out' (Schama 1995 : 7).

The idea of protected areas spread around the globe, often ironically linked to 'modernising' values imposed on colonised land. As part of this spread, 'World Congresses on National Parks' were staged every ten years. They provided a platform in which diverse positions could be debated, building a form of consensus within what increasingly came to be seen as a worldwide 'movement'.

1962: the first World Congress on National Parks

The first World Congress on National Parks in 1962 marked the beginning of a worldwide awareness of the role protected areas played with the ambition of establishing 'more effective international understanding and to encourage the national park movement on a worldwide basis' (Adams 1964 : xxxii), bringing together delegates from 63 different countries. The first World Conference stood at a crossroads between two conflicting views of what protected areas should be, referred to exclusively in this context as 'national parks'. The first suggested that they should be wilderness areas predominantly designated in view of their aesthetic value and for contemplation by human beings, the second that they should exist to protect what was then called 'fauna and flora'.

Illustrating the idea that protected areas are islands of wilderness in a sea of altered landscape, Stewart Udall, then US Secretary of the Interior, said in his keynote address that 'with few exceptions the places of superior scenic beauty, the unspoiled landscapes, the spacious refuges for wildlife, the nature parks and nature reserves of significant size and grandeur that our generation saves will be all that is preserved. We are the architects who must design the remaining temples; those who follow will have the mundane tasks of management and housekeeping' (Udall 1964 : 3). 'Parks' were both areas for experiencing the sublime, and instruments for preserving it. Using the familiar metaphor of Noah's Ark, he likened park managers to 'the Noahs of the 20th Century' (Udall 1964 : 7), locking up nature in specific places in order to carry it intact into the next century. Romantic and biblical language likened the destruction of nature to the rape of a pure creation. Parks were for people's enjoyment of nature, 'created by the people for the use of the people' (Wirth 1964 : 20) either in the romantic pristine wilderness experience, or in the more pragmatic American parkways 'which are elongated parks with studiously landscaped highways, designed for the pleasures of scenic travel' (Wirth 1964 : 15). Parks were places where there were 'opportunities for contemplation and regaining the almost forgotten sense of timelessness the world once knew' (Olson 1964 : 48), featuring the Eden-like and virginal quality of an untouched wilderness.

This aesthetic approach was contrasted by a more pragmatic 'scientific' position. Chasing wilderness was an illusion: 'in very few areas can we still refer to unspoiled nature and sound ecological units. Natural preserves have been interfered with to such an extent that balanced ecological units are very rare'

(Knobel 1964 : 165). A protected area, far from being only sublime scenery was 'an area set aside for the protection, propagation, and the preservation of wild animal life and wild vegetation and for the preservation of objects of aesthetic, geologic, prehistoric, archeologic, or other scientific interest for the benefit, advantage, and enjoyment of mankind' (Knobel 1964 : 160). In other words, it was an area of land not only for human contemplation, but also for the preservation of nature itself, fundamentally distinct from human existence.

Protected area boundaries

The introduction to the conference proceedings noted that 'the problem of conserving nature is not a local matter, because nature does not respect boundaries. The birds winging their way southward over Europe neither know, nor care, whether they are passing above a Common Market or a group of feudal duchies. (...) Nature takes no heed of political or social agreements, particularly those that seek to divide the world into compartments. It has been – and always will be – all-inclusive' (Adams 1964 : xxxi). Despite such a pronouncement, nobody present at the Congress questioned the notion that protected areas were necessary, or desirable, and therefore that it was useful to define an area in order to protect it by means of an outside boundary.

The actual planning of the areas designated as protected also underwent a change at this time. The one unique exterior boundary keeping humans out yet allowing them in to enjoy the site was reviewed. For although 'it sounds relatively easy to make laws prohibiting people to enter certain areas, to build strong fences or walls around such areas, to refuse to build roads to, and in, such areas and virtually to provide complete protection against man' (Knobel 1964 : 160), in reality it was not.

It was clear that humans were understood to live on the exterior, looking in across the boundary. They could travel through the area, but not stay for long. The idea that human populations could inhabit these parks was anathema to the basic idea of pristine wilderness. The terms used to describe these entities were in themselves revealing, including words like 'reserve' and 'sanctuary', indicating that humans were kept out yet selectively allowed in to contemplate the land. Hales noted that while the accepted principle was that Parks were for People, 'carefully excluded from the notion of "people" are those who would make "nonpark" use of the resources, those not oriented to the enjoyment of the values for which the unit was set aside' (Hales 1989 : 140). It was therefore accepted that 'permanent human settlements within the sanctuaries and reserves should not be permitted. Even existing settlers, if any, should be evacuated. Alternative sites outside the parks and reserves could be found for their occupation. Experience has shown that some settlers have been extremely unscrupulous, and their presence in the sanctuaries has been fraught with danger to wildlife' (Badshah 1964 : 28). A national park was a sanctum sanctorum, 'inviolate, as it often represents the last remnant of the original stand of the country' (Badshah 1964 : 30, see also Wirth 1964 : 16). The ultimate aim was to keep hostile humans out while the wilderness remained pristine for contemplation by those who could really appreciate it.

The image of a protected area as fortress with one large peripheral wall was recognised to be of limited use in combining the paradoxical challenges of conserving nature and providing an area for recreation and contemplation. A spatial solution was suggested to solve the problem: multiple boundaries designating specific areas for various uses (Beltran 1964 : 38; see also Monod 1964 : 263).

1972: the second World Congress on National Parks

A hundred years after the designation of Yellowstone and ten years after the first World Congress on National Parks, the second World Congress on National Parks was convened in Grand Teton National Park in the United States in 1972. The conflicting forces apparent in the first World Congress, balanced between a romantic ideal of wilderness and the scientific need for the 'preservation' of nature and natural resources no longer coexisted peacefully.

In a provocative statement at the beginning of the congress, Nicholson severely blamed the proponents of the romantic movement according to whom parks were 'still viewed as the living embodiment of romantic values, and therefore as an unashamed anachronism in the modern world' (Nicholson 1974 : 33). To move beyond such a vision, he suggested that parks could only be managed by scientific pragmatists, since allowing 'the compulsively emotional champions to continue to dictate policy and to handle tactics would be to condemn the movement to go down in limbo' (Nicholson 1974 : 33).

The position of science as arbiter was reinforced. Concepts such as carrying capacity, population control, ecological equilibrium and plant succession became widespread (Reed 1974 : 40). This did not mean that the biological sciences reigned unchallenged as new societal approaches emerged. Issues of local population involvement, economic value, visitor use management, and social and economic development also engaged park managers. No longer exclusively an idealist or a natural scientist, the ideal park manager was 'thought to be an ecologist with a strong social science capacity' (Erz 1974 : 154).

Protected areas were no longer fortresses. It was 'highly important that parks should not be treated as isolated reserves, but as integral parts of the complex economic, social, and ecological relationships of the region in which they exist' (Hartzog 1974 : 155). Hartzog argued against what he called the 'forester syndrome' which monopolized much of national park management, saying that 'it is high time that we recognize that sociologists are as important as natural scientists' (Hartzog 1974 : 158). Quite why sociologists to the exclusion of other social scientists were selected for this role was unclear. Nevertheless, the natural science hegemony was losing ground. Science and planning became tools to reconcile use with preservation. Humans were no longer on the outside, looking in, but were acting in the centre of the action, trying to simultaneously read and write the user manual.

Protected area boundaries

The main change in the ten years between the two congresses was the appearance of new models of protected areas, variously termed 'natural park', 'landscape park' or, in the case of Britain, confusingly labelled 'national park'. These were protected areas that no longer followed the wilderness ideal, but rather were areas 'in which agriculture and forestry, hunting and fishing can still be pursued but where urbanization and industrialization arc barred' (Harroy 1974 : 26). Protected areas as fortresses of encircled wilderness were increasingly questioned.

The idea of specific zonations within protected areas prevailed as one way of overcoming differing objectives. Nicholson noted that 'the existing boundaries of many parks need urgently to be reviewed, both to conform to ecological realities and to add buffer areas in cases where incompatible development just across the boundary would compromise the integrity of the park (Nicholson 1974 : 36). A pragmatic approach to boundaries gained standing, contrary to previous definitions of the outer boundary as inviolate: 'a too literal-minded and rigid insistence on the unalterability of every park boundary is almost certain to give reason to think that no boundary will ever be adjusted by reasonable means (. . .) some of which are well-known to have been hastily fixed for mistaken reasons in the past' (Nicholson 1974 : 36).

However, the actual criteria for defining the boundaries of a protected area were still open to debate. Boundaries should follow ecological features since 'instead of moving to acquire the smallest possible area, we must now consider the maximum feasible area, then delineate management boundaries with a full consideration toward maintaining ecosystem integrity' (Reed 1974 : 42). Likewise, 'in the past, national park boundaries have usually been drawn to delineate a fairly compact area of simple shape. There could be greater elasticity in the areas chosen for designation' (Crowe 1974 : 164). However, she also noted that 'the essential boundary of the area must be assessed, both for biotic reasons and visual integrity' (Crowe 1974 : 165).

In parallel to issues of local definition, the idea of a representative 'world network' that emerged in the first World Congress gained further ground (Curry-Lindahl 1974 : 93). Thus 'the process of land planning is a series of plans, progressively becoming more detailed and more localized, but each fitting into the wide, overall concept of a master plan. In this hierarchy, the planning of national parks should be seen as an ingredient of total, worldwide conservation of resources localized, in the first place, into a broad master plan for a whole country or region' (Crowe 1974 : 163).

A protected area boundary was no longer a high wall keeping people out, but rather could be compared to a filter letting selective influences through. Managers therefore had to insure through spatial planning and management that the boundaries of the protected area fulfilled this crucial filter role.

1982: the third World Congress on National Parks

The third World Congress was held in Bali, Indonesia, in 1982. Unlike the two previous ones dominated by North American and European participants, the Third Congress was attended by managers from many 'developing' countries, reflecting the fact that in the previous ten years 'more national parks have been established in the Third World than anywhere else' (Malik 1984 : 10). The Congress Proceedings reflected this worldwide representation, dividing the report into nine 'realms' representing different biogeographical provinces (Udvardy 1984 : 34), avoiding political units. Each 'realm' was divided up into 57 'biogeographical provinces', suggesting a new world map based on purely biophysical criteria. In addition, since the idea of a global network was accepted by all, and enshrined in programmes such as UNESCO's World Network of Biosphere Reserve initiated in 1976 and including 208 sites by this time, such a classification was meant to help in 'identifying major holes in the protected area network' (Harrison *et al.* 1984 : 25).[2]

The Third Congress reflected an increasingly pragmatic approach. While the importance of 'the wilderness and sacred areas on which so many draw for aesthetic, emotional, and religious nourishment' (McNeely 1982 : xi) was not diminished, the need to 'recognize the economic, cultural, and political contexts of protected areas' (McNeely 1982 : xi) was enshrined in the Declaration. Rather than applying one North American model around the world a diversity of approaches was needed in different situations within the limits of environmental 'sustainability', a term endorsed by the World Conservation Strategy in 1980 (IUCN 1980). For the first time, the Proceedings included a strict series of definitions of the different categories of protected areas, ranging from one to ten. Diversity was codified and stringent protection 'is not necessarily appropriate for all areas which should be kept in a natural or semi-natural state' (McNeely 1984 : 1).

The need for a change in management philosophy was identified. This was summarised as 'the approach that a park is being protected against people, to the approach that it is being protected for people' (Talbot 1984 : 15). Although such formulas were also used in the previous Congress, the meaning of the expression had changed. It encompassed the need to make protected areas contribute to development, making them 'responsive to the needs of development' (Talbot 1984 : 16). Consequently, 'far from being considered as "set aside", a park should be viewed as being "brought into" the main arena of human affairs' (Myers 1984 : 656), accepted as an established phenomenon in a crowded world.

Protected area boundaries

The idea that 'even if the boundaries are fenced, there is inevitable interchange between the area and the surrounding world' (Croze 1984 : 628) was accepted, and even if the area appeared to be a self-contained ecosystem 'there will inevitably be trickles of energy and nutrients across the boundaries' (Croze 1984 : 628). The view that parks had to be part of the wider landscape, including people and local communities, also made ecological sense: 'Whatever may have been desired for them, parks can never be "islands". (...) Across a park's boundary, as across its ecosystem frontier, there are all manner of dynamic fluxes' (Myers 1984 : 658). Yet hiding behind the discourse of anthropic action, Muir's 'boundless nature' lurked: 'when we draw a line on a map and declare that within that line is a park, we make a gross intrusion on the landscape: we try to demarcate two separated entities in nature's seamless web of affairs' (Myers 1984, p.658; see also Garratt 1984 : 66). Thus the idea that protected areas could be isolated from the rest of a human-dominated area dissolved: 'it is a mistake to suppose that a protected area can be isolated, through park manager's fiat, from its hinterland' (Myers 1984 : 658).

In many ways, as Hales noted, 'the perspective had changed. No longer was the view from the border inward; the debate was whether one should focus outward from the border, or whether borders existed at all' (Hales 1989 : 141). Boundaries were increasingly likened to filters, letting selective elements through. The spatial model endorsed was concentric zoning, fulfilling various objectives within one area. Thus, 'this multiple-use approach is to achieve all its goals by use of concentric zoning. The park core will be protected, human needs will be met, preservation and development will coexist across a series of barrier zones so designed that all the purposes of each will be attainable' (Hales 1989 : 142). The idea of a buffer zone was reinforced since 'regrettably, and to the great detriment of the park movement, the border zone strategy has not been fostered with a fraction of the enthusiasm it merits' (Myers 1984 : 659). Buffer zones – a surprisingly militaristic term – were an interesting element in the evolution of the concept: boundaries were no longer linear but zonal.

Integrated regional planning stemmed from this idea of filters, complicating the idea of zonation. It was endorsed as a physical link between protected areas, adjacent land and human relationships to such areas (Garratt 1984 : 71). The actual physical definition of the area to which such an integrated plan was to be applied was also important. Arguments relating to 'the extent and boundaries of the planning region in logical geographical, ecological or human terms' (Garratt 1984 : 66) were mentioned, although what constituted a 'logical' geographical term was not specified other than as a combination of criteria linked to geology and soils, hydrology and scenic quality.

Thus the boundaries of protected areas changed from walls and fences to filters, no longer necessarily keeping humans out but supposedly integrated into the human use of the land. While 'national parks' were still promoted, other forms of protected area gained increased recognition implying different boundaries to different types of protected areas. Some were designed to keep people out, some to keep some human uses outside an area and some to keep people in 'anthropological reserves' 'to allow the way of life of societies living in harmony with the environment to continue undisturbed by modern technology' (CNPPA 1984 : 52).

1992: the fourth World Congress on National Parks and Protected Areas

Reflecting changes in terminology, the Fourth Congress on National Parks and Protected Areas was held in Caracas, Venezuela, in 1992, the same year as the United Nations Conference on Environment and Development, termed the 'Earth Summit' in Rio de Janeiro (Appendix I). The diversity and quantity of material presented during the World Congress meant that no single report was produced but rather a series of workshop summaries, as well as the Caracas Declaration and the Caracas Action Plan, a series of objectives endorsed by the Congress.

The tremendous diversity of topics addressed reflected the increasing roles taken on by protected area managers, making it clear 'that the park guard and park naturalist are being joined by the park community affairs officer, and earning the support of local people is being seen as a management opportunity, as well as a challenge' (McNeely 1993 : 192). While the first protected area congresses used romantic language, subsequent ones turned to scientific terms. In 1992, a surprising new language appeared, in which users of protected areas were referred to as 'customers' or 'market', and protected area management was termed a 'business' (McNeely 1993 : 192). Social, cultural and political issues were central to the success of protected areas. The premise was that 'we need to be more aggressive in marketing the goods and services of protected areas' (McNeely 1993 : 192). The private sector was called in as a possible partner and funder, as were local communities, non-governmental organisations and ... women (McNeely 1993 : 193). Arguments with an economic flavour appeared more and more, and protected area managers were expected to 'use the park's assets as a base upon which to build customer satisfaction, investment and interest' (McNeely 1993 : 192).

Protected area boundaries

The concept of the protected area as island received further scorn, since 'such an "island mentality" is fatal in the long run' (McNeely 1993 : 8). The idea that protected areas needed to be integrated into 'broader regional approaches' (McNeely 1993 : 9) was endorsed by the appearance of the term 'bioregion', 'used to describe extensive areas of land and water which include protected areas and surrounding lands, preferably including complete watersheds, where all agencies and interested parties have agreed to collaborative management' (McNeely 1993 : 9). Arguments relating to natural boundaries for protected areas received wider support, in particular the idea that management should follow watersheds which provide 'a natural unit for land and water management' (McNeely 1993 : 9). Such ideas extended to widespread calls for 'transboundary' protected areas, illustrating the return to planning on the scale of nature, unbounded by political jurisdictions (Fall 1999). Additionally, buffer zones were joined by complex spatial corridors, physically joining up protected areas (see Chapter 4 for further discussion on the design of protected area boundaries).

2003: the fifth World Parks Congress

The fifth congress, rechristened the 'World Parks Congress', was held in September 2003 in Durban, South Africa, subtitled 'Benefits Beyond Boundaries'. This meeting was organised as a plethora of parallel sessions held within a vast conference centre, ironically – in view of the title – surrounded by barbed wire, high fences and tight security patrols. Maps of the town included blacked-out 'no-go areas' for delegates, contrasting with the congress centre power enclave. Like previous congresses, the sheer size of the gathering and the multitude of parallel sessions meant that despite plenary sessions it was difficult to discern clear trends. Perhaps the defining moment was Nelson Mandela's opening speech: his enthusiastic call for transboundary initiatives as vehicles for peace and development. 'Boundaries' were overwhelmingly seen to be 'bad': obstacles to be overcome with new partnerships, especially with the private sector and local communities. Reflecting the location of the meeting, economic development was taken as instrumental to conservation, mostly by selling nature to tourists, often as a certified good. Simultaneously, critical approaches to protected areas as agents of neo-colonialism were present, if not in the mainstream sessions.

Protected area boundaries

One 'stream' was entirely devoted to transboundary issues, coordinated by the softly-spoken but dynamic Trevor Sandwith, reflecting the fact that 'since 1990, the total number of transboundary protected areas doubled and many others are set to launch within the next few years' (IUCN 2003). Following the initial surge in enthusiasm for transboundary protected areas, there was a new recognition that 'protected area managers are confronted with entirely new issues that they are ill equipped to deal with effectively. Because these new issues are set within a context of international relations and global politics, the more localized communication systems by which protected area managers currently share expertise and knowledge are not adequate to meet their growing demands' (IUCN 2003). As a consequence, a 'Global Transboundary Protected Areas Network' was launched, designed to 'act as a clearinghouse for all TBPA information and would allow for communication across diverse audiences and vast distances' (IUCN 2003), further marking the institutionalisation of 'transboundary' issues as a field it itself, since 'the individuals most well versed in TBPA issues do not have full institutional support to answer inquiries or present at conferences, workshops and other meetings as they are fully committed to other projects' (IUCN 2003).

In addition to discussions during the congress, the web of support for transboundary work was extended on an international level. Research and management guidance, including the publication of many guidelines and case studies, was further sponsored by the Biodiversity Support Program, the International Tropical Timber Organization, Conservation International and the German development agency Inwent, in addition to the familiar clan of IUCN (originally the International Union for the Protection of Nature and

Natural Resources, rebranded the World Conservation Union in the 1990s), UNESCO, Europarc and the Peace Parks Foundation. This diversity of partners, although the sign of a dynamic increase in interest, also reflected a variety of different perspectives. One presentation by the Peace Parks Foundation, for instance, consisted of an aesthetically pleasing succession of images of 'wild Africa', consisting wholly of animals trampling an earth devoid of humans. Red sunsets and charismatic megafauna set the scene, branding the landscape as a product, accompanied by a soppy song in English. The malaise of some participants was palpable.

In stark contrast, more critical approaches were equally visible, including a suggestion to 'take power relations seriously since power dynamics cannot be wished away by naïve assumptions about good governance' (Wolmer, 2003, *pers.comm.*). Comments reflected that 'going transboundary' was 'not a purpose, but a tool' (Van der Linde, 2003, *pers.comm.*) and that any such process had to be understood as 'superimposed on complex institutional landscapes' (Wolmer, 2003, *pers.comm.*), marking a change from the initial uncritical embracement of transboundary spaces as unproblematic entities. Elephants crossed boundaries, tourists consumed and travelled within transboundary entities, yet local people continued to be removed from their land, continually described as 'poachers' (Mdluli, 2003, *pers.comm.*), and 'only settled people were thought of as people since everyone else is seen as outside' (Kothari, 2003, *pers.comm.*).

Thus the wheel had come full circle: like the protected area movement as a whole, the 'transboundary' subtopic had developed sufficiently to contain many contradictions, divergent approaches and coherent critiques, put forward by a wide variety of actors each pursuing their own specific agendas ranging from neo-colonial continent-wide initiatives to radical analyses.

Patterns from the review

In this short overview of the main trends within protected area boundaries, I discussed the coexisting and divergent spatial discourses that existed and continue to exist within the worldwide movement. The definition of these entities sheds light on the construction of spatiality by laying emphasis on the role of different arguments relating to the nature of boundaries. In this discussion, this was expressed in two main distinctions deriving from the modernist nature/culture dichotomy: the spatial dichotomy between Insiders needing protection and Outsiders posing a threat, and the ontological distinction between biophysical and societal conceptions of boundaries. The succession of discourses within the protected area movement defined various Insides and Outsides constructed around differing understandings of whom or what should figure in each. Initially, romantic visions of 'nature' as the ultimate Other were constructed around the notion of 'wilderness', separate from human culture. Nature was a tableau for human contemplation. Engaging with it aesthetically further entrenched the divide. The boundary between human and non-human was ontologically unbreachable. Protected areas were nothing other than vignettes of wilderness with humans on the outside looking in across a boundary defining the archetypal Other. The boundary was defined on the basis of (societal) aesthetic criteria.

Subsequently, in a series of more or less defined steps (Table 3.1), boundaries were taken to be concentric sieves attempting to blur the Inside and the Outside in a series of zones defining increasing levels of 'naturalness'. Certain people were considered more 'natural' than others and were allowed to be more or less permanent Insiders. Protected area managers were designated rational decision-makers in this process. Boundaries were defined around biophysical arguments with science as the 'objective' arbiter and definer. Concurrently, there was an increasing desire to include human activities in areas designated as 'protected' which appeared to be based on a less clear-cut dualism between nature and culture. Comprehensive wide-scale approaches including local communities and women as Insiders were promoted within an ideology of free-market capitalism and political devolution, entrenching the idea that the natural could be sold for profit as a commodity. This seemed to herald a new conception of nature.

Yet this merchandisation of nature did not lead to a fundamental rethink of the nature/culture dualism. Paradoxically, the attempt to incorporate protected and non-protected areas in the wider landscape, including through market processes, did not and could not lead to a rethinking of the dualism. The ontologically distinct biophysical and societal conceptions of boundaries could not be breached: rather than lead to a redefinition of nature/culture, the expansion of protected areas into 'networks' led to a return of the idea of boundless nature, to the idea that 'nature's seamless web of affairs' could not be divided. In fact, as a consequence of this, a return of the idea of 'natural boundaries' was apparent in notions such as bioregions and ecoregions, heralding a return to forms of biophysical determinism. Nature, the archetypal Other, was seen to inherently contain spatialised political scenarios.

The conservation or 'protection' of nature has been reduced to a question of boundary definition on a spatial level. Yet protected areas are spatial models constructed out of the struggle of people and organisations which remain overwhelmingly professionally separated along the nature/culture divide. There is therefore little understanding that such a process also entails the theoretical need to transcend these dualisms between nature and culture. Within even the most integrative protected area administrations, the natural and the social scientists and managers have not come up with ways of work that transcend this boundary, as a brief look at any protected area administration staff diagram will confirm. Until this happens, no amount of joined-up thinking or differentiated spatial scenarios will bring about new conceptions of protected areas that fully reflect the 'reinvention' of nature.

Biosphere reserves: a planning model based around boundaries

The debate on the design of spatial models in international conservation programmes would not be complete without a specific review of biosphere

Table 3.1 Boundaries and protected areas: changes in dominant discourses

Time frame	**Dominant discourses on nature/culture within protected areas**	
19th C to mid 20th C First national parks	Nature encircled, wild, sacred. Boundary as wall. Humans out, wilderness in. Spaces for human contemplation *Stark divide nature/culture: nature as Other to be contemplated*	
1962 First World Congress on National Parks	Romantic ideal Aesthetic value, contemplation Parks for people, but no inhabitants	Scientific preservation Islands, temples, reserves, sanctuaries, Ark First zonations
	Discourses upholding stark divide: nature as Other to be contemplated or preserved	
1972 Second World Congress on National Parks	Romantic ideal acknowledged but largely discarded	Rejection isolated reserves Reconcile use/preservation 'Scientific' decision-making Boundaries as filters
	Pervasiveness 'scientific' discourse: nature as Other to be managed	
1982 Third World Congress on National Parks	Nature protected for people but some people more natural than others 'Seamless web nature' Rejection 'island' discourse, integration and exchanges Concentric boundaries *First spatial attempts to reconcile nature/culture: nature as pervasive Other*	
1992 Fourth World Congress on National Parks and Protected Areas	Call for support from local people People as 'customers', 'market', business 'Natural units' of management, ecoregions, bioregions Corridors, networks *Failure spatial attempts to reconcile nature/culture: nature as pervasive arbiter*	
2003 Fifth World Parks Congress	Widespread discussion of 'boundaries' as elements to be overcome institutionally, internationally and physically. Surge enthusiasm for transboundary initiatives and large-scale projects. *Plethora of diverse spatial models: 'partnership, governance, co-management': nature as Other to be sold and certified.*	

reserves. This section can be seen both as an example of a specific planning model and as an illustration that spatial planning policies do not reflect only 'rational' choices, following uncontroversial scientific principles. Instead, using the example of the biosphere reserves model designed by the United Nations Educational, Scientific and Cultural Organisation (UNESCO), it is argued that the process of design is dynamic and contested through time, resulting from repeated negotiations situated within specific social and political contexts. Like protected areas as a whole, the biosphere reserve programme and model has undergone a series of changes since it first appeared in the early 1970s. The model now combines three different zones within one wider area, divided by a series of internal and external boundaries.

The Biosphere Conference

In 1968, an international conference was set up by UNESCO as a way of stimulating a larger undertaking of international scientific cooperation, in association with the Food and Agriculture Organisation of the United Nations (FAO) and in collaboration with IUCN. This became, in due course, the Intergovernmental Conference of Experts on the Scientific Basis for the Rational Use and Conservation of the Resources of the Biosphere, shortened, not surprisingly, to 'The Biosphere Conference' (Holdgate 1999 : 97).

In 1968, Michel Batisse, assisted by Waddington, the Secretary of the International Biological Programme, drafted a resolution during the Biosphere Conference creating the Man and the Biosphere (MAB) programme. The mandate was spread between use and conservation, separating the protection of genetic resources – traditionally the role of the FAO – from the protection of nature. Batisse said that 'one day, and here I don't know who nor when, someone started to talk about biosphere reserves. And usually people say that I did, and I don't have any memory of this. (. . .) In any case I'm not sure that it's a very good expression. The word "reserve" may not be so . . . Anyway, we didn't come up with an alternative'[3] (Batisse 2000, *pers.comm.*). The concept of biosphere reserves (BR) formally appeared in 1971, when the idea for a World Network of biosphere reserves that combined conservation and research was formalised (UNESCO 1971 : 21).[4] In practice, due to the objective of linking conservation with research, the first BRs were usually national parks in which there was some level of research.

1971–1982 Defining the first framework

Following the International Coordinating Council, biosphere reserves were addressed in a meeting on the 20–24 May 1974, in Paris, within an international panel of scientists, including representatives from American and Russian state departments, as well as representatives from FAO and IUCN. This working group produced the first spatial model for BRs (Figure 3.1). From the start, the idea of buffer zones, or buffer mechanisms, was regarded as crucial.[5] Participants decided that BRs should have one or several buffer zones, dependent on local conditions and locations. These were assumed

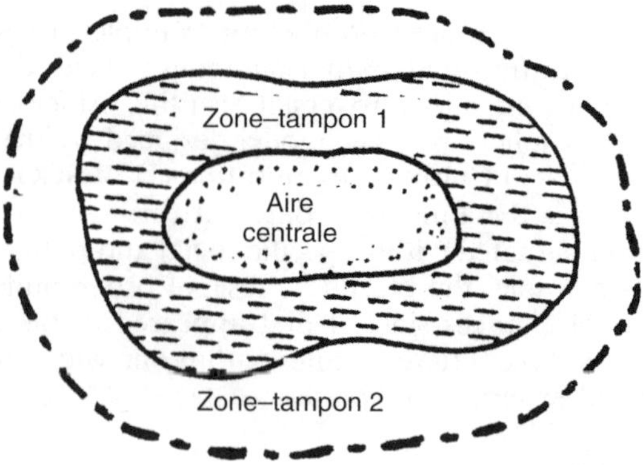

Figure 3.1 Design principles for biosphere reserves (UNESCO 1974)
Reproduced with kind permission of the UNESCO Secretariat in Paris.

to be concentric rings around core areas, as drawn in the accompanying figures.

The core areas were designated as 'sanctuaries', quasi-religious vocabulary implying areas free from all human intervention. However, defining a spatial configuration for BRs did not mean that this was easily applied or followed on the ground. In fact, 'most reserves had been superimposed on existing protected or research areas, and the idea of formal buffer zones involving other administrative entities had rarely been implemented' (Price 1996 : 647). There was no formal procedure for designating BRs and selection was left to individual countries. In practice, therefore, little was new in planning, design or management, notwithstanding the establishment of an international list. The precedence given to biophysical arguments implied that no populations would be allowed to settle in the buffer zone(s), reflecting the prevalent 'conservation dominant' (Price 1996 : 646).

Facing growing criticism

Despite the shortcomings, the list of BRs around the world continued to grow during the second half of the 1970s, with little changes in the basic philosophy. Between 1976 and 1981, 208 BRs were designated in 58 countries (Price 1996 : 647). Yet criticism of the model started to emerge. 'Some people say that "this is not the way it should be done, because that doesn't improve conservation in any way, zones should be chosen where there is nothing, where there is no protection". If there is no protection, then it becomes more complicated, there need to be zones where people are participating' (Batisse 2000, *pers.comm.*). Slowly, therefore, awareness of a new role for local populations emerged, in parallel to that within the wider protected area movement.

These growing pains also included combining different institutional traditions within the world of international conservation. At the meeting in

1971, IUCN and the FAO were present, but this did not mean that collaboration between the different organisations and United Nations institutions was always easy. Holdgate, a former Director-General of IUCN, noted that IUCN's participation in the BR model 'was less than had originally been expected, perhaps because the Commission on National Parks preferred to do things its own way' (Holdgate 1999 : 98). Batisse, from UNESCO, made a similar comment in saying that 'there is also some hostility in certain circles, notably IUCN, who perhaps grasps the programme's interest, but who says that if it's not IUCN doing it, then it's not good' (Batisse 2000, *pers.comm.*). Both individuals from different and sometimes rival organisations agreed that cooperation was not as straightforward as might have been expected.

1982–1994 Widening and clarifying the concept

In the early 1980s, UNESCO produced an attractive poster showing two opposing models for conservation: on one side a picture of animals and plants jammed in a bottle and on the other an open landscape, with people and nature interacting, reflecting the BR philosophy. This marked a clear departure from the previous paradigm. At the same time, in 1983, another international conference was organised in Minsk, Russia, in a climate of intense political tension following the gunning down on suspicion of spying of a South Korean civilian plane above Kamtchatka. Despite the logistical complications of getting international delegations to attend, often requiring travel by road and not by air, the conference took place.

Despite the adverse circumstances, an Action Plan for Biosphere Reserves was adopted, even if 'in reality we cheated a bit, because the conference adopted babble, and we made up the action plan afterwards as though it had been adopted by the conference ... a little bit later!' (Batisse 2000, *pers.comm*). This Action Plan was also endorsed by the United Nations Environment Programme (UNEP) and IUCN, although this did not mean that they committed any resources to implementing it. The result was a list of suggestions of what BRs could do, rather than a list of minimal fulfilments of what they should do: 'An action plan with no action' (Batisse, October 2000, *pers.comm.*). At the same time, the Scientific Advisory Panel on Biosphere Reserves was established with the mandate to clear up some of the confusion and lay some clear guidelines for the future.

Following this, a meeting in Czechoslovakia in 1985 further clarified the objectives of a BR. Batisse recalls that: 'there was a point when I said "this is all rather confused", and so I went to the board, and I drew a triangle. And that is the triangle of conservation, development, logistics. Before that, we didn't have a triangle. (...) There was no rigour between the main functions. So making a triangle was my main contribution. What have you done with your life? Me, I've designed a triangle' (Batisse 2000, *pers.comm.*). The birth of this conceptual triangle, separating yet connecting the three functions, was not enough to make BRs operational on the ground, even despite further clarifications of the outer buffer zone, defined as a 'transition area' or 'area of cooperation' (UNESCO 1986 : 73).

In 1992, once the Scientific Advisory Panel had been disbanded, a new Advisory Committee on Biosphere Reserves was established. This followed the recognition that the innovative planning principles were failing in practice and the discrepancies between what conservation-orientated academics dreamt up and what managers actually did on the ground was becoming something of an embarrassment. In 1993, a review mechanism was designed on the recommendation of the Advisory Committee, based on an expert assessment of the effectiveness of the concept's implementation on the ground (Price 1996 : 649). This was only formally adopted in Resolution 28 C/2.4 by the UNESCO General Conference, at its 28th session in 1995, after the Seville conference.

1995 The Seville Strategy for Biosphere Reserves

In March 1995, an International Conference on Biosphere Reserves was organised in Seville, Spain, to lay a more formal framework. Although the principles discussed were broadly similar to those raised in Minsk, the Seville conference brought together a much more representative set of people, both from the field and from national MAB committees. The result was the Seville Strategy, a set of recommendations for 'developing effective biosphere reserves' (UNESCO 1996) as well as the Statutory Framework of the World Network of Biosphere Reserves.

> *Each reserve is intended to fulfil three functions: a conservation function, to preserve genetic resources, species, ecosystems and landscapes; a development function, to foster sustainable development, and a logistic support function, to support demonstration projects, environmental education and training and research and monitoring related to local, national and global issues of conservation and sustainable development*
> *(UNESCO 1996 : 4).*

Of the three zones, only the core area required specific legal protection (Figure 3.2). Individual biosphere reserves remained under the sovereign jurisdiction of the countries in which they were situated. In certain cases, countries enacted legislation specifically to establish biosphere reserves.[6] A formal mechanism for a periodic review for BRs was established (Robertson Vernhes 1997 : 3), stimulating a revision of existing biosphere reserves in several countries.

When retracing the steps that lead to an increased formalisation of biosphere reserves, choices taken and policies adopted appear much more haphazard than any official institutional history might suggest. Driving ideas and concepts were dreamt up on blackboards, central principles governing the definitions of different zones were drafted after meetings took place and global policies followed the enthusiasms and choices of individuals within the Secretariat. Rather than a science-led initiative of 'rational' planning, the biosphere reserve programme has to be understood – like all international programmes – as the outcome of contested, politicised and dynamic processes, linked to individuals and socio-political contexts. Arguing that policies emerge in contested ways does not lessen their intrinsic value. The underlying design principles laid out in the BR model have undeniably contributed to contemporary protected areas

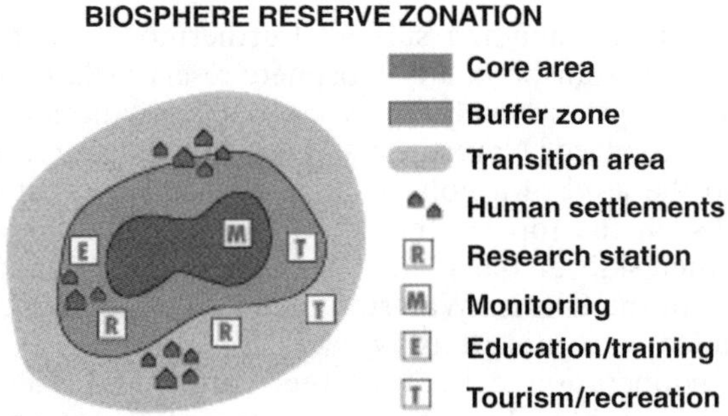

Figure 3.2 Design principles for biosphere reserves (UNESCO 1999)
Reproduced with kind permission of the UNESCO Secretariat in Paris

paradigms: 'the concept is accepted by all (. . .) people call it "bioregional approach", some people call it "corridors", others call it all sorts of things to avoid calling them biosphere reserves. So on a conceptual level, we've absolutely won' (Batisse 2000, *pers.comm.*).

Expanding protected areas across boundaries

The preceding short review related the changes related to protected area boundaries, both in general and within the specific biosphere reserve programme. This following section gives more emphasis to the formation of a transboundary agenda, pointing to key meetings and players while seeking to identify discourses on political boundaries. In an increasingly competitive world of organisations vying for a limited amount of funding, transboundary protected areas became a leading paradigm: the new Big Thing in nature conservation. This review is limited to publications produced by some of the international organisations and European non-governmental organisations mentioned earlier. It does not cover the wide variety of material published specifically on non-European parts of the world such as the wealth of writing on Southern Africa (see for example Wolmer 2003; Singh & van Houtum 2004) or Central and South America. While this section remains general, a more specific analysis of individual discourses on the construction of transboundary entities is developed in Chapter 6.

Definitions of transboundary entities

It is difficult to define a 'transboundary protected area' since it is not necessarily clear quite what this entails. It is therefore equally tricky to decide which the first one was. No legal definition will help, since this is a minefield in itself. Does a twinning agreement, for instance, constitute a transboundary protected area? Is a specific legal framework necessary or does something like a

Memorandum of Understanding suffice? Furthermore, what about international entities such as transboundary biosphere reserves that are recognised by the United Nations yet not necessarily by the legislation of the individual countries that are involved? Nevertheless, it is generally accepted – and perhaps even elevated to the level of a holy myth quoted in many an introduction to articles or books on the topic – that the first transboundary protected area appeared on either side of the United States/Canada boundary. In 1932, a 'friendship park' in the Glacier-Waterton area was established, although quite what this entailed was not terribly clear. In Europe, explicit forms of transboundary cooperation existed in the Tatras and Pieniny mountains between Poland and Slovakia at least since the Nineteen Fifties. These precedents have however not stopped individuals from other areas repeatedly claiming that their own project was 'the first', as Hanks recently did for Kgalagadi Transfrontier Park, in Southern Africa, in the Preface to a IUCN document, noting that 'a giant step was taken on 12 May 2000 (. . .) the world's first formally designated transfrontier park' (Hanks 2001 : ix).

A brief look at the literature further complicates the issue. The term 'transborder protected area' was chosen in the first IUCN report published in 1996, while the term 'peace park' also appeared, either applied to international boundaries where political tension subsisted, or else simply as a synonym. At the same time, different spatial scenarios of transboundary protected areas were identified and classified following several broad patterns, according to the shapes and disposition of protected areas along one or several boundaries (Shine, 1997 in Brunner 1998 : 16), ranging from protected areas scattered on both sides of a border to an integral multi-lateral unit.

A year later, Zbicz coined the phrase 'adjoining protected area complexes' to define 'all those places in the world where protected areas physically meet or nearly meet across international boundaries' (Zbicz 1999 : 106). This was an important distinction, also reflected in Brunner's choice of referring not to transboundary protected areas, but to transboundary cooperation since 'as far as things are affected by national sovereign competence there are no "real" transfrontier protected areas – or hardly any' (Brunner in EUROPARC 2000 : 51). In 2001, an IUCN publication finally came up with one – supposedly final – definition for transboundary protected areas as 'an area of land and/or sea that straddles one or more boundaries between states, sub-national units such as provinces or regions, autonomous areas and/or areas beyond the limits of national sovereignty or jurisdiction, whose constituent parts are especially dedicated to the protection and maintenance of biological diversity, and of natural and associated cultural resources, and managed cooperatively through legal or other effective means' (Sandwith *et al.* 2001 : 3).

This legalistic and rather contorted definition was partly the reflection of the background of one of the authors, Clare Shine, a freelance legal consultant specialising in environmental issues. This definition was interesting, not least because it banished the various 'transfrontier' or 'transborder protected areas' or the specifically Southern African 'transfrontier conservation areas' that appeared in the literature, but also because it marked the return of the notion of a transboundary entity within the limits of one sovereign state. This definition was substantially different from the position endorsed by UNESCO. That emerged after a somewhat acrimonious debate in the meeting in Pamplona at the 'Sevilla + 5' International Expert Meeting, when during a plenary debate six speakers took to the floor to present their views. Two people argued that internal boundaries should also be considered, particularly in countries with federal systems, while the others suggested instead that this would dilute the issue. Following an intervention of the Secretariat, in the person of the Director Peter Bridgewater, it was settled that the definition should only cover international transboundary sites because of UNESCO's status as an international United Nations organisation. Cooperation within countries was also taken to be a priority, but this was considered to be covered already within the Statutory Framework of the World Network of Biosphere Reserves.

Shifting paradigms across boundaries

At the first World Conference on National Parks in 1962, Walery Goetel of Poland presented a report on the transboundary national parks along the Polish-Czechoslovak boundary, reviewing other initiatives around the world. He noted that 'the political frontiers of countries often run across mountain chains and massifs, drawn in an entirely artificial way either along mountainous ridges, down the slopes or along the valleys' (Goetel 1964 : 288), and therefore that the creation of a national park on one side of the boundary could be extended to the neighbouring country. However, apart from Glacier-Waterton along the United States/Canada boundary, he noted that the other projects had 'not yet been definitely completed insofar as their legal status and other specific details are concerned' (Goetel 1964 : 289).

Similarly, at the second World Conference on National Parks in 1972, Krinitskii of Russia suggested building on the initial 'twinning' of protected areas in Eastern and Western Europe, as well as concluding 'bilateral or tripartite agreements concerning the regimes of appropriate groups of nature reserves in different countries; such agreements could also provide for setting up nature reserves jointly administered by neighboring countries under a single coordinated program' (Krinitskii 1974 : 67). Nicol also noted the need for such cooperation, stating that 'the international parks demonstrate that there is a growing recognition of the fact that management and cooperation must flow across international boundaries wherever and whenever possible' (Nicol 1974 : 383). A formal Recommendation was accepted, stating that the Congress 'requests governments to collaborate closely in the planning and management of neighbouring or contiguous national parks' (Elliott 1974 : 444, Recommendation #6).

While protected areas spanning international boundaries were mentioned in the second World Congress on National Parks, they were not formally discussed in the third World Congress in 1982, held in Bali. However, the tremendous political upheavals within Europe that started at the beginning of the Nineteen Nineties put boundaries back on the agenda. In 1990, the Protected Areas Programme of the World Conservation Union (IUCN) published a short report entitled 'Parks on the Borderline: Experience in Transfrontier Conservation'. This caught the imagination of many, offering promises of new beginnings for the European continent shaken up by boundary changes.

In 1991, a Mountain Theme working group was established as part of IUCN's Commission on National Parks and Protected Areas (CNPPA), led by Lawrence (Larry) Hamilton, a delightfully dynamic man fired by enthusiasm. This group organised a meeting in the Australian Alps on 12–21 November 1995, coordinated by the Australian Alps Liaison Committee and members of Australian parks services and funded by a variety of Australian conservation and development bodies. Following the meeting, a short publication on 'Transborder Protected Area Cooperation' (1996) was published by IUCN and the Australian Alps National Parks, laying the path for a surge of enthusiasm for the topic. The report was composed of a series of chapters on the benefits of cooperation, listing some elements of effective cooperation and examples of management activities. These recommendations, mostly written in the form of bullet points and lists in a rather random way, were followed by a series of case studies, in what became a popular combination of 'examples' and 'recommendations' adopted by many similar publications. The case studies reflected situations in Europe, North America and Australia, reflecting the participants' origins (35 individuals from 18 countries). Some of these people became the core members of the international 'transboundary scene', building reputations and creating a new niche of expertise for themselves in the process.

The following year, the fourth World Congress on National Parks and Protected Areas reflected the return of the idea, with a specific workshop on transboundary cooperation. The idea was further enriched by new bioregionalist jargon. Thus 'protected areas which occur along international boundaries call for international cooperation for which the bioregion approach also provides a framework' (McNeely 1993 : 9). The bioregional idea, defined rather vaguely at this stage, captured the imagination. It offered a basis for applying all the seemingly contradictory ideals of conservation, development, sustainable development, regional identity construction, political devolution and tourism promotion within one site, defined primarily along biophysical criteria. This was seized upon as the geographical equivalent of the Theory of Everything, the latest fashionable paradigm that would secure support and therefore new funding from untapped sources. It was, as noted in one report 'a concept whose time has come' (IUCN 1998 : 7). Despite the prevalent enthusiasm, the workshop rapporteur noted realistically that 'more sites have the potential to become protected areas but it appears that the majority of existing border protected areas have yet to work well in practice, even those sites recognized under international conventions' (Karpowicz 1993 : 154).

Creating transboundary experts

At this time of growing enthusiasm for supranational ideas, it quickly emerged that the main players in Europe promoting transboundary cooperation belonged to a relatively small clan, with certain names appearing with regularity in the various meetings organised by a variety of different international and non-governmental organisations. Within the space of a few years, the World Conservation Union (IUCN), the United Nations Educational, Scientific and Cultural Organisation (UNESCO) and EUROPARC had all formed specific taskforces or working groups on transboundary cooperation in protected areas. Within these groups, a few key figures quickly emerged, with 'experts' much in demand at the various meetings, ably promoting themselves on the newly-formed circuit and participating in elevating the topic on the international scene. Fluency in English was practically a pre-requisite for attaining this status, as few meetings had any substantial budget for interpretation. Field experience of cooperation was prized over academic experience, although key figures working on the policy side played a large part in convening the meetings, providing the funding and coordinating publications. A number of publications were produced following international meetings, with organisations often racing against each other to be the first, each looking for a specific slant that would guarantee a niche in what increasingly seemed like a competitive market.

Meetings explicitly focussing on transboundary protected areas were organised around the world. A large one was hosted by IUCN in Somerset West near Cape Town in South Africa, on 16–18 September 1997 under the title of 'Parks for Peace – International Conference on Transboundary Protected Areas as a Vehicle for International Cooperation'. The role of the South African 'Peace Parks Foundation' was instrumental in this event. This was a foundation founded by Anton Rupert in 1997, with his money from the tobacco industry. He obtained the support of the then-President in a brilliant public-relations coup, as Mandela stated that 'I know of no political movement, no philosophy, no ideology, which does not agree with the peace parks concept as we see it going into fruition today. It is a concept that can be embraced by all' (Wolmer 2003 : 261). This large meeting was followed by a more targeted meeting in 1998 entitled an 'International Symposium on Parks for Peace' in Bormio, Stelvio National Park in 18–21 May 1998, with the explicit aim of producing a set of guidelines for organising transboundary cooperation. At this time, key players were identified and a loose task force was set up.

Around this time, the Council of Europe started to add a 'transboundary' dimension to its European Diploma of Protected Areas programme. Diplomas were awarded to individual 'protected natural or semi-natural areas of exceptional European interest from the point of view of conservation of biological, geological or landscape diversity (. . .) by virtue of their scientific, cultural or aesthetic interest if they have an appropriate protection system, perhaps also in conjunction with programmes of action for sustainable development' (Council of Europe, Resolution (98) 29; 1999 : 2). Although such diplomas were awarded to individual protected areas, the programme specifically mentions 'transfrontier' areas, which 'shall not be granted without the consent of all the States concerned' (*op. cit.* 1999 : 3). This award took the form of a document certifying sponsorship by the Council of Europe, awarded for a period of five years and renewable for successive five-year periods. The shape of this sponsorship was unclear, and did not seem to go beyond being a form of label, with the brand name Council of Europe as the main advantage. This was viewed as important in Central and Eastern Europe, as managers were enthusiastic to be considered part of 'Europe' – implicitly or explicitly on the list for joining the European Union.

Transboundary biosphere reserves

Initially, biosphere reserves were designated within single countries, but in 1993 the Man and the Biosphere Programme of UNESCO designated the first two transboundary entities in the Tatra mountains between Poland and Slovakia and in the Krkonoše/Karkonosze mountains between what was then Czechoslovakia and Poland, in line with Objectives I.2 (1) and IV.2 (5; 6; 16) of the Statutory Framework (UNESCO 1996). Calls to 'encourage countries with biosphere reserves or potential biosphere reserves on each side of an international boundary to start exchanges to explore possibilities of creating a transboundary biosphere reserve' (Price 2000 : 93) were made within meetings in Europe and Asia (Kim 2000). These were further strengthened during the Seville + 5 International Expert Meeting on the Implementation of the Seville Strategy of the World Network of Biosphere Reserves 1995–2000. Held in Pamplona, Spain, this meeting offered a venue for the *ad hoc* Task Force that brought together representatives from Africa, North and South America, the Asia-Pacific Region and Europe. A set of 'Recommendations for the Establishment and Functioning of Transboundary Biosphere Reserves' (UNESCO 2000) was prepared on the basis of a draft report, partly drafted within a workshop and subsequently pulled together by the Secretariat, with two people working on a laptop in the corridor, cobbling together and drafting principles to be approved in the final plenary session. In July 2003, a book of case studies of the existing transboundary sites was published, based on field studies carried out by two consultants in 2000–2001 (Fall *et al.* 2003).

However despite this apparent increase of interest, the enthusiasm was somewhat short lived, at least on a global scale. The *ad hoc* Task Force never met again and no more global meetings on the topic were scheduled, despite an increase of projects on the ground, in Europe, Asia and Africa. Budget and time restrictions were invoked; other more pressing agendas took over. Attempts to link up UNESCO's work on transboundary biosphere reserves with other global initiatives such as that run by the IUCN Task Force on Transboundary Protected Areas failed. The MAB Secretariat, fearful of being absorbed by and confused with other initiatives, particularly IUCN, chose not to send formal delegates to international meetings, such as that held in La Maddalena, Sardinia, in 2004. Varying institutional logics, competing mandates and the need to carve out separate turfs within and between international organisations maintained the split between the biosphere reserve programme and other initiatives. With a Secretariat run by a handful of people, of whom no more than four or five senior staff, individual choices and priorities and responding to immediate deadlines, meant that transboundary issues no longer held centre stage.

The following year, EUROPARC, a federation of protected areas within Europe, organised a meeting in Zakopane, Poland, on 15–19 September 1999 on 'Transcending Borders – Parks for Europe'. UNESCO joined the club in 2000, creating a specific *ad hoc* Transboundary Biosphere Reserves Task Force that met during the EuroMaB meeting from 10–14 April 2000 in Cambridge, England. The same year, the IUCN group met again in Gland, Switzerland for a further meeting on 'Promoting a Global Partnership', and discussing the publication that eventually came out as a Best Practice Protected Area Guidelines manual in 2001, published jointly by IUCN and Cardiff University.

The question of funding was a course crucial and I have already hinted at the competitive nature of international conservation. Within this context of a fashion in transboundary projects, having a transboundary label – whether as a transboundary biosphere reserve, a European Diploma, a Eurosite label, a transboundary World Heritage Site or a twinning agreement – was seen as a definite advantage for securing a competitive edge over other sites, attracting more funding for projects. The Danube Delta in Romania, for instance, was a transboundary Biosphere Reserve, a Natural World Heritage Site, a Ramsar site, a member of a network of protected deltas ... Thus while such designations did not in themselves bring in money, they were seen to assist in securing it from other donor agencies.

Arguments for transboundary protected areas

From the start, the emphasis was that 'protected areas that share common borders share common problems' (Hamilton *et al.* 1996), and therefore that 'areas of natural or cultural significance shared by two or more countries or other resource-owning jurisdictions lend themselves to transborder protected areas' (Hamilton *et al.* 1996). At the time, cultural factors were seen as important, something that was subsequently dropped and given less importance in the literature, reflecting an emphasis on nature conservation. The fact that the first international meeting was held in Australia, where boundaries between federal states were an issue, no doubt contributed to the inclusion of 'other resource-owning jurisdictions' being added to the definition. In later international meetings, debates over whether 'transboundary' should only be applied to sovereign states became a leitmotiv, receiving different answers in different contexts.

In the meeting near Cape Town in 1997, the emphasis was on cooperation as a means to promote peace and regional stability, offering a comprehensive and irresistible package: 'a major contribution can be made to international cooperation, regional peace and stability by the creation of transfrontier conservation areas which promote biodiversity conservation, sustainable development and management of natural and cultural resources, noting that such areas can encompass the full range of IUCN protected area management categories' (IUCN 1998 : 15). The Declaration of Principles also stressed that such areas can be 'managed cooperatively, across international land or sea boundaries without compromising national sovereignty' (IUCN 1998 : 15), although it was not explained exactly how this should be done. As well as the emphasis on local community participation and the promotion of tourism – two unavoidable fashions at the time – the biophysical arguments for the creation of such areas were subsequently mentioned: 'such areas also have a vital part to play in the conservation of biodiversity, in particular by enabling natural systems to be managed as functional ecosystem units, for species conservation and ecologically sustainable development through bio-regional planning' (IUCN 1998 : 15). However, despite endorsing bioregional planning, there was no mention of how transboundary areas were to be defined on the ground. It was supposedly assumed that these would simply be created on the basis of existing protected areas that could be linked up.

Four main arguments were usually presented in this literature to argue for the necessity of transboundary protected areas, three ecological and one political. The first issue related to scale, since linking protected areas across political boundaries did help 'to achieve protected status for an area of sufficient size to insure its ecological integrity (e.g. to maintain an adequately diverse and sufficiently large gene pool, or to encompass the range necessary for large mammals)' (Westing 1998b; see also Hamilton *et al.* 1996; Fall 1999). The second issue, in direct consequence of the first, was the increasing importance of the network approach used in nature conservation, physically and administratively linking protected areas across the landscape. As such, a protected area spanning a boundary was seen to effectively create a key node in any continent-wide network. The third was that ecologically valuable sites that most repay conserving were seen to straddle the 220 000 km of land boundaries (Westing 1998a), for various historical and topographic reasons. The fourth argument concerned the opportunity of using the field of environmental management for fostering good neighbourly relations, cementing and reinforcing confidence between states through the joint management of protected areas (Hamilton 1996; Brunner 1998) (The construction of transboundary spaces is further examined in Chapter 6).

Critical voices

It took several years of overwhelming enthusiasm before critical voices began to be heard, questioning the near-universal endorsement of transboundary protected areas and the assumptions about cooperation that prevailed. It was fair to say that these critiques were initially mainly restricted to academic debates, slowly stretching into policy and becoming more mainstream, following the pattern set by other topics such as critiques of species conservation or community participation.

The first critical voices that emerged to feed the debate were not specifically aimed at transboundary planning but rather discussed some of the dubious assumptions prevalent within the literature on bioregionalism (Chapter 2). A revision of the assumption of 'naturalness' of spatially bounded entities inevitably flowed into the debate on transboundary protected areas. Specific critiques focussed on the assumption that the world could be divided into distinctive and discrete ecosystems and the corresponding argument that most contemporary territorial boundaries were artificial and therefore in discord with nature's ways (Olsen 2001 : 73). The role ascribed to state boundaries was interesting, since bioregionalists – and proponents of transboundary planning – did not identify the bioregions with existing states and nations. The argument was that 'inasmuch as state boundaries rarely reflect ecological considerations, the present world of nation-states is anti-ecological' (Olsen 2001 : 80). This was an important point, for certain critiques have drawn attention to some of the assumptions of bioregionalism, drawing disturbing parallels with right wing ecology, although assumptions about state boundaries are noticeably different.

Building on this idea, Wolmer stated that 'bioregionalism has been criticised for its reductionist understanding of natural regions and undifferentiated human societies, its frequently ahistorical analysis, the environmental determinism of its simplistic nature-culture causal linkage, and its romanticized representation of "traditional" indigenous cultures living in harmony with the environment. Yet elements of this bioregionalist philosophy have entered mainstream conservation thinking – particularly in the United States – from where they have been exported to underpin much thinking about TBNRM [Transboundary natural resource management] globally' (Wolmer 2003 : 264). However, he did qualify this by saying that the focus was not so much on the 'utopian and slightly New Age rhetoric' (Wolmer 2003 : 264) but rather on the more explicitly scientific and managerial discourse deriving mainly from the field of conservation biology. Furthermore, others noted the unequal distribution of power and the use of representation to argue that bioregional models are typically based upon a series of visualization methods that simultaneously establish the framework for environmental interventions while distancing 'bioplanners' from the effects of those interventions (American Anthropological Association 2001).

When it came to considerations of societal factors, the focus on a 'natural' scale of planning became more tenuous. For while proponents of transboundary planning drew on biophysical arguments, they also identified with the idea that each human 'culture' and the individuals who made it up, were the resulting expressions of a particular environment that defined the scale of planning (Olsen 2001 : 75). Olsen noted that such thinking built on the assumption that environmental protection encompassed both the protection of the human world – or in other words, the protection of human cultures – as well as the non-human world. He argued that this distinction ultimately led to a 'natural', determined scale of planning linked to a form of neo-racism 'whose central theme is no longer centered on biological characteristics, but rather the intractability of cultural difference' (Balibar 1990 in Olsen 2001 : 75). Thus arguments for appropriate scales of planning, while initially appealing, could easily slip into the protection of human diversity by protecting every culture 'distinct and unique as it is, from the non-native, the foreign' (Olsen 2001 : 76). Olsen noted that the far right did not have the exclusivity of such thinking which also found expression among the political left.

The second form of criticism currently emerging focuses more specifically on transboundary protected areas *per se*. Authors directly challenge the re-establishment of ecological integrity across 'artificial' frontiers and administrative boundaries (Wolmer 2003), noting that this concept has impacts far beyond the realms of biodiversity protection and 'natural resource management'. One article specifically centres on the link between regional debates on national sovereignty, land reform and poverty alleviation in Southern Africa. Wolmer particularly raises the question of the ideological, political and economic rationales behind the flowering of transboundary projects in Southern Africa and in particular in the Great Limpopo Transfrontier Park – a Kiplingesque name whose neo-imperialist significance is not lost on him. In this article, Wolmer coherently questions the 'unlikely grouping of concepts and philosophies including radical environmentalism, conservation biology, and neoliberal economic agendas, as well as donor and NGO funding prerogatives' (Wolmer 2003 : 261).

Focussing on the example of Southern Africa, he further notes that 'it appears very unlikely that southern African governments will be willing to cede any power or territory to the ethnic groups spanning their borders and (. . .) TBNRM may be as likely to cause inter-state disputes as to assuage them' (Wolmer 2003 : 266). In additional to many of the claims made for transboundary areas as 'peace parks', Wolmer more specifically attacks some of the economic assumptions behind these projects: 'the discourses of radical environmentalist bioregionalism and technical conservation biology somewhat surprisingly articulate with an explicitly neoliberal free trade agenda' (Wolmer 2003 : 267). He concludes by saying that these projects are no more than a form of green imperialism, 'driven by Cecil Rhodes clones – rather than seeing great expanses of red on the map they want to see a great wedge of green as their legacy to Africa' (International Conservation NGO staff member, quoted in Wolmer 2003 : 268).

Furthermore, Wolmer notes that in parallel to the promotion of tourism in these transboundary sites, forms of 'ethnotourism' are encouraged, in which 'indigenous communities – rather than being viewed as anathema to a "wilderness experience" and being removed from protected areas, are being reconceptualised as a useful "cultural" adjunct to wildlife as long as they are visually pleasing' (Wolmer 2003 : 274). While this appears particularly shocking in Africa, a similar comment could be made about sites in Central Europe, notably in the Polish Tatry mountains where populations expulsed in the Nineteen Fifties were 'brought back' in the Nineteen Nineties to herd sheep within the National Park/Biosphere Reserve on the condition that they wear traditional costume at all times. During a field visit with a park ranger in Spring 2000, we came upon a shepherd wearing jeans. He was duly chastised, and told to go back into his hut and change. He instantly did so.

Conclusions

The discussion above has charted the idea of expanding protected areas across boundaries, discussing how such spatial entities have been defined. The emergence of the concept within the worldwide protected area movement has been discussed, from the first ideas in the 1930s through to the establishment of an 'industry', with a core group of international experts, heralding the arrival of the 'Next Big Thing'. The continuing uncertainty about how to define transboundary protected areas reflects the difficulties in grasping what these imply both spatially and institutionally. Despite the appearance of critical voices, publications describing transboundary protected areas in largely unproblematic ways continue to appear, despite a nascent internal critique. Examples of uncritical reproduction of transboundary spaces as unproblematic entities include articles that reproduce lists of examples, thus taking their existence on the ground for granted rather than questioning what such a 'transboundary' spatial entity implies. Kliot, for instance (Kliot 2002 : 407), uses Zbicz and Green's (1999) list of transboundary protected areas, interspersed with a bit of theory on more or less 'integrated borderlands'. Perhaps slighting such publications is unfair, yet such unproblematic publications further the myth that transboundary protected areas are unproblematic spatial constructs that exist because they can be listed. This particular example illustrates the creation of a self-quoting clique, with key references systematically appearing unchallenged, quoted in a closed circle, with little or no fieldwork informing the discussion. It also leads to dramatically naïve pronouncements such as: 'many of the protected areas in Europe have benefited from geography' (Kliot 2002 : 432). What on earth is that supposed to mean?

Choosing an approach

Transboundary protected areas are part of a global phenomenon promoted and established around the world. One possible method for grasping this is carrying out a global survey at a distance, as Zbicz did in her thesis on transboundary protected areas, never actually setting foot in a protected area (Zbicz 1999). Another frequent approach is to carry out one in-depth case study and generalise from that particular standpoint. Both methods constitute extreme examples of the diverse choices facing researchers. At the start of this work, separate individuals each kindly recommended one of these approaches. Yet neither seemed appropriate. If the objective of the work was to explore why transboundary cooperation was so difficult, surely different scenarios needed confronting? Yet carrying out a variety of case studies, each in different places and contexts, could prove both logistically complex and methodologically fraught.

Having decided that a series of case studies would be most appropriate, I settled on carrying out five, with additional emphasis on two of them. This was based on the assumption that separate sites were involved in comparable socio-spatial processes. They could complement each other and shed light on elements that could be generalised. The final choice of sites was linked to a consultancy project for UNESCO; it made practical and financial sense yet was sufficiently ambitious to be interesting: it involved carrying out case studies in five transboundary areas spanning seven countries and eight different languages. It also limited me to Europe, stretching far from West to East, where the existing transboundary biosphere reserves were at the time. This was the continent I knew most about, narrowing down the amount of background information I would need.

As it turned out, the fieldwork was carried out in two rounds, stretching from 1998 to 2002. During this study, I carried out interviews with protected area managers, national MAB Committee members, employees of relevant ministries and non-governmental organisations, as well as taking part in the life of the administrations by sitting in on meetings and field trips. During the course of the interviews, it appeared that while many individuals were involved in transboundary cooperation, protected area managers were key players in the process. These were the people actually confronted with 'cooperation', dealing with the day-to-day difficulties and challenges of applying abstract ideals to their daily workloads. I decided to focus on this group: individuals employed by protected area administrations, including directors, senior managers, project managers, researchers as well as more field-oriented rangers and foresters. Because of the nature of the topic, most of the more formal interviews were carried out with more senior managers although time spent in the field meant that informal discussions involved a wider variety of employees. The interviews are listed in Appendix II.

The first round of fieldwork 1998–2001

Initially, a series of short case studies was carried out in four sites as part of a wider study of transboundary biosphere reserves, partly undertaken for UNESCO. These visits stretched from 1998 to 2001. The work consisted of a series of interviews of people involved in transboundary cooperation in various capacities. During the course of this fieldwork, 43 people were interviewed (Appendix IV), with additional interviews of three UNESCO staff members in Paris.

- East Carpathians Biosphere Reserve (Poland/Slovakia/Ukraine); July–August 1998 and May–June 2000
- Tatry/Tatra Biosphere Reserve (Poland/Slovakia); May–June 2000
- Pfälzerwald/Vosges du Nord Biosphere Reserve (Germany/France); September 2000
- Danube Delta Biosphere Reserve (Romania/Ukraine); May 2001.

The first round of fieldwork produced a wide panorama of experiences and situations, leading to an increased awareness of the issues and challenges of transboundary planning. It was also at times depressing, as the full challenge of transboundary cooperation appeared starkly in contrast to the glossy published reports appearing on an international level. However, while the wide variety of situations shed light on many points, it remained poor on others. Such a wide field experience also led to a certain number of theoretical rethinks that could not be answered with the information at hand. Therefore, in order to further develop some of these themes, I planned a second round of fieldwork, focussing more specifically on two sites.

The second round of fieldwork 2002

The two sites were chosen to reflect two distinct situations, including one transboundary site not currently applying the biosphere reserve model, but examining the possibility of doing so in the future:

- Pfälzerwald/Vosges du Nord Biosphere Reserve (Germany/France), May–June 2002
- Parc National du Mercantour/Parco Nazionale delle Alpi Marittimi (France/Italy); March–April 2002.

Both sites presented the advantage of being relatively easy to get to, keeping travel costs down. More to the point, I spoke the languages: French, German and Italian. This linguistic choice was the single most important factor in the choice of sites, since the research method called for extensive qualitative interviewing, a difficult exercise to carry out successfully when using interpreters.

Interviews are narratives, constructed in situ, a product of a talk between interview participants: knowledge is the result of the action taken to obtain it. Keats states that 'one of the more interesting, yet demanding, aspects of interviewing is the relationship between the interviewer and the respondent. It is a dynamic relationship which develops as the interview proceeds' (Keats 2000 : 21). How the researcher is perceived and what she is perceived to be working on and what she will use it for, influence responses obtained. However, while social scientists acknowledge that this must be taken into account, the actual practical implications are unclear. As Rose warns, this is not straightforward: 'I found this an extraordinarily difficult thing to do. Indeed, I think I found it impossible' (Rose 1997 : 305). Likewise, when asked to state explicitly the effect of positionality on results obtained, Gibson-Graham said 'Stuffed if I know!' (Gibson-Graham 1994 in Rose 1997 : 314). Thus the call for positionality and reflexivity assumes an alarming reflective and analytical power on behalf of the researcher who can somehow objectively edit out barriers to reveal a more objective truth. This was particularly difficult in this study, where different languages, modes of interviewing and quality of information varied widely. Furthermore, undertaking some of the fieldwork while being strongly identified with UNESCO was also fraught with difficulties. Further reflections on the methodological implications can be found in Fall (2003).

Notes

1 Elements from this chapter first appeared in Fall, J.J. Divide and rule : constructing human boundaries in 'boundless nature', *Geojournal*, (58), pp. 243–251, Kluwer Academic Publishers. Reproduced with kind permission of Springer Science and Business Media.
2 More details of the birth of the Biosphere Reserve programme in relation to protected area boundaries can be found in Fall, J.J. (2003) 'Drawing the Line : Boundaries, Identity and Cooperation in 'Transboundary' Protected Areas', PhD thesis, Département de géographie, Université de Genève, pp. 87–100.
3 All quotations from Michel Batisse, unless stated otherwise, are taken from an interview carried out on the 23rd February 2000, in his office in Paris. The quotations are freely translated by the author from the original French.
4 The terms used by UNESCO are all rather unfortunate by current standards, leading some to comment that 'biosphere reserves have perhaps suffered from a rather uncertain image – compounded no doubt by the user-unfriendly title' (Philips, 1998 : vii), and while 'reserve' still conjures up images of ostracized Indians, Man and the Biosphere hardly gains points for inclusiveness.
5 For more on protected areas and the design of boundaries, see Fall J.J. 2004, 'Divide and rule: constructing human boundaries in "boundless nature" ', *GeoJournal*, (58) 243–251.
6 This led to the designation of certain BRs not formally recognised by UNESCO and thus not part of the World Network, notably in India and Mexico.

Chapter 4

Science, Politics and Legitimacy in the Design of Protected Areas

(Re)drawing the line: boundaries as contested acts of power

Drawing a line is an act of power. Because boundaries participate in spatial socialization (Paasi 1996), structuring all socio-political and socio-economic projects by modifying territoriality (Raffestin 1980 : 153), their (re)definition inevitably heralds (re)territorialisation. In this chapter, the specific definition of protected area boundaries is analysed by drawing attention to the discourses informing boundary construction and maintenance. Various processes lead to the designation of protected areas following either biophysical criteria such as levels of high endemism or biodiversity, or societal criteria such as low population density, political opportunism or aesthetic value. Rather than being a benign – and 'rational' – act of defining an area for conserving charismatic fauna or flora, the process is analysed as a contest between different discourses. This redefinition of boundaries is set within an inevitable web of power relations that is renegotiated or forcefully modified with all boundary changes.

The establishment of protected areas is often presented as being framed by four questions: the aims of the protected area (why?), its nature and chosen spatial model (what?), its location (where?) and the actual planning or designing process (how?). In contrast to what appears to be a rational process, in fact 'whether in terms of process or the end result, park design is a puzzle! A puzzle because neither is predictable (. . .) the design of a park remains a speculative attempt at putting its pieces together' (Oneka 1996 : 2). Within the following sections, the arguments underpinning the definition of boundaries to and within these areas are discussed, drawing on interviews in the five case study areas. I argue that these cases illustrate the role boundaries play in territoriality, shedding light on the contested nature of boundaries and the resulting resistance to them.

This builds on the review of protected areas and boundaries that suggested two initial conclusions: firstly, that the idea of 'natural' boundaries is returning, secondly that the idea of a societal definition of boundaries linked to human spatial practices also has increasing support. In other words, protected areas managers have to deal with two radically different conceptions of boundaries. This assimilation of biophysical and societal boundaries has attained the status of a sacred, founding myth and is heralded as the main objective of successful planning. This is similar to what Massey has called 'the assumption of (. . .) an isomorphism between space and society' (Massey 2001 : 10) which 'was not simply "wrong" in the sense of not true, nor was it, in spite of nationalisms and exclusivist parochialisms for which it has performed such an essential legitimating function, always and only politically "reactionary". But it certainly did have, and has, powerful regulatory and political functions. It has been both an outcome of, and a support to, particular forms of power and politics' (Massey 2001 : 11). I argue that the conceptually doubtful and ontologically impossible reconciliation of boundaries of different natures – typical incarnations of *fiat* and *bona fide* objects – leads down politically suspect paths. The resulting tension between these two discourses is inherent in the resulting spatial entities and creates lasting management difficulties as individual actors promote particular conceptions of boundaries in line with their professional training. Furthermore, such a notion echoes the one-time popularity attained by the idea of 'natural boundaries', fostering politically conservative motivations behind seemingly benign or laudable objectives.

The role of 'science' in defining boundaries

Natural scientists, including biologists and ecologists, have traditionally been called on to develop techniques for identifying the most advantageous way of planning a protected area, especially regarding location, size and design. During the fieldwork, thirty of the people interviewed were involved in one way or another in drawing lines and establishing zonations, either as scientific advisors, final decision-makers or both. Of these thirty, twenty-four (80%) came from a natural science background. This reflected the importance given to biophysical criteria in the decision-making process. Although it would be naïve to suggest natural scientists were immune to other more political or social arguments, these were often initially set aside. Political complications were more likely to be mentioned when precise questions were asked, or when practical difficulties were evoked. Other elements such as administrative or jurisdictional boundaries were in practice taken into account in deciding 'where to draw the line'. Nevertheless, the process as a whole was dominated by the natural sciences, reflecting the prevalence of such arguments in the protected area literature. The spatial discourses clearly postulated a pre-existing biophysical spatial coherence, an ideal spatial scenario.

The construction of scientific knowledge has been explicitly studied as a process of construction, no longer solely as a 'discovery' of pre-existing facts. Latour, for example, described following pedologists taking soil samples in the Amazon forest as they slowly built up a map. The construction of a soil diagram was not unlike the process of establishing a zonation. 'Is the diagram a construction, a discovery, an invention, or a convention? All four, as always. The diagram is constructed by the labours of five people and by passing through successive geometrical constructions. We are well aware that we have *invented* it and that, without us and the pedologists, it would never have appeared. Still, it *discovers* a form that was until now hidden but that we retrospectively feel was already there beneath the visible features of the soil. At the same time, we know that without the *conventional* coding of judgements, forms, tags, and words, all we could see in this diagram drawn from the earth would be formless scribbles' (Latour 1999 : 67) (Emphasis in original). He noted that the end diagram was not in itself realistic, rather it did more than resemble 'reality' since it took the place of the original situation, truthful 'only on condition that it allows for *passage* between what precedes and what follows it' (Latour 1999 : 67) (Emphasis in original).

Latour's critical approach to the construction of scientific knowledge sheds light on processes involving a variety of actors and forms of knowledge. The parallel with protected area zonation is easily made. However, unlike scientists whose only (declared) aim is to explain phenomena, managers are involved in a profoundly political process. Their choices and recommendations have practical, political consequences. Yet in order to maintain the privileged position and increased legitimacy guaranteed by this access to rationalised knowledge, the political dimension of zonation is negated. Due to the limited size of protected area administrations, one individual may be involved in the two stages of the process: the initial zonation (taken to be value-free) and the subsequent intrinsically-political implementation. The initial declaration of scientific objectivity thus directly serves to legitimise subsequent political choices, de-legitimising resistance by other actors.

Protected area design

The assumption throughout the literature on protected areas is that they are one of the best, if not the best, tool for conserving biodiversity. Peck, for instance, notes that protected area design 'is one of the primary ways in which planners influence biodiversity preservation' (Peck 1998 : 89; see also Bridgewater and Cresswell, 1996 : 3). There is a rich literature on techniques for the most efficient location of protected areas. Given notes that protected areas 'must take into account the patchy distribution of both species and the communities they form. Biological diversity is not evenly distributed but tends to be concentrated into areas of local endemism, the hot spots of biodiversity' (Given 1994 : 87). In order to address this reality, he notes that a variety of techniques have been developed for definition (he uses the term 'delineation') of protected areas. As examples, Given cites broad-scale ecological surveys, pattern analysis to define species assemblages, computer modelling to interpolate geographic patterns of species, as well as field sampling (Given 1994 : 88, see also Primack 1993 : 310–325). These techniques combine concepts developed within landscape ecology, ecology and conservation biology, using terms such as patch, matrix and corridor to describe spatial patterns and configurations. The arguments developed are almost exclusively biophysical, addressing issues such as ecological scale, ecosystem dynamics and protected area design.

The actual design of each protected area is naturally seen to depend on its purpose, be it the protection of a particular community, species or genetically distinct population (Given 1994 : 85). The scenarios and considerations refer to ideal approaches to planning, albeit accepting that the application of theory to real situations is always constrained by the practicalities of what can be achieved. There are in effect two main arguments:

The first argument suggests locating a protected area in order to maximise the representation of biodiversity within all existing protected areas. In other words, if grassland or alpine ecosystems are already protected in one site, then an additional one will be located in an area of woodland or mudflats. The actual scale at which such representativity is to be achieved is of course open to question and representativeness is often given as a criterion for protected areas at local, regional, national, and international levels. The concept of representativeness has been applied, for example, as a fundamental criterion for choosing biosphere reserves throughout the world.

The second argument takes a different line and argues for the establishment of protected areas in locations of high ecological value. Species richness, total biodiversity or unique endemic species are thus some of the elements taken into account. This approach has led to sophisticated methods and techniques for estimating the total biodiversity in a given area, since counting every organism is often not an option. Such techniques assist in identifying and subsequently defining the most appropriate or valuable sites to protect, for example based on a review of the number of endemic species, calculating the location of maximum endemism (Terborgh and Winter 1983 : 45) or biodiversity hotspots (Myers 1986).

Once a site has been chosen as the likely location of a protected area, other issues appear linked to design and in particular to boundary definition. Oneka makes a clear distinction between planning and designing a protected area in stating that 'whereas both planning and designing deal with the structuring of functional and aesthetic relationships amongst components of a system (. . .) they are distinguished from each other in the extent of their details. If the ideas expressed exceed basic concepts and tend towards operational details then "design" is suggested. In contrast, if the iven information is strategic rather than operational then "plan" is suggested. Planning and designing are two inseparable processes, the design defining the operational structure for the strategic and tactical plans' (Oneka 1996 : 2).

The design – the shape and size or spatial configuration – of a protected area has been specifically subjected to analysis and arguments. The model referred to most often with regard to spatial configuration is derived from MacArthur and Wilson's seminal theory of island biogeography (MacArthur 1972), the most well-known 'science-led' approach to design. Their research based on species distribution on oceanic islands led to theories of population dynamics in terrestrial ecosystems, linking area and species richness. MacArthur and Wilson's model is based on a dynamic balance between colonisation and extinction rates on oceanic islands. Since natural resources are limited, the more species already present on a given island, the harder it is for new species to become established, since when competition increases, the extinction rate does also.

The metaphor therefore suggests that isolated habitats, such as mountain tops, lakes, forest fragments and nature reserves, might also be viewed as 'islands' surrounded by a 'sea' of unfavourable habitat: an 'area where species can exist, surrounded by an area in which the species can survive poorly or not at all and which consequently represents a distributional barrier' (Diamond,

1975 : 129). This theory has stimulated a large body of research into the consequences of habitat fragmentation and isolation of species (Diamond 1975 : 135; Pickett 1978 : 28, see also Shafer 1990 : 12–13). The debate branching off from it has been termed the SLOSS debate, or Single Large Or Several Small, leading to a certain number of geometric principles (for a clear review of the debate, see Shafer 1990 : 79–80), although the assumption that big is best has been increasingly questioned (Zuidema, Sayer & Dijkman 1996 in Wolmer 2003). It is not surprising then that theory of island biogeography has somewhat lost favour among conservation biologists, although more for technical than for fundamental reasons (see for e.g. Peck 1998 : 92). The island metaphor has furthermore been critiqued for promoting a dichotomous approach where there are areas that are considered 'natural' and others that are not (Ingram 1999 : 5), reinforcing the perceived 'hostility' of humanised land surrounding protected areas.

Constructing bounded entities using biophysical arguments

Drawing on theories such as 'island biogeography', the literature on protected area design deals explicitly or implicitly with the definition of boundaries since 'the precise location of a boundary can affect a range of ecological processes, including species movement and hydrology' (Peck 1998 : 89). The discourses founding these techniques and rationales follow arguments related to the definition of 'natural' boundaries developed earlier (Chapter 2). Lucas, for instance, suggests that 'it is desirable (...) to have boundaries which are sensible, logical and identifiable in terms of nature, people and management' (Lucas 1992 : 43). His use of the terms 'sensible' and 'logical' is interesting, since it reflects the rationalisation of the process. In the same line, Given suggests that 'the shape of these areas should be compact, and the boundaries should be biologically meaningful (e.g. including whole watersheds, ecotones, and buffer zones)' (Given 1994 : 87; see also Balmford 1998 : 23). Lucas further notes that 'because administrative boundaries may not always follow logical geographic or physical boundaries, it may not always be practical to achieve the ideal' (Lucas 1992 : 43). Thus, for him as for many others, there is an ideal, optimal scenario dictated by nature that must be identified by planners.

Lucas' 'guidelines for boundaries' includes three principles directly relating to such a vision: 'Boundaries should encompass complete landscape units and ecosystems; (...) Boundaries should be readily identifiable and should desirably follow natural physical features as these are unlikely to change and are more easily understood on the ground than straight line boundaries drawn on maps (...); Boundaries must be recorded clearly in writing and on maps including both a topographical map showing physical features and a cadastral map showing land title boundaries (...)' (Lucas 1992 : 44). It is not clearly argued why the choice of 'natural' boundaries is necessary the most opportune choice, as – on a purely biophysical level – the reverse might well be true in some cases. Discussing plant conservation, Given notes that 'many rare species occupy disturbed edge habitats and rely on a shifting mosaic of disturbance for their continued existence. In such instances, preservation and management of an edge has to be an integral part of the protected area concept' (Given 1994 : 89). However, on a conceptual level, this is less revealing than the unsaid suggestion that 'nature knows best' and inherently contains a political programme. This programme is seen to find coherent expression in a series of spatial scenarios and entities, with 'buffer zones' and 'networks' as the recognised favourites.

Buffer zones

A buffer zone implies creating an additional boundary, either around the original core of a protected area or within the wider unit in order to maintain specific areas in a desired state while minimizing edge effects of stark boundaries. Buffer zones are a form of extended boundary, in effect a boundary zone, between the inside and the outside, between the wilderness that needs protection and the humanised landscape beyond. Given, for instance, identifies two forms of buffer zones: *extension buffering* which 'extends habitats that are in the protected area into the buffer zone, allowing much larger breeding populations to survive than may be possible in the protected area alone' (Given 1994 : 89) and *socio-buffering* which 'manages a buffer zone to concentrate activities such as raising crops or harvesting. Such use should be compatible with the objective of the core itself' (Given 1994 : 90). Other classifications for types of buffer zones have been suggested, notably identifying the nature and permeability of the boundary-zone between the protected core and the surrounding land. The *negative moat* concept, for instance, suggests that the primary purpose of a buffer zone is to defend natural areas against human onslaught. The *isolation* concept sees buffer zones encircling the protected area and isolating it from the surrounding human communities, even if some are allowed to live within it. Finally, the *protectionist management* approaches are aimed at the protected area only in isolation from the lands and the communities surrounding or in the vicinity of it (IUCN 1986, quoted in Given 1994 : 92) (the issue of defining insiders and outsiders is further discussed in Chapter 8). In all these terms, the explicit use of words with strong military and gendered overtones should not be ignored, as briefly mentioned earlier (Chapter 3). Passive, (feminine) nature is managed by the imposition of (implicitly male) reason. Yet paradoxically, at the same time, nature is framed as containing a plan for its own management: a political project that is intrinsically rational, based around 'natural boundaries'.[1]

Networks

Taking the idea of buffer zone one step further leads conservationists to promote the idea of network, in which the land is no longer only divided into protected area plots isolated from the surrounding matrix. Rather, the whole landscape is connected by a series of geometric shapes involving more complex boundaries defining corridors, core areas and a variety of buffer zones. This follows the growing view among conservation biologists 'that a reserve-based approach, on its own, will not be adequate to ensure long-term conservation requirements (. . .). Substantially increasing the number and extent of reserves is an important step, but even this (if it can be achieved) will not be sufficient in many regions' (Bennett 1999 : 157; see also Given 1994). Instead, the emphasis for conservation is seen to be on the inter-relationships of sites and no longer on space and general characteristics, such as area.

Nevertheless, some authors have noted that it might still be necessary to conserve at least some very large areas with relatively limited human access since managing a network of postage-stamp sized reserves is both expensive and difficult (Balmford 1998 : 22). This is seen to ensure that some areas contain whole metapopulations or an entire range of elevational migrants, maintaining large enough populations that retain their evolutionary potential, conserving areas large enough to contain complete landscape-wide processes such as fire and flooding unchecked within area of protected area (Balmford 1998 : 23). There is widespread recognition that it is necessary in many regions to extend the protected-area approach and find ways of enhancing biological diversity conservation through management of the whole landscape system. Protected areas are therefore no longer seen to be necessarily isolated from influences from a neutral surrounding matrix, but rather are 'subject to a host of pressures from their surrounding environments' (Bennett 1999 : 162). This means that the underlying principle of an integrated landscape approach to planning conservation extends beyond the boundaries of protected areas.

When habitats are highly fragmented, conservation efforts are seen to need to focus primarily on the matrix (Balmford 1998 : 25), since its condition affects the status of biological communities within the patches of protected areas. Other suggestions include moving to a broader scale by using networks of protected areas (Bridgewell and Cresswell, 1996 : 3; Glowka *et al.* 1994 : 39 in IUCN Information Paper May 1998; Bennett 1999), an approach endorsed by the Global Biodiversity Strategy. In the Convention on Biological Diversity, Article 8 implicitly recommends that signatory countries set up such a network when referring to establishing a 'system of protected areas' (UNTS 3069, 1992), viewing protected areas as part of the landscape matrix, rather than as 'islands'. This is seen to lead to the creation of a highly complex set of geometric entities, introducing the need to interact with the human populations living in and around the defined protected areas.

Confronting prescribed patterns to case studies

In order to understand precisely to what extent boundaries were defined and drawn in line with biophysical principles, I asked managers to describe the process, indicating in particular who was involved at which stage. Where appropriate, the same question was asked of the internal zonation and in particular the specific zonation carried out when establishing a biosphere reserve. While answers varied, initially, all did stress the scientific, rational nature of the process. Formal responsibility largely rested with natural scientists or foresters:

> *The zonation was done on the basis of a biodiversity assessment. This was done by specialists, scientists, according to a set of biological and physical criteria. It was scientific. This new concept was part of the new law adopted to manage the Danube Delta biosphere reserve*
>
> *(Gheorghe, Danube Delta, Romania).*

The process was seen as rational, with specialists revealing what was already there. It assumed pre-existing natural boundaries: *bona fide* boundaries revealed by a rationalised process. The belief in the objectivity of the scientific method granted tremendous power to the decision-makers. Because any resulting zonation was 'scientifically' established, resistance by other actors was often seen as irrational or untenable. As Latour said: 'Yes, scientists master the world, but only if the world comes to them in the form of two-dimensional, superposable, combinable inscriptions' (Latour 1999 : 29). A belief in the validity and legitimacy of such 'inscriptions', or mediated scenarios or representations of the world such as zonations, conferred substantial authority.

The power relationship between the expert and other actors was a key to understanding the discourse of many managers, since they considered themselves imbued with a non-negotiable authority rooted in rationality. Initial oral descriptions of the zonation process rarely mentioned other actors, or possible local unease about the zonation process. Such issues could only be touched upon in subsequent questions. Initially, the process was largely described as straightforward, with 'experts' deciding where lines should be drawn:

> *In their zonation [in Ukraine], there is a move towards having a larger biosphere reserve with more strict protection. Now it is a national park. A professor pushed for the creation of the Nadsianski Protected Landscape Area which is managed by the Forest Authorities*
>
> *(Andrzej, Bieszczady, Poland).*

> *The Forestry commission carried it [the zonation] out, it was a common project between the foresters, that is me the environmental manager, and a woman S.H. who has just had a baby and is on holiday now for maternity leave, otherwise she could have done it with me, I can imagine that, but at the moment I am doing it alone. But it is, I think, important to say that, not the foresters, not alone ... but rather foresters, because of conservation. We worked together and also ... all the ... towns and communities and councils and nature protection agencies were informed and had the opportunity to take part, so we made a proposal and put this forward and we went around and presented ourselves and people could respond to this, say whether they were favourable or whether changes were still necessary, and so, as far as it went, bearing it all in mind, this is now the result*[2]
>
> *(Alexander, Pfälzerwald, Germany).*

The second quotation mentions the possibility of local participation. However, here this was limited to contributing comments after the initial rationalised process had been carried out. Thus participation came after the scientific process had been completed, and was largely restricted to commenting on the coherence of the proposal. In order for the process to have scientific validity and subsequent legal status, it was considered necessary for 'experts' to carry it out from start to finish, with a limited amount of political input from local communities. 'Experts' did not always make decisions on blank slates. The zonations of biosphere reserves were all carried out on the basis of pre-existing protected areas. Yet, surprisingly, these different zonations were not always carried out in collaboration with the people working within the pre-existing protected areas. Furthermore, the resulting biosphere reserve zonation was not always compatible with a pre-existing scenario. Thus various combinations of boundaries constructed different spatial scenarios of the same area, set up for differing purposes and implicitly for different audiences.

In the case of the Slovak East Carpathians it did not matter to managers that the zonation of the national park and biosphere reserve appeared contradictory. Actual management interventions were carried out in any case by a different agency under the control of the Forestry Ministry. Management was supposed to be based on the national park plan, although this was not always the case. Thus the same area was covered by three conflicting scenarios produced by overlapping administrations. It was instructive to examine which zonation appeared on published material or on signposts, targeted to different audiences. Thus the work of 'experts' could take on almost ethereal characteristics: sometimes reduced to a purely intellectual exercise destined for a limited audience, removed from political processes almost to the point of absurdity.

Not all situations flirted with the surreal to quite such an extent, although getting conflicting scenarios to agree was rarely straightforward. In the Vosges du Nord, the process of negotiating a suitable fit between two conflicting zonations took over ten years:

> *The zonation of this part ... yes ... in the park Charter there was a plan that came with it that said more or less where the park was going to intervene, what the objectives were and so on. And when we applied to be a biosphere reserve, we did a second more specific zonation. With three zones, core, buffer and transition, and with criteria that were subsequently strongly revised because in fact initially the core areas were restricted to the nature reserves and the Bitche forest, the buffer zone was the perimeter that covered protected areas and natural areas, and then after that all the rest of the park was in the development zone*[3]
>
> *(Théo, Vosges du Nord, France).*

Coming from the director of the park, this sounded perfectly straightforward. However, the person actually in charge of the zonation said that behind this simile of rationality, there were both haphazard decisions and large doses of improvisation. The following extract highlights the subjective nature of decisions that were often made in haste, and with incomplete knowledge of both the situation on the ground and the desired result. This was interesting in that it demonstrated that even seemingly 'scientific' design procedures in practice relied on arbitrary individual judgements:

> *The first zonation was centred on the northern part of the territory of the park, which is here, and where, I'm not sure quite why, we had decided to put the core area. (...)*

Why? A zonation was needed, we were taken a bit short, I didn't know anything about the biosphere reserve programme. When I was told about the zonation in this programme, I thought a bit about it with the scientists who were the instigators of the scientific council and we said to ourselves: 'there is a high concentration of protected areas and sensitive zones and places of remarkable heritage interest here, so let's take this zone. We had other ones in other places but here there is a high concentration, a golden triangle. So that's what we took, and very quickly, when we set up the first version in 97/98, when we set out the table and looked at the method, we realised that it didn't really work[4]

(Hugo, Vosges du Nord, France).

Here, the biosphere reserve zonation was established on the basis of existing boundaries, both protected areas and jurisdictional entities. The decisions were made by a small *ad hoc* committee of scientists, led by the manager who first launched the idea of establishing a biosphere reserve. Yet while all these individuals upheld the scientific nature of their expertise, the actual decision-making process was fuzzier. This particular interview was quite unique in describing the details, complications, negotiations and arbitrariness of the process. Overall, managers were loath to question the pertinence of their zonation, and the idea that rationality and science prevailed throughout the decision-making process was very widespread.

Conclusions on biophysical arguments for boundary definition

In the literature reviewed above, it was assumed that entities should be defined essentially by biophysical criteria, contained within a bona fide boundary. In the literature and on the ground, however, because of the need to integrate local communities, other arguments were surfacing, moving beyond the strict division between areas that were defined and protected by boundaries and those that were not. The introduction of buffer zones was the first step in making boundaries more permeable, but in certain cases these were considered to be little more than extended fortress walls, buffering core zones from outside hostile influences. However, with the shift towards allowing human activities within buffer zones, these became more than just walls and turned into lived-in, cared-for territories. In practice, planners often had little choice in locating protected areas. As Peck realistically noted, 'there may be little open space remaining in the region. Public pressure to create the reserve may also be based on protecting certain landscapes or species habitats' (Peck 1998 : 93). The complexity of the questions and the need for a pragmatic approach meant that choices had to be made. As Given noted 'social and economic issues are particularly sensitive ones, especially in regions of dense human habitation. Final boundary placement for reserves may have to be a compromise among financial and political factors, and existing patterns of land-ownership and use' (Given 1994 : 85).

Frankel and Soulé have argued that the issue was not in fact design but subsequent management. They stated that 'we contend that the issue of reserve design, *per se*, is something of a red herring; design is important, but too much emphasis on design alone is highly myopic. The issue for the future is *maintenance*' (Frankel and Soulé 1981 : 125). They argued that the 'the main issue, as we see it, however, is the absolute necessity of careful and continuous scientific management. Unfortunately, the past preoccupations with physical design features inadvertently promote the false notion that proper size and distribution of reserves will, in itself, guarantee the success of conservation' (Frankel and Soulé 1981 : 125). They continued raising the crucial question of the feasibility in saying that 'politicians easily gain approval of the international community by drawing some lines on a map in a sparsely populated region, and proclaiming the establishment of a new national park. This costs virtually nothing, but without an equal commitment to reserve maintenance, it guarantees the preservation of virtually nothing. It is more difficult to direct the attention of politicians, planners and conservationists to the necessity of careful and continuous management, but it is essential' (Frankel and Soulé 1981 : 126). Thus having reviewed the literature on biophysical principles to protected area design and confronted it with situations emerging on the ground, a pragmatic truth emerged: optimal spatial configuration was useless if it was not operationalised and made to work within some sort of coherent institutional and managerial structure. Spatial ideals and scenarios were one thing, but experience on the ground suggested a much more *ad hoc* approach to design, despite the prevalent framing of the process within a scientific, rational discourse.

Constructing bounded entities using societal arguments

Boundary definition is an intrinsically political process. In contrast to the rich literature on the actual management of protected areas in collaboration with or directly by local people, the use of 'societal' or 'political' arguments in designing them is surprisingly little discussed. Somehow in the enthusiasm of finding management solutions to complex situations, the first step of coherently setting up protected areas according to societal concerns has been largely forgotten. One problem is that when dealing with human populations, a whole set of complex factors appear that are much harder to identify than those discussed in the section on biophysical arguments. Social, economic, spiritual, legal, historical and political arguments coexist with purely aesthetic ones, promoted by a variety of groups and individuals.

In the previous section, the natural scientist or the biodiversity expert proposed an ideal scenario based on 'rational' data. However, finding the right site to conserve was far from the whole issue. Even the enthusiasts of rationalised sampling methods pointed out that 'it is a relatively easy matter to identify a location on a map and decide that is where there should be a reserve, but because of conflicting cultural, political or economic considerations, it is certain to be far more difficult to translate the recommendation into concrete action' (Terborgh & Winter 1983 : 51). Similarly, Given noted that 'few protected areas occur in totally uninhabited regions and many are in regions where people have lived for many generations' (Given 1994 : 95).

Correspondingly, local people had long-established patterns of living, harvesting regimes and resource use which could be changed dramatically by the establishment of a protected area.

Participation in protected area design

The idea of involving local people and residents in the design of protected areas is a relatively recent idea, as noted previously in the brief history of the protected area movement. This refers to participation in the first steps of the process, leaving aside the vast issue of subsequent 'participatory' or 'co-management' on which there is a much more extensive literature base and corresponding debates. Paradigms change quickly in volatile international conservation and there is already a backlash against participatory approaches and a return of the 'protection paradigm' in protected area design and management (for a discussion of the trend see Wildhusen *et al.* 2002 : 17–40). Nevertheless, participatory paradigms still abound and remain the norm, at least on an official level, despite a very wide and differing understanding of what these imply. In many cases, the issue of local participation in the first steps of design is dealt with in very general recommendations. The IUCN Guidelines on National System Planning for Protected Areas, for example, state that 'the needs of local communities should be assessed and information arising from these consultations should be used in protected area planning and management' (Davey 1998 : 30). However, this still implies that the information collected is subsequently processed by an expert group, rather than relying on local usage to define boundaries from the start.

In a book on community-based conservation, one author notes that 'reference is seldom made to site-selection criteria as such. Instead, particular circumstances often lead to a project's development' (Seymour 1994 : 477), adding that biophysical and societal elements are inextricably linked in the process. She notes that sites are usually chosen along two countervailing tendencies: for their intrinsic value or for their potential as models for dispute resolution and community resource management. In a number of studied sites, 'despite their initial concern with biological conservation objectives, a systematic site-selection process based on biological criteria was not necessarily employed. In only a few instances were sites selected according to some national or international framework of conservation priorities. (...) The interest and advocacy of a particular individual with personal or professional ties to the area appears to have been the key factor differentiating project sites from other similar sites with equal or greater biodiversity value' (Seymour 1994 : 477).

The territory a group inhabits is a spatial entity to which it feels it belongs, towards which local people feel an attachment linked to the need for a sense of identity. However, whether this is always a clearly bounded entity of space, sufficiently well-defined and therefore useful to planners, is not always clear. In many cases, a sense-of-belonging is loosely attached to an area, without there necessarily being sharp contours, and clear-cut boundaries are unlikely to appear forthright (Entrikin 1994 : 113). Given notes a series of general principles that should be adhered to as far as possible, endorsing the idea that local populations should remain within designated protected areas and be part of the management process. These principles imply that boundaries should take account of the existence and livelihood of the residents. Within these, there is a strong undertone of insider/outsider, with the former being seen as a 'legitimate' partner in the process, implicitly 'more natural' than outsiders, and deserving 'protection' and 'non-contamination' as such:

- 'Resettlement should be avoided whenever possible, since an indigenous culture will remain intact only in its home territory, where productive capacity of the environment is intimately understood.
- The protected area should be sufficiently large to accommodate its dual function – a reserve for nature with lands for indigenous people. The creation of reserves of reduced area only serves as a symbolic end and begins a process of cultural devolution and ecological degradation if the indigenous people do not have access to the resources they require.
- Protected area planning must accommodate population increase and cultural change. It is unrealistic to expect a group to atrophy, or worse, to return to some traditional technology long ago discarded in favour of a more modern alternative.
- Park staff should be traditional residents. Threats to integrity chiefly originate from outside. This is sensible use of resources, but also helps retain the necessary goodwill of the people in the area' (Given 1994 : 96).

The biophysical arguments developed by Given concerning buffer zones were mentioned previously. However, he goes beyond simply discussing biophysical arguments and dwells a little further on societal aspects in what he calls *socio-buffering*, arguing that when such areas are defined, local practices and use must be taken into account. The implicit leitmotiv is that buffer zones mediate between 'wild' and 'humanised' expanses, facilitating the integration of protected areas into the wider landscape. Given argues that buffer zones should in part be defined by local people, since 'the rating system should take account of how buffer zones are viewed by local people. It is important that buffer zones be viewed positively and not as yet another imposition on people's rights' (Given 1994 : 92).

Given notes that 'there is considerable scope for protected areas and buffer zones to relate to local people in a very practical way (...) it is vitally important that buffer zones benefit more than just a small number of people; they must be perceived to be of value to the whole community' (Given 1994 : 90). Better buffer zones are meant to help foster a new conception of boundaries, since 'managers of larger protected areas can develop a siege mentality, feeling encroachment from all sides. This attitude often heightens feelings of "them and us" and of conflict between preservationist advocates of the protected area and land-users outside the boundary. It can perpetuate a romantic vision of parks, which, in itself, may actually threaten parks by reinforcing the siege attitude' (Given 1994 : 92).

Putting participation into practice

Like many trends in protected area planning, despite the principles laid out in the literature, the practical implications of 'local people's' or 'public' participation remains elusive and somewhat undefined on the ground. The crucial question is to identify who is ultimately taking decisions when defining boundaries. One additional central issue is identifying who is in a legitimate position to speak on behalf of a heterogeneous population.

Although various forms of public participation were present within the case study areas, no administration had successfully set up a mechanism for involving people in the initial zonation. What participation existed was limited to informal consultation of 'key actors' (usually elected persons, lobbies or non-governmental organisations) for the subsequent management of the area. Other more strident voices got themselves heard through demonstrations. These, for example, involved fishermen demonstrating in the Danube Delta or hunters organising rallies in the Alpi Marittime (see Chapter 5). Determining who was a legitimate partner in public participation was the first hurdle as competing insider/outsider discourses abounded. During the fieldwork, elected officials were most often mentioned as the most appropriate partners. Although non-governmental organisations played some role in subsequent management, they were rarely if ever involved in initial zonation. One of the difficulties identified by managers was getting people who supported or were indifferent to the actions of protected areas to get their voices heard.

> *In these cases here it is always the case that even those who are in favour and aren't opposed [to extending the park] are not as active and don't go to the meetings because all in all because they don't give a damn whether there is a park or not. Instead those who do go to the meetings are those who are against, and that is why the meetings went, I'm not saying badly, but, well …*[5]
>
> *(Alessia, Parco Naturale Alpi Marittime, Italy).*

In the absence of real extended participation, the role of the expert-scientist as sole decision-maker was completed by that of elected persons, seen as legitimate ambassadors for the wider community. This position was seen to be different from that of an impartial expert speaking on behalf of 'science'. However, the similarity was found in the notion that certain individuals were imbued with decision-making power beyond their position as individuals:

> *You see, we in Aisone have voted, or rather we have joined the communal council and have voted unanimously, we approved the enlargement of the Alpi Marittime park in our territory, although we already had a bit. We will add an extremely characteristic site, extremely appropriate, which is that of the archaeological caves, we did it because we accept that including such a particular habitat that belongs to the commune could represent an important safeguard'*[6]
>
> *(Local mayor, Alpi Marittime, Italy).*

In this case, the decision to extend the boundaries of the park to include a historical site was made in order to safeguard it. As the land already belonged to the commune, this only required the vote of the elected local councillors. As elected officials, they were considered (by themselves!) to be in the best position to make this decision. However, this was not enough to justify the decision to others. The quotation also indicated an attempt to legitimise the decision through science and rationality. It was interesting to note here the use of the term 'characteristic' (caratteristico) and 'qualified' (qualificato) in the description of the place. The inclusion was justified on its intrinsic societal value (cultural and historic). Without specifying on what basis this judgement was made, the coherence and rationality of the inclusion was put forward, legitimised by the decision-makers' position within the community. Individual opinions were replaced by a declaration of objectivity. As in the use of science as arbiter, this recourse to the status of the decision-maker compounded by objectivity or 'obviousness' occluded the political dimensions of the decision-making process.

This illustrated combinations between forms of arguments. In practice, 'expert' opinions were rarely fully removed from political processes. It was sometimes politically impossible, even for the most foolhardy manager, to follow a scenario that appeared most desirable from a biophysical point of view. The redefinition of boundaries along biophysical lines was therefore not always possible for societal reasons – two spatial discourses could not be made to concur. Rather than accomplish this by force and provoke fierce resistance, the strategy of managers was to attempt consciously to modify the territoriality of the resistant populations. The ultimate aim was for the situation not to be perceived as a constraint, with contested boundaries, but rather to be embraced as a desired spatial scenario. It was thus an attempt to influence the process of spatial socialisation, actively seeking to define the process of internalising a collective territorial identity:

> *For example, it is necessary to extend the national park to limit hunting and logging but there are many constraints linked to finance and the local population. There are also pressures between the State Forests and the national park: some forests are being transferred to the park, and the people managing the land will also be transferred. Within the national park, park employees manage the forests, but within the landscape protected areas, it is State Forests employees who do this work. But salaries within the national park are 60% lower than within the State Forests! So this means that if the park is extended, then this area should be part of the active protection zone, not under strict protection, because if not we will have problems with the local people. So there is the need for some sort of commercial success before we can extend the park to show that the national park can make use of its existing real estate, good public relations showing that the national park is linked to local development"*
>
> *(Andrzej, Bieszczady National Park, Poland).*

Although this appeared sound, such persuasion was far from easy on the ground. The manager was assuming a substantial increase in salaries, something that seemed unlikely under the existing circumstances. Yet this example was interesting for at least two further reasons. It represented an attempt to reconcile two different rationales: the park's and that of the local people. The need to extend the park was carried out by 'experts' and was taken for granted ('it is necessary'). The manager suggested a direct engagement with local opinion to align it to the need to redefine the boundaries. However, rather than put forward this proposal using biophysical arguments (such as 'save the forest') a strategy seen as destined to fail, he chose to directly engage the population using economic arguments: 'the park can make you money'. There was never any question of public participation in the zonation process. Public opinion – no doubt realistically – was assumed from the start to be hostile to the park's intention to expand. Rather than seek participation in the planning process and thus possible compromise, the only solution identified was to change the existing local attitude directly, through judicious economic persuasion. The role of expert knowledge was thus maintained and reinforced, despite an approach that initially appeared more politically aware.

Legitimating political decisions

In the Vosges du Nord, despite the prevalence of natural scientists, a subtle shift from purely biophysical arguments towards more societally-based spatial divisions was noticeable. The discourse on 'protection' was extended to cultural heritage (historical sites). The areas of high historical interest – mostly castles and ruins – were incorporated into the biosphere reserve core areas. Interestingly, this reflected a change in the way nature was considered, moving from a discourse based around 'wilderness' to considerations of culturally determined landscapes:

> *What is new is that we took of course the natural areas, and we put all the historical monuments, especially the castles that are in the middle of the forest. We assumed that all around, in any case, the zone as well as the castle were carriers of nature (imbued with nature), after all this is Europe, a continent strongly transformed by man [sic], where nature and culture are strongly bound together. (...) We added the traditional orchards, which are absolutely nature-culture sites, disappearing rapidly and for which we have a specific programme. We thought it was logical to have these as part of the buffer zone, I would say areas that are a bit natural that still have high heritage value but that, in any case, are managed by man [sic], on which we must specifically experiment alternative techniques (...). That's it for the zonation*[7]
>
> *(Hugo, Vosges du Nord, France).*

This highlighted the importance of anthropic elements in the landscape, reflecting a changing conception of nature and nature protection in line with international trends. This was unique among the case studies, and this conception was certainly not shared by the protected area administration across the boundary. The German managers were absolutely horrified to discover that the French included castles with millions of visitors each year in their core areas, illustrating a very different conception of what was 'natural' and therefore worth conserving (Chapter 10).

In this case, varying forms of negotiation took place during the zonation. The first instance of negotiation involved determining the management objectives of the various zones within the park's administration. Subsequently, the process was extended to other bodies: the National Forestry Office, private land owners, as well as the myriad of (non-specified) people involved in urban areas and centres of economic activity. Listening to the terms chosen, it was difficult to imagine that the decision-making process involved these other actors from the start of the process. On the contrary, the impression was that the senior managers (the plural 'nous' or 'on' was used) made the decisions, and presented the result to others ('it makes the foresters more accountable'). However, the plan had to be endorsed by local communities before it was applied, despite only having indicative, non-legal standing.[8] At the time of the interview, this had not yet happened. Thus the role of the experts drawing lines was described according to their own spatial discourses before any subsequent political process had taken place. The power of these omnipotent experts, exercised through the imposition of rationalised knowledge, was reflected in the words chosen to describe the various decisions: 'tout naturellement'; 'c'est clair et net'; 'on trouvait logique'.

I mentioned earlier that Raffestin described territorialisation as a social and spatial process through which space is organised. In one article, he argues that domestication and simulation are key processes of this, producing territories through the application of science and technology. The overwhelming role of scientists and scientific methods in the zonation of protected areas illustrated this, as natural ecosystems were 'domesticated': 'the story of human ecosystems is in fact a long story conditioned by domestication and simulation applied to natural ecosystems in order to construct territories. A territory compared to a natural ecosystem is, in fact, nothing else than the projection of human work mediated by practices and knowledge that are rooted in science and technology. Technico-scientific practices and knowledge systems have never ceased to be mobilised and used to reorganise natural ecosystems in order to transform them into human ecosystems. This reorganisation of natural ecosystems leads to a territorialisation that can itself be questioned – deterritorialisation – by new scientific or technical mediators that construct new territories, thereby illustrating reterritorialisation'[9] (Raffestin 1997 : 101).

The metaphor of domestication is interesting and appropriate here. It is a paradoxical metaphor since the actors who adopted 'scientific', rationalised arguments for zonation, would no doubt have strongly rejected the idea that they were involved in domesticating nature. Despite the imposition of human management on the landscape, the modernist division between nature and culture was rarely questioned. Rather, the prevalent discourse and rationale was linked to the opposite end of the spectrum: managers saw themselves as defining areas of wilderness, setting nature free from human intervention. Yet this was clearly an act of territorialisation and institutionalisation: restricting human impact or access within an intrinsically political and societal process.

Conclusions

The literature review on the criteria used to define the boundaries of protected areas illustrated that while biophysical and ecological criteria had traditionally been used to define protected areas, other societal concerns were increasingly being explored. Nevertheless, despite the pragmatic coexistence of both spatial discourses, the idea of an isomorphism between space and society remained, both in the literature and on the ground. The temptation was to look for particular pre-existing spatial scenarios inscribed in the landscape. Yet 'nature' does not dictate particular politics, nor does it inherently contain spatial management units. There are no 'patterns given by nature itself' (Natter & Schultz 2003). Yet such an idea prevailed both within the protected area movement as a whole and among managers on the ground.

To a large extent, the realisation that an 'ideal' situation from an ecological point of view did not necessarily produce the most effective unit from a management point of view coexisted with the firm belief in biophysical arguments Societal criteria were therefore increasingly being used to create entities that, rather than being ecologically perfect, could be much more effectively managed. However, rather than be accepted as progress, this was viewed as a compromise, a less-than-optimal scenario that diverged from the biophysical 'ideal'. Throughout the interviews, protected area managers were keen to position themselves as 'experts' and privileged decision-makers. Scientific methods were taken to be value-free, objective and rational tools for decision-making, removed from political influences. The social construction of spatiality underpinning the managers' conceptions was rooted in their training as natural scientists as well as their professional environment in the natural sciences, premised by notions of pre-existing biophysical boundaries revealed by objective enquiry.

Despite the prevalence of notions of 'wilderness', the definition of protected area boundaries was a politicised domestication of nature, implying an institutionalisation (spatial socialisation) of spatial discourses. As such, it was far from value-free. The attempts to include societal arguments for zonation brought about a clash between ontologically distinct conceptions of boundaries. Although there were burgeoning attempts by managers to engage with societal arguments, these were still carried out by omniscient 'experts' upstream from subsequent participatory and politicised processes. Yet the institutionalisation of territoriality that underpinned all zonation of protected areas could only be understood in reference to spatialised political projects. Decision-making and negotiation between a variety of actors and the inevitable power struggles these involved further reflected the contested nature of the process. The apparent fear of engaging with the political nature of zonation was not only a fear of losing control of the process, but rather stemmed from an ontological difficulty: the prevalence among managers of a professional background rooted in the natural sciences. Managers were therefore unused to considering boundaries as subjective, constructed phenomena that were intrinsically contested.

Notes

1 The link between protected areas and the military is certainly worth exploring further, perhaps building on Woodward's work on Military Geographies (2004) and the environment. Other authors worth looking at include Van Schaik & Kramer (1997) in Wilshusen 2002 : 34 who actually argue for a direct military enforcement of protected area boundaries, with soldiers patrolling their boundaries and keeping people out, arguing that protecting such areas could be a strategic objective of a state.

2 Personal translation from: '(...) und gemacht haben es die Forstarbeiter, also Zusammenarbeiten zwischen Forstarbeitern, das war, bin ich und Naturschutzverwalter, das war eine Frau, S.H., die ist im Moment nicht da, die hat ein Baby bekommen und hat Urlaub jetzt und hat Mutterschaftsurlaub. Sonst könnte sie auch mit mir zusammen ich könnte mir das vorstellen, aber jetzt mache ich es alleine. Aber es ist, denk ich, wichtig zu sagen, also nicht Forster, weil nun alleine, sondern Forster wegen Naturschutz. Wir haben zusammen gearbeitet und auch ... alle ... ... Städte und Gemeinden und Landkreise und Naturschutzverbände wurden informiert und hatten Gelegenheit, also wir hatten einen Entwurf gemacht und diesen vorgestellt, überall sind wir hingereist und haben uns vorgestellt und alle konnten sich äussern dazu, ob sie einverstanden sind, oder Änderungen noch nötig wären, und wir haben, so gut es ging, alles berücksichtigt, und das ist dann das Ergebnis jetzt'.

3 Personal translation from: 'Le zonage de cette partie ... oui ... dans les chartes de parc, il y avait un plan qui accompagnait qui disait grosso modo à quelle endroit le parc allait intervenir, les objectifs etc. Et quand on a fait l'acte de candidature pour être réserve de biosphère, on a fait un deuxième zonage, spécifique. Avec les trois zones, centrale, tampon et transition, et avec des critères de zonage qu'on a révisés fortement car en fait ils réservaient des zones centrale au zones de la réserve naturelle et de la forêt de Bitche, la zone tampon c'était une zone qui était le périmètre qui englobait ces espaces protégés et naturels, et puis après tout le reste du parc c'était en zone de développement'.

4 Personal translation from: 'Le premier zonage était centré sur la partie nord du territoire du parc, qui est ici, où, je ne sais pas vraiment pourquoi, on avait retenu des aires centrales. (...) Pourquoi ? Il fallait un zonage, on a un peu été pris de cours, je ne connaissais rien du programme des réserves de biosphère. Quand on m'a parlé de ce zonage dans le dossier, j'ai réfléchi un peu avec les scientifiques qui étaient les premiers pionniers du conseil et on s'est dit : "il y a une forte concentration d'espaces protégés, de zones sensibles et de caractère remarquable de grand intérêt patrimonial ici, alors prenons cette zone. Nous en avons d'autres ailleurs mais il y a vraiment ici une concentration, un triangle d'or". On a pris donc ça et, très vite, on s'est rendu compte, quand on a fait la première version en 97/98, qu'on a fait le tableau et déroulé la méthode, que ça ne tenait pas debout'.

5 Personal translation from: 'In queste occasioni qui succede sempre che magari quelli che sono favorevoli e non sono contrari, ma non sono così attivi non vanno alle riunioni perché tutto sommato che ci sia il parco o non ci sia non gliene frega niente. Invece quelli che vanno alle riunioni sono quelli che sono contro, per cui le riunioni sono andate, non dico male, perché comunque ...'

6 Personal translation from: 'Guardate, noi a Aisone abbiamo votato, o abbiamo aderito al consiglio comunale e votata all'unanimità, approvato l'ampliamento del parco Alpi Marittime nel nostro territorio, nonostante ne avessimo già un pezzo. Inseriremo un sito estremamente caratteristico, estremamente qualificato che è

quello delle grotte archeologiche, lo abbiamo fatto perché riteniamo che inserire un ambito così particolare del comune potesse rappresentare una allenta importante'.

7 Personal translation from: ' La nouveauté est aussi que l'on a pris les zones évidemment naturelles, et on a mis tous les monuments historiques, notamment des châteaux qui sont en pleine forêt. On a estimé qu'autour, de toute manière, la zone qui s'y trouve et le château lui-même étant porteurs de nature, on est quand même en Europe, qui est fortement modifiée par l'homme, où nature et culture sont fortement imbriquées. (...) Donc, on a rajouté les monuments historiques (...). Si on rajouté les vergers traditionnels, qui sont des zones tout à fait nature-culture aussi, en fort déclin, pour lequel on a un programme d'action. On trouvait logique donc de les avoir en zones tampon, je dirais les zones un peu naturelles qui ont encore du patrimoine fort mais qui, de toutes façons, font l'objet d'une gestion par l'homme, sur lequel on doit justement expérimenter des pratiques alternatives (...). Voilà pour le zonage'.

8 In Europe, biosphere reserves only had direct legal status in Germany, Romania and Ukraine. In other countries, they only gained legal standing when they were made to concur with other existing legal designations, requiring substantial negotiation.

9 Personal translation from: 'l'histoire des écosystèmes humains est en fait une longue histoire conditionnée par la domestication et la simulation appliqués aux écosystèmes naturels pour produire des territoires. Un territoire par rapport à un écosystème naturel n'est, en fait, rien d'autre que la projection de travail humain à l'aide de médiateurs – pratiques et connaissances – qui s'enracinent dans les sciences et les techniques. Les pratiques et les connaissances technico-scientifiques n'ont pas cessé d'être mobilisées et utilisées pour réordonner les écosystèmes naturels pour les transformer en écosystèmes humains. Ce réagencement des écosystèmes naturels débouche sur une territorialisation qui peut elle-même être remise en cause – déterritorialisation – par de nouveaux médiateurs scientifiques et techniques qui postulent de nouveaux territoires qui constituent des exemples de reterritorialisation'.

PART II
CONTESTED BOUNDARIES

Chapter 5

Contested Boundaries and Complex Spatial Scenarios

Boundaries crystallising conflicts

Boundaries are always contested. It was stated earlier that boundaries are ontologically parasitic and cannot exist in isolation from the entities they bound (Chapter 2). Because they are resisted as forms of reified power, the contest often focuses on what the boundary enshrines and creates, that is to say the spatial entity itself. Conflicts about boundaries are wider contests about the control of space by different actors: a conflict about differing spatial discourses. 'For territory to be meaningful it has to be reproduced by the enactment of challenges to it, by questionings and erasures of boundaries as markers of space, but also through the inscription of new boundaries' (Albert 1998 : 61). The conflicts and questionings can be summarised as struggles for legitimacy: contrasting understandings of who is a legitimate insider or outsider, linked both to spatial discourses and conceptions of nature. Building on the previous chapter, I continue to argue this by drawing from extracts of speeches taped during a hunters' demonstration in the Alpi Marittime, in Italy. Subsequently the argument is enriched by quotations from interviews.

Contrasting spatial discourses: insider and outsider

In 2002, vocal resistance was organised by the hunting lobby following the announcement of a project to enlarge the existing protected area in the Alpi Marittime. This series of demonstrations contributed to postponing the establishment of a biosphere reserve beyond the territory of the existing protected area. Using rousing language, the president of the provincial hunting lobby captured the idea of resisting an imposed spatial project:

> *I just want to remind you of one thing, for you this should be an alarm bell, and we should remember that as it can happen in this area – and that is that our management is not recognised – it can happen in other places, because if we chop off a piece here and we chop off another piece in another place, maybe one year a mayor in another place decides to do something similar, and our territory is reduced a little each time. That's it*[1]
> *(President Hunter's Association, Alpi Marittime, Italy).*

The very populistic appeal to defending the remaining 'territory' from the onslaught of protected area managers was very powerful. His language conjured up geopolitical images of 'balance of power' and the need to control strategic territory by occupying it. The redefinition of boundaries was depicted as illegitimate. Any project of 'protection' was described as less valid than the control gained through the 'management' carried out by the hunters themselves. Implicit in this discourse was the idea of defining an insider (Us) from an outsider (Them) seen as less legitimate. Thus any increased control was taken to be imposed by outsiders, as an aggression and invasion. Such views were similar to Ardrey's 'territorial imperative' (Ardrey 1967 in Storey 2001 : 10), harking back to crude deterministic positions in which territorial behaviour was seen as a natural and therefore legitimate behaviour linked to the control of space. Additional legitimacy was gained through an appeal to labour, in a quasi-calvinistic way. The selected culling of animals throughout the year was seen to imply moral and practical rights over the territory as a whole. Rather than a discourse of 'wilderness', the prevailing conception was that of man[2] as organiser, as rational domesticator of nature:

> *So, what results do I retain? The results are only those to manage as much as, no, not as much as, I would say better than, because according to me wherever man [sic] is and man [sic] lives, it's impossible to think of enclosing the territory in a glass casket, because man [sic] lives and therefore must live also within the territory, and manage the fauna and the territory as best he can*[3]
>
> *(President Hunter's Association, Alpi Marittime, Italy).*

The extension of the boundary of the protected area was thus likened to an artificial act. The protected area was compared to a glass casket, not a space of wilderness but rather an unnatural object that had no rightful place. The very idea of boundaries (and therefore control) was rejected since the territorial ideology it was based on was rejected. Thus paradoxically both the protected area managers and the hunters were brandishing conflicting version of 'nature': on one hand various degrees of 'naturalness' bounded within protected areas, on the other the idea that such boundaries were intrinsically 'unnatural'.

Much of the discussion was implicitly or explicitly concerned with defining insiders and outsiders within both the park and the wider region. It was not only a question of rejecting the extension of the park ('Them coming to a place that belongs to Us'), but also rejecting the implicit definition of hunters as 'outsiders' within the existing park:

> *Considering that we don't want, as they have written in the newspaper, to go and destroy everything in free communities, in places where there isn't a park, because that is an absurd thing, we are not destroyers. I should like us to start thinking about it and start saying 'no' to the park that takes its responsibilities to be unique, and let us give space to the hunter's management, maybe specifically inside the parks, and where we enlarge, let us try to maintain them, because these are also important things*[4]
>
> *(President Hunter's Association, Alpi Marittime, Italy).*

The term 'free community' was instructive in revealing the paradoxical attitude to boundaries: it described an area defined by political boundaries (acceptable boundaries), that was not part of the protected area (unacceptable boundaries). Furthermore, the President complicated the discussion of insider / outsider by arguing that even in spaces where hunters were defined as outsiders (inside the park), they should have a role. Thus as well as 'defending' the remaining 'free territory', he called for an expansion into spaces previously out of bounds. Opposed to this very populistic rejection of the status quo, other voices attempted to express themselves, seeking to break up the stark attempt to create insiders and outsiders. Yet, as Alessia noted later on, who the insiders and the outsiders were was not entirely clear and depended on points of view:

> *Those who don't want it [the enlargement] are the hunters who are non-residents, that is to say those who come from outside but who are listed as hunters but who come from Genova, from Torino or also only from Cuneo or from Borgo San Dalmazzo. They will see the zone they can hunt in diminish and so they are not at all happy. So it becomes, how can I put it ... Their right to hunt is diminished in favour of resident hunters, which seemed quite correct because well the resident population are always those who have limitations put on them, aren't they?*[5]
>
> *(Alessia, Parco Naturale Alpi Marittime, Italy).*

> *If we only consider the park as a hindrance ('momento di vincolo'), as the closure of a territory where you shouldn't go because something is fixed, then this is a losing park. (...) So I would think that within the hypothesis of an enlargement, of the park or of the buffer zone ('pre-parco') this needs to be studied, defined together, it could be a way of ... not cheating the hunters, but it could be a way ... [whistles and shouts do not allow him to continue]*[6]
>
> *(Local mayor, Alpi Marittime, Italy).*

This conciliatory language, seeking discussion and negotiation while implying that boundaries were not exclusionary walls, was however less effective in mobilising high emotions. It failed to confront the real problem: a staunch rejection of the idea of boundaries controlling behaviour. The mayor was silenced by aggressive whistles and shouts. This particular meeting, although formally a 'discussion' between opposing views, was in fact organised as a show of force on behalf of the hunting lobby.

Contrasting conceptions of nature: insider and outsider

The discourse also focussed on opposing conceptions of nature, nature protection and appropriate behaviour. The Us and Them debate extended to consider conflicting claims of territorial legitimacy. Following an attempt by the hunters' lobby to portray themselves historically as 'the first environmentalists', a local elected person stood up, picking up on the issue of insider and outsider:

> *The first environmentalists are not the hunters. Do you know who the first environmentalists were? They are the people who live in Argentera, all year round, the people who in order to take their kids to school get up at six in the morning, they are those who go up to the 'alpe' with four sheep (...). The environmentalists, or better, the hunters from Cuneo who come and hunt on our land are only half environmentalists,*

> *we are the first environmentalists, and if we are also hunters then that is fine too. This is a correction that was due to all those who live with the mountains all year round*[7]
> *(Local mayor, Alpi Marittime, Italy).*

Thus the debate became one of contrasting belonging, of opposing claims to spatialised legitimacy. The position of insider was seen to convey increased – or sole – decision-making power. Here, the mayor indicated the varying degrees of legitimacy between the different groups of actors: inhabitants who had the strongest claim to being insiders were more legitimate. For him, legitimacy was not rooted in work (hunters) or management (park) but in a commitment to long-term belonging. This emotive argument was alien to the protected area managers attending the meeting. They were equally sure of their own position of insiders, rooted in several decades of environmental management. They perceived themselves as legitimate managers of the land, hampered in the rational need to extend the park by political fanatics.

This Italian example illustrated how definitions of Self and Other were constructed and negotiated by different groups. Each group used a variety of arguments linked to territoriality in order to advance claims linked to the definition of boundaries. The conflict was particularly visible in the Alpi Marittime due to the very well-organised and vocal hunting lobby. Such a conflict was however far from unique. All protected areas visited during the fieldwork faced conflicts of varying intensity with local groups over the (re)definition of their internal and external boundaries. The feeling of being excluded from the process was a recurrent fear. This was expressed for example in the zoning of the biosphere reserve in the Pfälzerwald:

> *Within communities there is always the fear that 'we will be limited in our freedom to plan; will we still be able to build housing projects or will we be, will we have ... will our choices be reduced?' That used to be the greatest fear*[8]
> *(Alexander, Pfälzerwald, Germany).*

Although Alexander partly used the past tense to describe this conflict, the matter was not resolved. Again, fears focussed here on the notion of 'outsiders' dictating a course of action, imposing their particular conception of nature and space on the resident population. The new boundaries were seen here to literally limit or bound the local communities, controlling their future planning options.

This struggle between nature conservation and local development was expressed repeatedly in the different case studies. Conflicts crystallised around the notion that boundaries incarnated power, power from outside the area (outsiders) or from above ('the State'). The location of the boundaries became a crucial strategic issue, with pressing arguments for their definition being put forward by different interest groups. In the Polish Tatras, for instance, the tension between high-impact tourism in the town of Zakopane and strictly enforced nature protection within the national park clashed along the park's boundary:

> *The Tatras is the only national park in Poland without a buffer zone, so there is a very sharp border between the park and the surrounding land*
> *(Tadeusz, Tatras National Park, Poland).*

> *Talking in this moment, there is a small buffer zone for animals around the national park, up to the town, but it is not definite*
> *(Jurek, Tatras National Park, Poland).*

The conflict in the Tatras centred on the park manager's identified need and desire to establish a 'buffer zone' as enshrined in Polish law. Concurrently, developers wishing to extend infrastructure around the town had managed to get some land reclassified for development. The park's outer boundary had therefore been redefined with tacit support of the Ministry of the Environment. The language used by managers to express this struggle reflected the diverse spatial strategies adopted: attempting to gain strategic land for a buffer zone (advancing) while protecting the status quo in other areas (retreating). The military term 'buffer zone' thus became poignantly pertinent, expressing the strategic issues of land control perceived to be at stake. The feeling of entrenchment was further compounded by the perceived treason of the Minister of the Environment, accused of towing the line of big business.

In other areas, conflicts did not centre on redefining boundaries so much as changing their function and role. Thus rather than seek physical changes, local authorities sought to extend their authority over protected areas that were often managed by other governmental bodies:

> *With the recent political changes in Poland, we now have stronger local governments. They want to extend their area of authority over the land in the national park. For example, now the money from the local entrance fees comes to the national park, but the local government wants it*
> *(Andrzej, Bieszczady National Park, Poland).*

The strategy of the local government was not so much to seek to modify boundaries but rather to change the prevailing institutionalisation of the existing bounded spatial entities. Resistance was more indirect, yet by no means less effective. Arguments for increased local management of the area held strong appeal to local people, keen to identify the park with a distant centralised power: the central State constructed as an outsider removing resources from their rightful local owners.

Conclusions

These various examples have illustrated two possible resistance strategies to territoriality perceived as imposed or constrained: direct resistance to boundaries through attempts to redefine them by force or indirect resistance by attempting to change the nature, meaning or effects of these boundaries. The latter led to increasingly complex situations with decision-making power shared by a variety of administrations and individuals. Increasingly complex spatial scenarios emerged in which competing interests vied for legitimacy over an area, using a selection of arguments linked to spatialised identity and insider/outsider. Building on this idea of indirect resistance, the following section explores whether the coined term 'New Medievalism' is useful in grasping the spatial complexity of such situations.

New Medievalism: coextensive boundaries and complex spatial patterns

The term New Medievalism first appeared in a book by Hedley Bull in 1977 describing the 'reincarnation of the vertically segmented and overlapping forms of authority which existed in Europe before the rise of the modern state' (Bull 1977 in Anderson 1996 : 134). A couple of decades later, Anderson resuscitated the term linking it to 'postmodern territorialities' in which 'traditional "one-level" thinking is completely inappropriate for assessing contemporary political transformations. We have to deal with multileveled and multifaceted processes which span global regulatory regimes, global regions, world cities, substate regions and localities as well as states' (Anderson 1996 : 140). In addition to being multi-scalar, the term crucially describes an overlapping of various authorities on the same territory (Albert 1998 : 56). These postmodern or 'neo-medieval' spatial scenarios reflect 'changes in the world are of such a character that a reliance on established vocabularies to explain them would not do them justice, while the new vocabularies needed to provide the proper explanation are as yet only in the process of being devised and searched for' (Albert 1998 : 56). Yet is the metaphor of the Middle Ages useful in describing a current situation, or does it reflect both a failure to understand that particular epoch as much as our own?

Frame tells us that when examining the Middle Ages in Europe, the historian 'has learnt to live in a world characterised by permeable frontiers across which the interests of the nobility readily spread, and by powers of varying types whose fields of dominance ceaselessly expanded, contracted, and overlapped. He *[sic]* must adjust his focus to take in vast multi-national empires, kingships marked by great unevenness of control from region to region and period to period, and principalities that cannot be categorized as consistently within or without the control of superior authorities' (Frame 1995 : 3). However, Anderson cautiously notes that 'the present is far from being a simple reversal to a long-lost past' (Anderson 1996 : 150), hinting at the danger of metaphoric simplification. The initial transition from 'medieval' to 'modern' territoriality involved changes in territorial control: 'the nested hierarchies and different levels of authority in medieval Europe – the vertically segmented overlapping sovereignties defined in terms of functional obligations as well as loosely territorial terms – gave way to sovereignty delimited only and much more precisely by territory' (Anderson 1996 : 141). The following section explores to what extent this suggestion of a return to the former scenario is pertinent, examining if this description of 'neo-medieval' boundaries has contemporary relevance. In particular, I explore cases of coexisting and coextensive forms of protected areas.

Contested boundaries and complex spatial scenarios

The simplest scenario for a new designation – such as a biosphere reserve established on the basis on a pre-existing protected area – is for both to be coextensive. In the field, this direct overlapping was however not always reflected in management structures. In the French Vosges du Nord and the Polish Tatras, for example, both designations covered the same territory and were directly overseen by the same administration:

> *Another link between NRP [Natural Regional Park] and BR [biosphere reserve] is that, for us, when I look at the French network, all the park is classified as a biosphere reserve, not only part of it. (. . .) We were able to really calibrate a perfect match between the park and the biosphere reserve because the territories are the same. So, the overlapping is complete, you can see for example in our publications, we have a fusion of the two logos*[9]
>
> *(Hugo, Parc Naturel Régional des Vosges du Nord, France).*

> *The boundary of the biosphere reserve is the same as the boundary of the national park. We cannot extend this because of the unfavourable political situation. So it is not possible to have a legally protected transition zone outside the park*
>
> *(Kazimierz, Tatras National Park, Poland).*

Yet the varying territories did not always overlap in such a direct manner. In these cases, the metaphor of neo-medievalism appears particularly pertinent. Scenarios were not fixed but changed and fluctuated, both spatially and institutionally. In the Bieczszady, in Poland, there were à variety of overlapping and contrasting boundaries, reflecting the complex combination of various forms of protected areas:

> *There is a gradual move towards having a larger biosphere reserve with more strict protection. At the moment, the buffer zone of the national park is smaller than the buffer zone of the biosphere reserve*
>
> *(Andrzej, Bieczszady National Park, Poland).*

Such a scenario was the result of a decision-making process involving a variety of actors. Rather than being straightforward, this was the result of various levels of negotiation involving actors wielding various spatial discourses. In the Vosges du Nord, the first biosphere reserve zonation was carried out by a small group of 'experts' within the pre-existing natural regional park. During the interview, the director reproduced this process of drawing lines by sketching the scenario on a sheet of blank paper:

> *It is a scenario something like this, we can get a copy of it. [Draws picture] Here is the perimeter of the park, so here we have a certain number of protected areas concentrated in the Pays de Bitche, here as well, well in fact, they are all over the place, you know, and so we said in the first MAB zonation: the central zone is this and this and this and selected a certain number of things, and we put the buffer zone sort of here, more or less, and then all the rest in the development zone. It was a bit dubious according to the*

criteria. But it was accepted as a biosphere reserve. But then next to this, and this is completely different, we had the park's plan. The park was less important at the time. The park's plans have only had real meaning since 1994, in fact[10]

(Théo, Parc Naturel Régional des Vosges du Nord, France).

However, it was not always straightforward to combine different protected areas and administrative structures. These were often seen to exist for different reasons or else were associated with specific key individuals. In the Slovak Tatras, there was an acute conflict between the two administrations following institutional restructuring, leading to a complete breakdown of communication between the two resulting entities. This was justified by one manager – the ex-director of the national park, subsequently involved only in the biosphere reserve – as being because of the radically different objectives of both. Here, the different locations of boundaries were taken to illustrate the inherent differences and varying objectives of the two structures:

The national park is based on the idea of a patchwork of zones, but the biosphere reserve has a concentric structure. The biosphere reserve zonation is not for management, but only for research. The only interest for a biosphere reserve is research

(Michal, Tatras National Park, Slovakia).

It is interesting to note that this manager, himself a natural scientist, saw the biosphere reserve as a purely scientific tool, reflecting his own interest and orientation. Thus managers attached meaning to territorial entities in line with their own particular objectives. This often meant that what constituted a biosphere reserve was understood very differently by various actors, even leading to a complete divorce between involved administrations, although both territories were coextensive. Yet, during interviews, such difficulties did not always appear, and the conceptual differences were ironed over, with the biosphere reserve being seen as simply an international quality label:

The task of a Naturpark in Germany, or in Rheinlandpfalz, is to develop the region for the recreation of people, big cities around, in the county of Saarland, with many people, the big towns, big cities like Frankfurt, Wiesbaden, Mainz, Heidelberg, and other agglomerations, these are the big cities where there is no nature, and these big forests, the Palatinate forests is the area which has to be developed to an area for recreation for that people in that big cities around, that is the main task of the Naturpark which was founded in the Fifties. And since 1992 we have another state, on a higher level, and that is the biosphere reserve

(Lukas, Naturpark Pfälzerwald, Germany).

This active construction and negotiation of meaning was particularly prevalent in situations where the situation was formally unclear or else where legislation did not recognise certain protected areas:

So far, what we had in the old system which was not generally accepted, was just a preliminary idea, we had the strict nature reserves, and the nature forest reserves being recognised as cluster-like core areas. This has been changed now. This is under

> *discussion. This is a proposition coming from the Ministry [He shows the second map]. It is under the authority of the Ministry to set up the zonation*
> *(Daniel, Naturpark Pfälzerwald, Germany).*

> *The biosphere reserve is mostly used as an argument when talking to local people in order to explain the international level of protection*
> *(Tadeusz, Tatras National Park, Poland).*

In the case mentioned in the second quote, the biosphere reserve was an extra designation, recognised on an international level, but with no legal grounding, implying no additional national funding. In this case, the biosphere reserve did not fulfil the Seville Strategy criteria and was little more than an additional label. It did not lead to any changes in existing practices. The additional spatial scenario only had a voluntary status, relying on individuals defining objectives on an informal level and taking them on. This was less selfless than it may sound, as it often made professional sense, involving trips abroad and increased international exposure for the individuals involved. When national legislation did not recognise biosphere reserves, the lack of legal grounding was repeatedly identified by managers as a crucial shortcoming, particularly regarding its effect on securing national funding.

Contested boundaries and institutional complexity

The neo-medieval metaphor suggests both spatial and institutional complexity. This seems particularly pertinent in cases where a variety of different Ministries (Environment, Agriculture, Foreign Affairs) and differing Committees and other decision-making or consultative bodies were involved with administrating the various spatial entities, often competing for limited funding:

> *The biosphere reserve still has no legal status, and in Poland we have a conflict with the Deputy Minister. In 1997, I asked about possible ministerial support for the work of the biosphere reserve but he told me that it was 'not within the interest of the Ministry of the Environment'. The relationship between the MAB Committee and the Ministry is difficult, and the MAB Committee is seen as a source of possible funding*
> *(Andrzej, Tatras National Park, Poland).*

In such cases, it was not surprising that different protected area designations were not used interchangeably. There was often a subtle difference in the use of names for the entities, reflected in varying positions amongst staff members in each administration. It was instructive, for instance, to listen to the name given when managers answered the telephone ('Hello, this is the ...'). Often, different staff members used different names, expressing varying enthusiasm for the complex structure. Some would never use the term biosphere reserve, for instance, while others consciously did or, more specifically, chose to do so in specific situations. This flexibility was reflected in published material and correspondence, reflecting complex overlapping territorial and institutional identities:

> *Sometimes, when I write to Catherine [the coordinator of the French biosphere reserve network], I put 'biosphere reserve'. Another time, I will put 'park', but, still, mostly, I use both together. So people know that we are talking about the same territories and practically the same action programmes. The difference is only that, for us, the biosphere reserve adds something: it brought or reinforced objectives that might not have happened had we only remained a 'park'*[11]
>
> *(Hugo, Parc Naturel Régional des Vosges du Nord, France).*

Communicating these complex spatial scenarios to local people was always challenging, particularly when situations were not clear to the managers themselves. In many ways the need to communicate spatial scenarios also highlighted the inherent contradictions of these increasingly complex entities. The meaning of these was constructed and negotiated both on an individual and shared level:

> *For many people, it is difficult to understand that in the biosphere reserve, there should be at least 3% core area; that you shouldn't impact too much on the ecosystem. (. . .) Because there is also a misunderstanding I think we can say, many people think a biosphere reserve is not like a national park an area where people in large parts do not have to interfere, but a biosphere reserve is for the people themselves, for sustainable development, and the understanding is also that they think they can go everywhere, it is difficult for them to accept that there will be 2 or 3 % of areas where they cannot do what they want*
>
> *(Daniel, Naturpark Pfälzerwald, Germany).*

In this case, modifying the status quo following the designation of a Naturpark as a biosphere reserve meant informing the local population of changes in legal status and corresponding changes in access. As in the Slovak Tatras, the coexistence of various spatial scenarios within the German Pfälzerwald led to an institutional split, with separate administrations taking over the Naturpark and the biosphere reserve, although these were coextensive. Furthermore, following the Seville Strategy, biosphere reserve criteria were revised. Thus, added to all the complexities, differing versions of what such a zonation entailed coexisted:

> *Biosphere reserves should be evaluated every ten years by UNESCO, and so that was different again so it was said that the old zonation no longer fit the criteria, you can see now, both international criteria and national criteria should ... Therefore a new zonation needed to be made. And my opinion is that this work should be carried out by the state. The association (Verein), the nature protectors shouldn't do this, but instead the administration should*[12]
>
> *(Alexander, Pfälzerwald, Germany).*

These difficulties were observed by the protected area in the neighbouring country. One manager in France suggested that not all spatial and institutional scenarios were feasible, hinting that in the adjacent German case the coextensive protected areas were too different to be reconciled:

> *I think differences are too entrenched between the biosphere reserve objectives and those of the Naturpark [in the Pfälzerwald] for these to be indeed overcome like that, very quickly, if you want, it needn't take much to say to people 'now you get down to work'. We [in the Vosges du Nord] have the MAB spirit, the spirit of a biosphere reserve unlike before when they only worked as a regional protected area. I think the difference is very big, so on the level of the management structure in charge of the project, it's obviously more difficult, that is obvious*[13]
>
> *(Hugo, Parc Naturel Régional des Vosges du Nord, France).*

These different conceptions of what a biosphere reserve should be conflicted. The debate was extended to the other side of the international boundary, with neighbouring protected areas adding voices to the discussion. Such different interpretations of zonation naturally compounded the complexity of attempting transboundary zonation, as discussed in Chapter 9.

The paradox of complex spatial scenarios

In order to overcome contradictions and conflicts between overlapping or coextensive protected areas, biosphere reserves were sometimes conceived as 'umbrellas' combining other existing entities. This created entities with complex combinations of boundaries on different levels serving different functions. However, on the ground, this did not always translate into anything concrete on an administrative or decision-making level. What was perhaps most surprising in these complex scenarios was that the spatial and the administrative structures did not always concur. A spatial entity was paradoxically not always matched and 'managed' by an equivalent administration. In certain cases, this directly resulted from resistance by members of other administrations who feared a loss of control. Thus any changes in spatial and institutional scenarios were actively resisted, as in the Bieszczady:

> *Within the national park, 100% of the forests are owned by the State. The protected landscape areas and the national park make up the biosphere reserve so the area covered by the biosphere reserve is protected according to Polish law. In a way, the biosphere reserve is an extra designation but with little added value, according to the Deputy Director. He sees the biosphere reserve as a direct threat to his own authority within the national park*
>
> *(Andrzej, Bieczszady National Park, Poland).*

Likewise, in other cases, it remained largely unclear who was responsible for what. Spatial scenarios remained undefined, matched by equivalent institutional uncertainty, despite the formal existence of a biosphere reserve designated by UNESCO:

> *The zonation of the biosphere reserve is not yet finished because many different institutions are responsible. The result will come out of scientific studies that are being carried out. There is no one person responsible. This is new in Ukraine since before land planning was divided into branches like forestry or agriculture*
>
> *(Ivan, Uzhansky National Park, Ukraine).*

When we became [i.e. obtained] the biosphere reserve, the Federal State did not want to manage it, and they said you do it, the Verein. And now we are asking, well, we started asking two years ago, we said we need an official contract telling us that we are the managing body, we get sufficient funding from the Federal State, etcetera etcetera. And we haven't had so far and we are looking forward to a new, 'ein Verordnung', 'a décret', you know what I want to say, in Rheinland Palatinate they are going to get new legislation for all the nature parks and this will also include the biosphere reserve zonation and will also include the role of the biosphere

(Daniel, Naturpark Pfälzerwald, Germany).

In the countries formerly under Socialist rule, this institutional complexity was further compounded by unresolved issues of land ownership. Following the change in regime, former collectivised land was increasingly privatised. This meant either returning it, where possible, to its former owner, or else finding ways of redistributing it to members of a community. In Slovakia, Ukraine and Romania, this process was ongoing, leading to situations of great uncertainty, further complicating the already complex spatial scenarios:

The Romanian biosphere reserve extends to the natural borders of the delta. The Ministry asked us to establish the exact cadastral limits and demarcate these on the land. This is important because we must define land ownership ... there have been changes in land ownership because of the changes in the political situation

(Nicolae, Danube Delta Biosphere Reserve, Romania).

Thus despite the apparent biophysical 'objectivity' in the design of the biosphere reserve that was meant to follow the 'natural' limits of the delta, uncertainty remained. In Ukraine, on the other bank of the river, similar uncertainties about the extent of the protected area's administration remained; the same happened in Slovakia, in a more terrestrial context:

At the moment, the borders of the northern part of the biosphere reserve are very unclear. So some of the management we do is carried out unofficially (Nikolaï, Danube Biosphere Reserve, Ukraine). The problem of land ownership makes zonation difficult. (...) Sometimes it is unclear because it is difficult to find the people who owned the land before Socialist times. So we would like this land to be part of the national park

(Dominik, Poloniny National Park, Slovakia).

Both these examples illustrated the difficulty of defining the exact boundaries of a protected area and therefore the spatial extent of the authority of the corresponding administration. In both cases, the managers concerned wanted the boundaries defined in order to secure their own role in controlling the land, moving away from 'unofficial' management. Uncertainty was seen as being best addressed by an increase in their authority over the contested land.

Conclusions

This review of the spatial and institutional complexity in a selection of protected areas has attempted to discuss the pertinence of the metaphor of 'New Medievalism' in this context. These examples have illustrated the multifaceted nature of boundaries to protected areas which appear much more complex than simple encircled areas. New Medievalism, or 'postmodern territoriality' is therefore about more than nested hierarchies, but instead hybrid, intermediate, ambiguous and uncertain territorialities. Territory has been defined as a 'distinction, indeed a separation, from adjacent territories that are under different jurisdictions' (Gottman 1973 in Storey 2001 : 14), implying the expression of power and control over space. Yet in this context, such a simple definition is no longer pertinent when spatial complexity does not allow for the identification of distinct, separate units. Thus the metaphor of New Medievalism assists in grasping the difficulty of ascribing one administration or institution to one space, illustrating the contested nature of boundaries in protected areas.

The boundaries of protected areas reflected this attempt at territorial control. They came to incarnate the power exercised over prescribed space. This reified power manifested itself in territoriality, in power made visible, deflecting the attention from the power relationship to the territory, away from the controller and controlled. The creation, and where applicable, the internal zonation of such spatial entities implied the institutionalisation of the spatial entity. In order to cope with the uncertainties inherent in such spatially complex scenarios, managers of protected areas sought to reinforce decision-making structures and administrations and thereby their own authority over the land. Managers searched for an increased institutionalisation of protected areas, leading to the clear emergence of spatial units that became established and were clearly identified as part of the spatial structure of a society. This attempt to provoke a geohistorical process was inevitably contested in a variety of ways, reflecting the inherently political nature of boundaries.

Building on this discussion of the intrinsically contested nature of boundaries, the next chapter further examines reterritorialisation and boundaries. The case studies provide material within which to examine the construction of 'transboundary' protected areas. The chapter is structured around three discourses promoting the creation of 'transboundary entities', dwelling on the arguments put forward by a variety of actors.

Notes

1 Personal translation from: 'Poi voglio solo ricordare una cosa, per voi questo dovrebbe essere un campanello d'allarme, ricordarci che come può succedere in questa zona, e quindi che non venga riconosciuta la nostra gestione, può succedere anche in altre zone, perché togliamo di qua, togliamo da un'altra parte, magari un altro anno un sindaco di un'altra zona decide di fare una cosa analoga, e il nostro territorio poco per volta si riduce, ecco'.

2 'Man' is indeed the most pertinent word here. The use of this term does not reflect a failure of political inclusiveness on my part, although the hunters made the point that their association of over 200 hunters included ... two women!

3 Personal translation from: 'Allora, quali risultati io ritengo? I risultati sia solo quelli di gestire, quindi altrettanto, no altrettanto, no, io direi meglio, perché secondo me l'uomo comunque c'è e l'uomo vive, non si può pensare di chiudere dentro in una boccia di vetro il territorio, l'uomo vive e quindi deve vivere anche nel territorio, quindi, gestire al meglio la fauna e il territorio'.

4 Personal translation from: 'Visto che noi non vogliamo, come hanno scritto sul giornale, andare a distruggere tutto di un comune libero, dove non c'è il parco, perché è una cosa assurda, noi non siamo dei distruttori, vorrei che cominciassimo a pensarci e cominciassimo a dire non 'no' al parco così fine a se stesso, diamo spazio alla gestione dei cacciatori, forse addirittura all'interno dei parchi, quindi dove allarghiamo, cerchiamo di mantenerli, perché sono delle cose anche importanti'.

5 Personal translation from: 'Quelli che non lo vogliono sono i cacciatori non residenti, cioè quelli che vengono da fuori, che però sono iscritti nelle liste del comprensorio di caccia e vengono da Genova, da Torino, anche solo da Cuneo o da Borgo San Dalmazzo. Che quindi questi qui si vedono ristretta la loro zona cacciabile e quindi non sono contenti per niente. Cioè diventa, come dire ... Si diminuisce il loro diritto di caccia a favore dei cacciatori residenti, che sembrava abbastanza corretto perché comunque le popolazioni residenti sono sempre quelle che hanno delle limitazioni, no?'

6 Personal translation from: 'E se consideriamo il parco solo come un momento di vincolo, di chiusura di un territorio dove non bisogna andare perché c'è qualcosa di teso, è un parco perdente. (...) Allora io penserei che nell'ipotesi di un ampliamento, parco o pre-parco, da studiare, da definire insieme, potrebbe essere un modo per ... non per fare le ali ai cacciatori, ma potrebbe essere un modo ... (Fischi e urla ...)'.

7 Personal translation from: 'I primi ambientalisti non sono i cacciatori, i primi ambientalista sapete che sono? Sono quelli che vivono ad Argentera, tutto l'anno, quelli che per portare i bambini a scuola li svegliano alle 6 del mattino, sono quelli che con 4 pecore vanno in alpeggio, (...). Gli ambientalisti, o, meglio, i cacciatori di Cuneo che vengono a cacciare da noi sono ambientalisti a metà, siamo prima noi ambientalisti, se poi siamo anche cacciatori va bene lo stesso. Era una correzione che era dovuta a chi la vive la montagna tutto l'anno'.

8 Personal translation from: 'Bei den Gemeinden ist immer, immer die Befürchtung, 'werden wir eingeschränkt in unserer Planungsfreiheit, können wir noch ein Baugebiet Häuser bauen oder werden wir, haben Sie, werden wir eingeschränkt?', das war einmal die grösste Befürchtung'.

9 Personal translation from: 'Autre lien PNR/RB, c'est que, chez nous, quand je regarde le réseau français, tout l'ensemble du parc est classé réserve de biosphère, et pas seulement une partie. Donc ça, si tu veux, c'est une chose intéressante, parce que au départ de la prise de conscience qu'une réserve de biosphère est un label international, une sorte de médaille en chocolat que nous remet l'UNESCO, on a pu glisser vraiment dans l'adéquation complète entre parc et réserve de biosphère parce que les deux territoires sont les mêmes. Donc, comme la superposition des territoires est complète, tu vois peut-être dans nos communications, on a une fusion des deux logos'.

10 Personal translation from: 'C'est un schéma du genre comme ça, on peut le retrouver d'ailleurs. (Draws picture) Voilà le périmètre du parc, et on a ici donc un

certain nombre d'espaces concentrés dans le pays de Bitche, aussi là, enfin il y en a partout, quoi, et on avait dit dans le premier zonage MaB la zone centrale c'est ça, ça, ça, et sélectionnés un certain nombre de trucs, et on avait mis la zone tampon comme ça, grosso modo, et puis zone de développement tout le reste. C'était un peu limite du point de vue des critères, quoi. Mais ça avait passé en réserve de biosphère. Mais, et à côté de ça et ça c'était tout à fait différent, le plan du parc, quoi. Le parc avait moins d'importance à cette époque. Les plans du parc on une réelle importance depuis 94, en fait'.

11 Personal translation from: 'Quelquefois, si j'écris un texte pour Catherine, je mets 'réserve de biosphère'. Une autre fois, je vais mettre 'parc', mais, la plupart du temps quand même, j'associe les deux. Les gens savent donc de quoi on parle, les mêmes territoires et quasiment les mêmes programmes d'action, à ceci près quand même que, pour nous, la réserve de biosphère est un plus : ça a apporté ou renforcé des missions qui n'auraient peut-être pas été autrement si on était simplement resté 'parc''.

12 Personal translation from: 'Biosphärenreservate sollen alle 10 Jahre evaluiert werden von der UNESCO und das war dann wieder anders, dass man sagte, die alte Zonierung für den Pfälzerwald entspricht nicht mehr den Kriterien, wie sie jetzt stehen, sowohl internationale Kriterien, als auch nationale Kriterien sollten ... Deswegen sollte eine neue Zonierung gemacht werden. Und die Auffassung hier meinenfalls war die, diese Aufgabe ist eine staatliche Aufgabe, also das soll nicht der Träger machen, der Verein, der Naturwart, sondern die Verwaltung'.

13 Personal translation from: 'Je pense que le fossé est trop grand entre les objectifs des RB et ceux d'un Naturpark pour que, effectivement, ça soit franchi comme ça très vite, si tu veux, il ne suffisait pas de grand chose pour dire aux gens 'maintenant vous bossez'. On a l'esprit du programme MaB, l'esprit d'une réserve de biosphère par rapport à avant où ils travaillaient simplement dans l'esprit d'un PNR. Je pense que plus la différence est grande, au niveau de la structure qui porte le projet, plus c'est difficile, c'est évident'.

Chapter 6

Constructing Transboundary Entities

Something there is that doesn't love a wall
He is all pine and I am apple orchard.
My apple trees will never get across
And eat the cones under his pines, I tell him.
He only says: Good fences make good neighbours
(Robert Frost)[1]

(Re)territorialisation in a transboundary context

Robert Frost's poem, set in rural New England, is a narrative about boundaries and walls in nature, culture and human minds. It suggests that boundaries exist in a metaphorical, as well as a physical, sense, building identities and practices between people by marking out clear divisions in space. However, unlike the oft-quoted line 'good fences make good neighbours', the poem may also be taken to imply that such fences are far less permanent than the neighbour would like. As Freyfogle notes, 'our culture has latched on to this proverb, no doubt because it captures so well a number of our foundational tendencies and assumptions. We like fences and erect them often, routinely separating mine from yours' (Freyfogle 1998 : 15).

In the past decade, the fashion for transboundary protected areas has gained tremendous momentum. This has led to the establishment of a variety of transboundary entities around the world including among others: transborder protected areas, transboundary biosphere reserves, transboundary national parks (a contradiction in terms!), peace parks, international friendship parks, and transboundary world heritage sites. Constructing a protected area is complicated enough. Carrying this complex spatial and institutional process across several sovereign countries further complicates the complex balancing act between competing interests. The definition of the boundaries of these transboundary protected areas is likewise additionally problematic, and further complicates the discussion of insider/outsider detailed earlier. I argue that the establishment of a transboundary entity on the basis of several existing protected areas implies a process of reterritorialisation implying negotiation between an increasingly complex set of local, national and international actors. Until now, when analysing transboundary protected areas, researchers have drawn on theories of international cooperation from the fields of international relations and political science. This has led to a failure to identify the complex nature of reterritorialisation that implies a common reinvention and redefinition of both social and spatial practices. In this chapter, I explicitly do not deal with 'cooperation' as such, preferring instead to examine the specifically spatial aspects of the process. 'Cooperation' as a process is examined more comprehensively in Chapter 7.

In the second part of the chapter, the discussion is divided into three sections portraying three differing discourses that are contrasted in order to shed light on their different assumptions and arguments related to boundaries, identity and cooperation. These three discourses are: the publications of international organisations, the publications of local protected areas and the declarations of individual managers. Throughout, I argue that these three levels of discourse reflect radically different assumptions about boundaries, leading to methodological problems when applying concrete projects.

Establishing a transboundary entity

The earlier discussion dealt with (re)territorialisation and the construction of boundaries, arguing that the drawing of a boundary is a contested process that reifies power. It has been noted that resistance to imposed internal boundaries is even more observable in peripheral boundary regions (Sletto 2002), something that has relevance in this case. In parallel to the recognition of the problematic nature of defining internal boundaries within boundary regions, it appears that 'across many disciplines there has been an 'epistemological celebration' of spaces and positions astride boundaries, and between nations and identities' (Hocknell 2000 : 53). This celebration of such ambiguous spaces has led to an unproblematized celebration of transboundary cooperation, often vaunted as the new glue that will participate in integrating and reterritorialising entities such as the European Union or Southern Africa, reinvented as spaces where distinct (state and non-state) spatial entities interact happily, cementing themselves together as a whole.

Within political geography, this celebration often replaces conceptualisation, as if cooperation were a universal 'good thing' which everyone can define and carry out instinctively. There is a plethora of monographies dealing with individual boundary areas, or comparing different cases. At most, cooperation is theorised as the modification of the functions of boundaries (Kratochwil 1986 : 46), an interesting idea, but not sufficient to explain the ongoing process of cooperation, especially when it goes wrong. Instead, a more analytical framework for conceptualising boundaries is suggested on the basis of Paasi's model of the symbolic construction of space. The scale considered here is therefore very different from that of international law or international relations. While these consider the actors to be principally states or institutions, the approach discussed here addresses much more localised phenomena.

The conception of boundaries as socio-spatial phenomena has important implications for the way cooperation is considered, since any transboundary activity has an impact on collective territorial identities within the process of spatial socialization. Hocknell notes that boundaries continue to play a key role as a distinct location for cooperation or conflict (Hocknell 2000 : 54). In approaching cooperation as a process, I use Paasi's discussion of 'difference' and 'integration' (Paasi 1996 : 15). From the idea that all political and territorial identities are, in a sense, fictional, and connected with imagined communities (Hassner 1993 in Paasi 1996 : 12), Paasi argues that the symbolic construction of space is based on a dialectic between two languages: the language of difference and the language of integration. Since the latter aims at homogenizing the contents of spatial experience, this is relevant to the idea of political integration.

Paasi's strength is in associating such ideas with what he terms 'we and the Other', through a discussion of how the latter is created as an external entity against which 'we' and 'our' identity are mobilized (Paasi 1996 : 12). The concept of 'otherness' spread beyond the context of anthropology largely as a result of Edward Said's ground-breaking 'Orientalism'. He described how the Orient had become the archetypal figure of otherness (Said 1978), the description of which sheds more light on those creating the discourse than on those described. The other reality or person is reified. Hence, says Paasi, the other must be mythologized and labelled; the label must be legitimised. Through sedimentation, word and myth come to have lives of their own, detached from the original act of mythmaking, and evolve into autonomous components of the everyday stock of knowledge which is taken for granted by a society (Aho 1990 : 22 in Paasi 1996 : 13). The definition of the other necessarily includes a spatial dimension, 'in the fact that the Other typically lives somewhere else, *there*' (Paasi 1996 : 13). Thus 'we' are necessarily different from it, leading to stereotypes which are not confined to any particular scale. The Other can be assimilated to supra-state entities ('Europeans'), neighbouring nationals ('Italians', 'French'), just as it can cover more local categories ('Piedmontese', 'those from the neighbouring valley', or 'the other park's employees').

This process of othering is central to understanding the process of cooperation. As Paasi argues, the process of distinction is inseparable from that of integration. Thus, 'it is reasonable to argue that the symbolic construction of space, territoriality and boundaries is based on a dialectic between two languages, the language of *difference*, and the language of *integration*' (Paasi 1996 : 15) (original emphasis). This is not a theory of cooperation, but instead offers the framework for understanding processes of confrontation with the Other, a central aspect of the construction of transboundary spaces. The Other is on the other side of the boundary and any process that leads to common work implies combining the language of integration ('building bridges across the boundary', 'cooperating') with coming to terms with the language of difference, of self-definition and identity. It is important to stress that these two aspects are necessarily intertwined, since boundaries not only separate groups and social communities from each other, but also mediate contacts between them. There is a paradox here: this approach risks equating the creation of a transboundary entity with cooperation, yet surely cooperation can only exist if two or more spaces are distinct. For this reason, I initially consider the creation of transboundary entities, only subsequently dealing with what cooperation means for those involved once these spaces have been defined (Chapter 7).

This approach leads to initially analysing the emergence of new territorial units within the process of regional transformation (Paasi 1996 : 15). Paasi describes this process as the institutionalisation of regions which refers to 'the process during which specific territorial units – on various spatial scales – emerge and become established as parts of the regional system in question and the socio-spatial consciousness prevailing in society' (Paasi 1996 : 32). Subsequently, I argue that any activity taking place across the boundary – usually described as 'transboundary cooperation' but rarely problematised – will necessarily partake in this institutionalisation of the (new) territorial unit.

Boundary areas and transboundary entities

The desire to divide and identify Self from Other is intrinsic to our human way of thinking, and Velasco-Graciet similarily makes a link between this social and the spatial process. She notes that 'every society, by a universal set of oppositions, constructs a world of realities through a classification system. Thus the construction by each group of its own identity enters into this logic of classification and according to the current research into ethnicity, it appears that the existence of the signifier "Us/Others" is fundamental to the organisation of the world within each society. (. . .) Therefore, a delimitation indicated on the ground allows the social opposition between "Us/Others" to be reinforced by a geographical opposition between "Here and There"'[2] (Velasco-Graciet 1998 : 119). This implies that boundaries are inherent to the notion of identity, founding the intimate difference with the Other.

Because interactions appear between groups on either side of a boundary, the concept of it as a linear phenomenon is no longer adequate and is often extended to cover a boundary area. The latter, unlike a simple line separating differing practices, takes on specific functions, since 'this boundary zone's function is to maintain cultural differences between the groups in contact. In this way, it is the place where organisational systems of relations are established allowing two present identities divided by a boundary line to be affirmed. This boundary zone is, therefore, the natural auxiliary to the boundary itself'[3] (Velasco-Graciet 1998 : 126). Thus it is because groups come into contact with each other that the concept of a linear boundary is inappropriate, since exchanges create the need for a specific space where they should occur. 'This zone defines the rigid spatial boundary and is the theatre of relations of interdependence, a necessary condition for mutual knowledge between groups and for reinforcing feelings of identity'[4] (Velasco-Graciet 1998 : 126).

As Hocknell notes, with time, awareness of a more dynamic functional role of 'border regions' has grown, leading to the context of 'geographically delineated regions that cooperate across a sovereign boundary' (Hocknell 2000 : 60). This is not necessarily a fixed and clearly defined area, but can be a 'concrete zone, a place of alliances and social innovation that far from being institutionalised and institutionalising, wanders on the margins of the certainties of two worlds that appear too bounded to be in touch with each other'[5] (Velasco-Graciet 1998 : 101). This is not to say that boundary areas become homogenised, and Hocknell argues that the contrary is often true: 'far from obliterating local cultural differences, international influences may in fact shape the "particularities of place" by stimulating and encouraging distinctive local and regional spaces' (Hocknell 2000 : 60). This is an important point, and can be linked up to Paasi's idea of difference/integration, since neither process is exclusive. Thus confrontation with the Other, far from leading to alienation, leads to a clearer vision of Self.

This idea of boundary area, frontier zone (Ricq 1998), border regions (Goonerate & Mosselman in Hocknell 2000; Hocknell 2000), zone frontière (Velasco-Graciet 1998), or borderlands (Martinez 1994) has become fashionable in geographical writing. The term border landscape has also appeared although there seems to have been little conceptual debate, even if its use can be traced back to 1937 (Jones 1937 in Rumley and Minghi 1991 : 1). Rumley and Minghi in fact note that 'many human geographers have only a vague and hazy notion of what the concept might entail' (Rumley & Minghi 1991 : 1). They note in particular that the 'specific definition of border areas as opposed to boundaries as the objects of analysis remains unclear in much of the literature, although the focus on disputed areas in conjunction with national boundaries provides concrete examples of an areal or regional milieu as opposed to a linear one' (Rumley & Minghi 1991 : 2). Furthermore, they note a lack of real concern with the development of 'border landscape theory', since 'the implicit assumption of uniqueness and even a general disinterest in theoretical and conceptual questions. Coupled with these problems has been a lack of concern with explanation and a consideration of process' (Rumley & Minghi 1991 : 4). This thesis chooses to use the more general term of boundary area, bearing in mind that these various terms have in fact very little conceptual difference.

In addition to the geographical aspects, boundary areas have also been more formally defined following legal criteria. Lapradelle used the French term 'le voisinage', translated rather curiously by Prescott into 'border landscapes' 'calling for the need to study the legal realities of the border and to avoid the belief that the entire state is subject to uniform internal boundaries right up to the boundary' (Lapradelle cited in Prescott 1987 : 13). Similarly, Ricq uses the term 'frontier zone' (Ricq 1998 : 11), identifying a certain number of legally approved definitions for this area. He cites, for example, that the former EEC 'thought in terms of a 20km radius, though 50km had originally been mooted' (Ricq 1998 : 11). Today, both the Council of Europe and the European Union 'regard a frontier region in the strict sense as a public territorial entity situated immediately below state level and having a common land frontier with one or several entities of the same type situated in a neighbouring state' (Ricq 1998 : 14). However, Ricq also notes that it is possible to define such zones 'pursuant to a variable geometry by the kind and importance of problems that they face' (Ricq 1998 : 14). These could be seen to relate to sectors of economic, social, environmental or cultural activities.

The extent of transboundary contact

Having established that boundaries as lines lead to the creation of boundary areas or zones, the question is usually framed as how and how much contact there is between the different sides. On this practical level, Martinez identifies three approaches to the study of interactions which he terms 'borderlands interaction' (Martinez 1994 : 1). Bearing in mind that although individual situations vary and in particular the 'size of nation-states, their political relationship, their levels of development, and their ethnic, cultural and linguistic configurations' (Martinez 1994 : 1), he suggests that generalisations on the dynamics of transboundary interaction – assimilated to 'cooperation' – can be made. Martinez bases his model on the amount of transboundary movement and the forces that produce it, leading him to identify four 'paradigms' of interaction between areas on either side of a boundary: alienated borderlands, co-existent borderlands, interdependent borderlands and integrated borderlands. Returning to the terms used in this paper, but conserving Martinez's key words, thus leads to a definition of four levels of transboundary interaction, taken to involve increasing levels of contact: alienated, co-existent, interdependent and integrated boundary areas.

Thus a boundary dividing two *alienated* areas is defined as a situation in which 'day-to-day, routine cross-boundary interchange is practically non-existent owing to extremely unfavourable conditions. (. . .) International strife leads to militarization and the establishment of rigid controls over cross-border traffic' (Martinez 1994 : 2). Obviously in such a tension-filled climate, cooperation between the two areas is non-existent and substantive people-to-people contact is very difficult if not impossible. A boundary dividing two *co-existent* areas, however, shows a slightly higher level of interaction. Co-existence arises when nation-states 'reduce extant international border-related conflicts to a manageable level or, in cases where unfavourable internal conditions in one of both countries preclude binational cooperation, when such problems are resolved to the degree that minimal border stability can prevail.' (Martinez 1994 : 2). *Interdependence* exists between two areas on either side of a boundary when the area in one country is symbiotically linked with the border area of an adjoining country. 'Interdependence implies that two more or less equal partners willingly agree to contribute and extract from their relationship in approximately equal amounts'' (Martinez 1994 : 4), contingent upon policies pertaining to the national interests of the two partners. The most intensive form of interaction is finally between *integrated* areas, at which stage 'neighbouring nations eliminate all major political differences between them and existing barriers to trade and human movement across their mutual boundary' (Martinez 1994 : 5).

This is not the only such classification present in the literature, and Hocknell has produced a comparative figure comparing four such classifications. It is reproduced here (Table 6.1) as an illustration of other possible approaches. The main idea that emerges from this exercise is that cooperation is almost universally considered as a phenomenon that happens at varying degrees, with different levels of integration. It can be argued that the principle of studying the levels of interaction between two areas can also be applied to internal boundaries within a given country. Although arguments like freedom of movement or trade restrictions are not directly pertinent in this case, other considerations concerning levels of interaction are. In particular, when it comes to discussing management models within a country that stretch across different administrative jurisdictions, or areas under the management of different Ministries, then such a classification does become pertinent. However, it is likely that differences between levels of interaction will fall into the 'interdependent' and 'integrated' categories, but at varying positions within their spectrum.

The extent of cooperation follows different patterns, and in practice is unlikely to follow the linear progression suggested above, but rather is made up of ups and downs, depending on both internal and external factors, respectively linked to the dynamics within each organisation (enthusiasm, allocated staff time, priority level, political commitment, budget, etc.) and outside the organisation (external funding, political acceptability, ease in transboundary communication etc.). All of these factors contribute to the success or failure of attempted transboundary cooperation. In the case studies, this process is analysed as a form of social construction of space, through the process of spatial socialization.

Implementing transboundary cooperation implies dealing with the dynamic consequences of boundaries, moving beyond the separation of all spheres of activities along national boundaries. The process of spatial socialization stretches to include issues of institutional structure, design and organisation – not a topic usually addressed by geographers, but rather by managers and business people. In many ways, this stems from the background of many managers, well-versed and skilled in activities specific to their field of expertise – be they natural or social scientists – but not well acquainted with what are in fact specific intercultural management skills. Any transboundary activity, however technical or practical, necessarily involves communication between people. Thus before either stage of planning can be addressed, basic management processes are needed, including determining objectives, identifying actors and resources concerned, as well as constraints and potential conflicts that might arise. Such a process is never neutral, and is inevitably hampered by a series of cultural assumptions made by both parties. In order to grasp the complexity of this process, the discussion leads on to an analysis of various forms of discourse dealing with issues of boundaries, identity and cooperation with the aim of identifying the links between integration and distinction in the creation of transboundary spaces.

Integration and distinction: discourses on transboundary cooperation

Paasi's categories of distinction and integration are not two separate processes that should be discussed one after the other but rather are taken to be intertwined and co-dependent. They are two facets of the same process of (re)territorialisation. This discussion builds on the Foucaldian notion that 'intellectuals, institutions, and ideologies constitute discursive structures that

Table 6.1 'Borderlands' and 'border regions': a comparison of conceptual frameworks based on an integration continuum. (Inspired by Hocknell 2000 : 62)

Intergration continuum	'Borderlands' (Martinez)		'Borderlands' (Momoh)		'Barrier networks' in 'border regions' (Suarez-Villa *et al.*)		'Border regions' (Goonerate & Mosselman)	
Non integration	*Alienated*	Tension. Border functionally closed; interaction totally or nearly totally absent.	*Zero-borderland*	Borderland coterminous with boundary but sides ideologically or religiously opposed. No cultural or ethnic affinity, and borderland space is minimal.	*Isolated*	Total boundary closure, although informal links may exist.		
	Co-existent	Stability on/off; border slightly open, limited binational interaction.					*Resource based*	Presence of a key resource necessitates cooperative exploitation and management.
	Inter-dependant	Stability prevails, increased interaction.	*Minimum-borderland*		*Inter-connected*	'Gateway node' exists, functional advantage of which in any border region is inversely related to number and size of competing nodes.	*Active*	Economically and socially integrated, offering a high potential' for coordinated regional development.
	Integrated	Stability, functionally-merged economies, unrestricted movement.	*Maximum-borderland*	Citizens on both sides have ancestral ethnic and linguistic affinities and links spanning millennia.	*Jointly controlled*	The network need not be official, but merely functional. Contiguity diseconomies are greatly reduced.	*Development*	Combination of 'active' and 'resource based', linkages and complementarities strengthened.

shape how we think about and act on relations between state and territories' (Foucault 1980 in Sletto 2002 : 183). Therefore an analysis of language and of its use in the creation of social representations and images is crucial to understanding the process. The main focus of this approach is the process of identity construction and othering.

The discussion is divided into three sections: the normative discourse of international organisations as expressed in a variety of 'guidelines' and 'best practice' publications; the discourse of protected area administrations within published material and official speeches; and finally the individual protected area managers themselves speaking during the interviews and discussions conducted in the field. These three discourses are contrasted in order to shed light on their different assumptions and arguments related to boundaries, identity and cooperation. Throughout, I argue that these three levels of discourse reflect radically different assumptions about boundaries, leading to methodological problems when applying concrete projects. Furthermore, I argue that any analysis of cooperation must take into account this conflicting diversity. A useful approach must move beyond the descriptions of 'cooperation' developed in political science and summarised above and instead must lean towards an understanding of the multi-faceted dimensions of (re)territorialisation.

Normative discourses: guidelines and best practices

There was an increasing wealth of literature produced by such bodies as the World Conservation Union (IUCN) Protected Areas Programme, the United Nations Organisation for Education, Science and Culture (UNESCO) or non-governmental bodies such as the EUROPARC Federation relating to transboundary protected areas, mirroring the enthusiasm for the concept. All sought to produce recipes for effective cooperation, often illustrating their recommendations by sets of examples from around the world. The aim was to create more and better transboundary protected areas, thereby spreading the approach effectively.

Publications produced by organisations such as IUCN and UNESCO were often Guidelines or Recommendations, the bulk of which emerged during meetings and were subsequently edited either internally or by an outside 'expert'. This process accounted for some of the pronouncements which seemed unbelievably naïve, or simply too prescriptive to be true. The terms used were normative, with flowerings of 'must', 'should', and 'it is vital that', interspaced with sweeping conclusions. Generalisations covered diverse topics including social, economic and political issues. Statements within these three fields concluded respectively, for instance, that 'improved staff morale seems to go hand in hand with transboundary cooperation' (IUCN 1998 : 29), 'joint approach in marketing is more likely to attract tour operators' (IUCN 1998 : 29), 'customs and immigration officials are more easily encouraged to cooperate by a joint effort of transboundary parks' (IUCN 1998 : 30).

Normative discourses and boundaries

The underpinning ideology behind such pronouncements drew on the idea that political boundaries were 'bad', specifically because they were not 'natural' but rather were against nature. Such documents often started with a sweeping statement referring to the 'anti-nature' dimension of political boundaries:

> *Since political boundaries between States have usually been drawn for demographic, geographic or security reasons, they may take no account of the parameters of an ecological unit: important watersheds or internationally significant natural areas are often transected by national boundaries*
>
> *(Shine in IUCN 1998 : 38).*

Thus the fact that political boundaries were defined through a political process was viewed as negative and required fixing through appropriate spatial management. These discourses implicitly hinted that it would be more appropriate for political boundaries to be based around biophysical features. Yet paradoxically, while 'natural boundaries' were alluded to, the realisation that biophysical features should not be divided prevailed. Somehow, it was assumed that the best political boundary would skirt round biophysical features, echoing the calls for a redefinition of boundaries in line with 'bioregions'. Political boundaries were taken to be 'artificial', a term that appeared almost systematically. If political boundaries were bad from an ecological point of view, they were equally taken to be bad from a societal standpoint.

> *You cannot divide a river, a mountain, a forest, a wetland in two or more pieces, following an artificial political boundary. They are unique ecosystems that should be managed as a whole*
>
> *(Rossi 2000 : 21).*

> *Frontiers are a convention: the product of international agreements, historical events and wars. (. . .) Normally borders coincide with a physical boundary, the ridge of a mountain range or the center line of a river, because they are easy to plot on a map, but from a geographical and naturalistic point of view the mountain or the river should be considered as single entities or at least as two complementary halves which require a common form of management and coordinated interventions*
>
> *(Rossi in Hamilton 1996 : 53).*

Yet while from an environmental point of view political boundaries were merely 'a convention', such conventions had indisputable concrete political effects. Initially, a least, this was set aside, replaced by further assertions about boundaries at other levels. 'Artificial' boundaries were bad and should thus be ignored, not only on an international level but also locally, transcending both international boundaries and the boundaries of protected areas themselves:

> *Important species, such as the Ibex in Europe and the Cougar in Central America are not interested in artificial boundaries drawn on maps. Protected areas, reflecting this*

perspective, must broaden their outlook beyond traditional boundaries if they are to survive in the next century

(Sheppard 2000 : 25)

The dimensions of the problem defined the scale at which management activities were to be carried out. Such arguments repeatedly called on flagship species as illustrations of the 'boundless' status of nature. In Europe, references to ibexes cropped up in many places, notably because Patricia Rossi, EUROPARC's then president, was involved in successful projects for reintroducing them in a transboundary context. Chosen species attained the status of icons, incarnating the issue of transcending boundaries. The stress therefore was on the dichotomy between nature's boundaries on one hand and political boundaries on the other. Furthermore, since protected areas were designed to protect biophysical objects, then these were seen to determine the management issues. Protected areas adjoined and shared a common space, so they consequently were seen to share common problems. These problems were taken to be independent of political context or political choices and were seen as rationally and objectively determined in a universal manner.

Regardless of the political jurisdiction in which the parks lie, or their political and legal differences, natural resource issues do not ordinarily confine themselves to boundaries described by humans

(Mihalic in Hamilton et al. 1996 : 39).

Protected areas that share common borders share common problems. (. . .) Areas of natural or cultural significance shared by two or more countries or other resource owning jurisdiction lend themselves to transborder protected areas

(Hamilton et al. 1996 : 1).

Such assertions ignored the fact that environmental problems needing redress were the result of human action. Naturally, such action was not universal but context-dependent, and thus in the throes of political processes that may well have been particular to one state. If, for example, a species needed conserving because its scarcity was the result of forest over-exploitation, this might not be mirrored on the other side of an international boundary. Furthermore, identification of what constituted a problem was often substantially different within national contexts, as in the case of the wolf in the Maritime Alps, where the return of the wolf was considered a problem in France but not in Italy.

These quotations also overlooked 'national' conceptions of nature, including the definition of what was considered natural – values and beliefs which were undeniably culturally constructed. To ignore such matters, as many natural scientists did, was to miss an important dimension of what environmental protection and 'management' was about. It was as though, while seeking universality, the writers of such documents preferred to ignore cultural differences since these were seen to be divisive elements, which like political boundaries introduced divisions into what was constructed as naturalised, homogenous space. Yet paradoxically, cultural elements were called upon when seeking to argue for the unity of chosen transboundary areas. Shying from the cultural and the political, focussing on a certain conception of the natural, such discourses displayed a naïve conception of what constituted a highly politicised process of establishing cooperation, inevitably inserted in complex networks of power.

Normative discourses on cooperation

This literature built on the idea that nature within protected areas was still pristine, wild and remained untouched by human hands and political processes. This was paradoxical, since protected areas often rested on the very idea that nature needed 'managing' within a defined area, barring perhaps carefully-defined core zones – itself a form of management. This was not to say that environmental issues were never shared in boundary situations, but rather that the idea that such areas automatically 'lend themselves' to cooperation was politically naïve. In parallel to the idea that boundaries were inherently *bad*, rested the idea that cooperation was always *good*. It was in fact taken to be a value almost universally promoted:

> *Cooperation in some form or sense is a common goal almost everywhere where two park areas share a common boundary. (. . .) If there is anything in common among all it is likely the desire on both sides to cooperatively work together to effectively solve natural resource management issues, regardless of whether or not they are common to the two park areas. Even in those countries where citizens of one nation poach protected resources across the border, the most optimal resource management solution lies in cooperation, rather than other alternatives*
>
> *(Mihalic in Hamilton et al. 1996 : 40).*

The statement that it was obvious everyone wanted 'to cooperatively work together to effectively solve' any problems was dubious. It ignored dimensions of power and political processes leading not only to the definition of problems but also to the definition of Self and Other. Assuming that cooperation was universally aspired to assumed a pre-existing knowledge of the Other. On the ground, in politically complex situations, it was unlikely to be clear who the Other was, or to what extent this Other shared political power, accountability or mandate. In the case of the Vosges du Nord / Pfälzerwald, despite a high level of contact between protected area managers in both countries, the respective powers and mandate of each administration were not initially clear to the other. Likewise, in the Tatras, it was unclear to the Polish administration who the Other actually was on the Slovak side following administrative restructuring. The same could be said of the Ukrainian entity in the East Carpathians.

Such caution in accepting cooperation as a 'given' might seem unnecessarily cynical, overlooking the inherent goodness of people and the dedication of protected area managers. This is definitely not the objective of my critique and I remain firmly convinced that dedicated individuals do aspire to cooperation, often against seemingly-horrendous odds. However, I believe that by pointing out some of the weaknesses of generalising assumptions, the political dimensions of cooperation will be identified from the start. Bearing in mind that people do not necessarily always want to cooperate, and may indeed feel threatened by such an idea, is a more helpful starting point than naïveté if cooperation is to be planned effectively.

The supposed effects of biophysical factors on the success of cooperation, drawing on images of biophysical determinism, cropped up repeatedly, and linked up to some of the reflections on the appropriateness of natural boundaries:

> *Mountain ranges form a natural barrier along a ridge, where ecosystems sometimes differ on both sides of the border, but there are also river valleys or flat areas where no significant natural borderline divides two neighbouring landscapes. In this case, cooperative management seems to be much more necessary or possibly effective than in mountain nature reserves or cultural landscapes*
>
> *(Brunner in Synge 1997 : 79).*

Quite why it would be easier to cooperate in a plain was unclear, although presumably this was building on the assumption that sameness fostered increased integration. It was obvious that certain quotes must be set in context, and become meaningless removed from the surrounding text. The following extract could only really be understood in reference to the setting: the opening remarks of a speech by IUCN's former Director General during an international meeting in South Africa. Despite its rather demagogic appeal to a rejection of colonial boundaries, it made certain interesting assumptions:

> *I read a statistic somewhere which is that over 50 per cent of the present national boundaries of the world were drawn up by six colonial powers. The boundaries wander whimsically over the face of the globe, the product often of the arbitrary actions of lost and lonely colonial surveyors with very vague briefs. Occasionally they used physical features to define the boundaries – drawing their lines down the thalweg (the middle of the navigable channel) of large rivers, for example. Apart from the fact that such lines tend to be a trifle insecure (the navigable channel shifts at flood time) they are a nightmare to ecosystem managers because they split river basins and watersheds precisely down the middle. They are also a nightmare to social scientists and community leaders and government administrators because they tend also to split human groups down the middle*
>
> *(McDowell 1998 : 23).*

The assumption seemed to be that colonial boundaries were even worse than boundaries defined in other ways – through wars for example – since they were even more 'artificial' than others. Not only were they seen to be unnatural but also 'antisocial'. McDowell's response to this was to promote transboundary protected areas, redressing the biophysical and societal shortfalls of existing international boundaries, stopping just short of a wholesale call for the redefinition of all political boundaries along theoretical bioregional or ethnic lines. Such a vision naturally required a legal framework and this was provided by Shine, a freelance advisor to IUCN's legal office based in Bonn:

> *In management terms, it would be preferable for the whole area to be administered as a single unit by one institutional body (the highest being a joint international commission established by treaty) in accordance with a single management plan. The international border would become purely symbolic, with immigration and customs controls being moved back to the park boundaries and uniform regulations being applicable throughout the TBPA. Such an 'ideal' will often, though not always, be politically impossible*
> *(Shine in IUCN 1998 : 38).*

> *Nature conservation and the fulfilment of the missions of protected areas as uniform entities is of primary importance. The corollary is that, in the long term, national interests should take second place*
> *(Brunner 2000 : 59).*

Thus transboundary protected areas would become utopian green spaces removed from state control, perhaps along the jurisdictional lines of joint off-shore trade zones, although the actual jurisdictional scenarios were not specifically suggested. The 'bad' boundaries would be pushed back, replaced with the wholesale creation of a new political entity. This, not surprisingly, had never happened, except – it could be argued – in the case of Antarctica. With ideas of abandoning the sovereignty principle openly promoted, it was not surprising that some governments were wary of entering into formal legal agreements. The role assigned to states was diminished, transcended by the higher moral right of environmental protection. Nature, therefore, became a force in itself that determined a political agenda. Protected areas were seen as the means of enforcing this agenda by transcending existing international boundaries through the creation of new homogenous spatial entities. Protected area administrations were assumed to be coherent and uniform actors in this process.

Conclusions on normative discourses

This section has reviewed some of the emerging understandings of boundaries and cooperation that appeared in the publications of international non-governmental and governmental bodies, broadly building on conceptions of bad 'non-natural' political boundaries versus good cooperation aimed at restoring naturalised spatial entities. I argued earlier (Chapter 2) that conceptions of boundaries could be divided into two separate ontologies: *bona fide* boundaries representing a realist position postulating a reality 'out there' than only needs revealing, and *fiat* boundaries, on the other hand, related to human cognition and decision. Thus the argument for basing political spatial entities on 'natural', biophysical boundaries rested on the assumption that *bona fide* boundaries logically entailed the formation of *fiat* boundaries, i.e. that 'natural' boundaries existed. As well as being politically suspect, this was also ontologically untenable.

In addition to the arguably naïve understanding of political processes this entailed, one of the main weaknesses was that this discourse assumed protected area administrations to be coherent entities acting as one integrated body. This failed to acknowledge that any administrative structure was composed of a selection of individuals, each struggling to put forward his or her particular conception of space. This myth of homogeneity created a naïve understanding of politics by ignoring the often intense power struggles within one structure. A full understanding of this was replaced by calls for 'ensuring that local staff and communities are more closely and effectively involved in the establishment and management of transboundary protected areas' (Sheppard in Europarc 2000 : 37), although how this was to be done remained unspecified.

These normative discourses on 'bad boundaries' and 'good' cooperation were produced on an international level. At the same time, local protected areas produced their own documents. These partially mirrored some of the ideas of the global discourse, while simultaneously building on other assumptions of the role and nature of boundaries.

Protected area publications: constructing homogenous spaces

In this section, I present a collection of views appearing within the published material of the case study areas, reflecting diverse and contradictory conceptions of boundaries, identity and cooperation. My purpose here is not to decide which of these two approaches is more prevalent, but rather to present a broad overview of the dominant themes appearing in local protected area publications in order to contrast these 'official' localised discourses with the opinions expressed orally by individuals.

There is an incredible wealth of material emerging from the five case study areas, and even after having set aside texts in the languages I cannot read, a significant amount remains. I therefore make no claims of thoroughness in this analysis, preferring instead to pick out subjectively those elements that appear the most pertinent. This includes identifying recurring themes, as well as pointing to unusual elements that are explored more in depth in subsequent chapters. This means that while no particular case study is fully developed here, an understanding of each individual situation informs the wider picture. Many of the quotes are extracted from protected area publications, either giving information about a particular feature or listing activities and projects for visitors. There are also extracts from key official speeches made during public events, inaugurations and conferences, subsequently published by protected area administrations.

Boundaries and social space

One particular intervention stood out amongst the official speeches made during a colloquium on the creation of a transboundary entity uniting the French Parc National du Mercantour and the Italian Parco Naturale Alpi Marittime. The intervention by the then-Maire of Menton referred to authors such as Augustin Berque and Martin Heidegger, stressing the societal dimensions of defining boundaries in the landscape:

> *Setting up an area of particular status within a society and in accordance with this particular society assumes the preliminary identification of the symbolic meaning*

> *ascribed to the territory. As Augustin Berque underlines, 'the stronger the symbolic charge, the stronger the appropriation'. The delimitation of the territory that will gain this particular status will allow eco-symbolic processes to happen, reactivating archetypes. (...) But before such a new transboundary space can be created, we must ask ourselves about the symbolic meaning that is linked to it. Perhaps it would be necessary to start with each existing park and grasp whether there is a symbolic meaning attached to other territories than the 'Vallée des Merveilles'. (...) The multiplication of actors involved in the establishment of a natural space hints that beyond the simple establishment of an administrative structure and the delimitation of a particular territory, the real issue is the design by all groups involved of a social space*[6]
>
> *(Guibal 1999 : 46) (Parc National du Mercantour, France).*

Although this appeared in a joint publication of the Parc National du Mercantour and the Parco Naturale delle Alpi Marittime, it did not constitute a representative view shared by protected area managers. It was striking, however, to discover an overt reference to ideas such as the symbolic meaning of shared space and the need for the explicit social construction of spatial entities defined by negotiated boundaries.

Other ideas circulating included a romantic visions of boundaries and protected areas as romantic spaces, in the North-American sense of frontier lands, similar to some of the arguments discussed earlier describing protected areas as the 'last frontier'. Protected areas were here symbolic frontiers of wilderness set aside for human contemplation and experience, mirroring romantic images of man confronting nature's splendours:

> *Like many men [sic] of my generation, because we had been taught that we were born into the age of the bounded world, I dreamt of finding its lost paradises, striving to maintain these privileged places where wild nature is still present and preserved for the benefit of all*[7]
>
> *(Servat 1999 : 13) (Parc National du Mercantour, France).*

The image of romanticised natural space that ignored human boundaries replaced that of a territory divided between two states, giving an alternate meaning to 'boundary zone'. This echoed the myth of boundless nature that recurrently appeared in this literature.

Biophysical determinism: inevitability of cooperation

The idea of nature without boundaries, stretching to the outer reaches of the earth ignoring human political divisions, appeared as a leitmotiv in transboundary protected area publications. This strong image very often popped up early on in texts and speeches, often in the first paragraph, leading on to the idea of an inevitability, a 'naturalness' of subsequent political cooperation. This almost exactly mirrored the images used in the international publications, constructing icons out of locally-relevant fauna and flora. Animals and nature thus showed the way, indicating the inevitability of cooperation between administrations:

Ibex and chamois were the first proponents of the need to collaborate: they have never taken boundaries into account and those tagged in Italy were observed in France and vice versa[8]

(Rossi 1998 : 7) (Parco Alpi Marittime, Italy).

Nature and animals have shown us the way and despite the many difficulties that we have had together there are remarkable results emerging from this common work[9]
(Westphal 2002 : 23) (Parc Naturel Régional des Vosges du Nord, France).

Shared fauna and flora were joined by endemic species, further creating an image of exclusivity, of homogenous space uniquely different from the surrounding landscape:

Not only do the parks complete each other on a geographical level, but they also do so with their fauna and flora. While they share exceptional biodiversity on either side of the boundary including such species as eagles, wolves, ibex, chamois or else marmots and stoats, each also possesses specific endemic species, such as the saxifrage[10]
(Anon in Monts & Merveilles 1998 : 9) (Parc National du Mercantour, France).

While these facts were uncontroversial, the subtle manipulation of such icons into a political vision was more problematic. Here, the biophysical conditions were seen to bring about a spatialised political project in a step seen as perfectly evident and natural:

What we have just described (...) shows that we are a natural unit. Moving towards one unique natural park is simply a natural consequence of what has been observable for years. Considering this massif as one homogenous entity is a logical consequence[11]
(Malausa 1999 : 28) (Parc National du Mercantour, France).

Likewise, the combination of the different spaces within different sovereign states was seen as inevitable for topographical and geometric reasons. They were two natural halves (or thirds ...) of one pre-existing natural unit:

Just examining the two parks on a map is enough to show how both complete each other geographically: one contributes the extent of its territory and its diverse geological, climatic and ecological situations, the other adds to the merging of these two spaces the indispensable thickness and depth necessary for the survival of the different wild species, something that the extended shape of the French park was unable to offer[12]
(Anon in 'Charte de Jumelage', 1998 : 14)
(Parco National du Mercantour/Parco Naturale Alpi Marittime).

More subtly, this shared nature was seen to imply a shared responsibility and a duty to work together despite political and cultural differences. Furthermore, nature not only had a space of its own, but also a unique timeframe, a permanence that needed to be taken into account:

It would be appropriate to ask whether nature never knows boundaries, or to put it differently, whether these boundaries have been created by humans and determined

> *culturally? Is nature showing us a path we have to overcome these boundaries and live together in order to survive? That means that if nature knows no boundaries, then it becomes our shared responsibility. Not a responsibility limited to states, communities or circumstances but rather a common responsibility that concerns all humans*[13]
>
> (*Duppré 2002 : 4*) (*Naturpark Pfälzerwald, Germany*).

The creation of a homogenised space was seen to be a return to a former entity, rejoining elements that had been artificially divided by history. Even cultural landscapes were seen to be naturally shared creations built on the same biophysical substrate:

> *Because this is a symbol that something has grown here together, nature and geology – in this area coloured sandstone – the forest, the cultural landscape, shared similar historical and economic development have meant that the area is already strongly united. So this means a lot for us today*[14]
>
> (*Conrad 2002 : 6*) (*Naturpark Pfälzerwald, Germany*).

The similarity of the substrate was seen as a more defining feature than history and images of mountains as barriers were replaced by mountains as unifying elements determining particular lifestyles, echoing the 'mode-de-vie' descriptions beloved of classical French geographers. Paradoxically while cooperation was universally taken to be natural and therefore always positive, there was some indication that it might be not entirely straightforward in practice.

> *On either side of the boundary, a theoretical line drawn by contingent policies, many intimate links have been woven throughout the centuries, strong links between the populations of the high valleys (both on a commercial and a cultural level). The mountain never was an obstacle when it was crossed on foot and the similarity in lifestyle among men [sic] in this territory is easy to understand*[15]
>
> (*Servat 1999 : 14*) (*Parc National du Mercantour, France*).

> *It is evident now, as in the past, that working in the same field of nature conservation of the same natural unit – the Tatra Mountains – with their similar characteristics and history – the two national parks have to cooperate very closely. Experience shows that cooperation always produces good results, even if the cooperation is not always satisfactory for both sides*
>
> (*Krzan 2000 : 69*) (*Tatras National Park, Poland*).

The second quote seemed to contradict itself, with the reality and complexities of day-to-day cooperation between separate administrations shattering the myth of 'good results'. There was an obvious hierarchy of interests, with nature having priority. For how else could results be always good if they were not satisfactory for both sides? There was a hidden third player: nature containing a political agenda. Politics, power struggles, and inequalities between administrations shook up myths of integrated fauna and flora.

Reference to other scales of spatial integration

In addition to compelling visions of shared nature, discourses of integration were often explicitly situated within wider spatial contexts. Thus explicit calls to 'Construire l'Europe' and 'Surmonter les réticences historiques' were linked to more local projects of creating integrated spaces. The idea that transboundary protected areas were laboratories for wider integration was repeatedly alluded to, often with the implicit idea that cooperation must be made to work lest the European project be weakened. Successful cooperation took on meaning beyond the boundaries of the protected areas, leaning on the sense of a responsibility to produce success. This leitmotiv was particularly present in the many speeches made during inaugurations and ceremonies, skimming out of the mouths of practised politicians, setting local situations in wider historical trends:

> *They are of high importance for the increased togetherness within Europe and therefore have special responsibility*[16]
>
> *(Conrad 2002 : 8) (Naturpark Pfälzerwald, Germany).*

Thus in a world with no more material boundaries, where everything was technically possible, political boundaries could be overcome with willpower. The idea of a European challenge or even crisis, rather than provoking a rethink, became an opportunity for proving spatial integration on a local scale, thus erasing the fear of a failure of the wider continental project:

> *The construction of Europe, in the field of protected areas, will undergo a rapid acceleration. At a time when the political integration of states remains as distant as ever, two parks, one Italian and the other French, are committing themselves to building one common entity in the Southern Alps*[17]
>
> *(Anon in Monts & Merveilles 1998 : 9) (Parc National du Mercantour, France).*

Nevertheless, if it was unclear what European integration involved, then the creation of local spaces was equally problematic. Such uncertainty was identified as a positive challenge requiring active (re)territorialisation. Thus the objective was not only integration at all costs, but rather the sharing of experiences linked to an acknowledgement of the specificities of the Other. In certain cases, this knowledge was recognised as an agent of change integrated into the process:

> *The creation of a transboundary biosphere reserve will bring about further opportunities to establish common projects enriched by diverse approaches in different fields of work*[18]
>
> *(Anon 1996) (Parc Naturel Régional des Vosges du Nord/Naturpark Pfälzerwald).*

These references to other scales of integration, in this case overwhelmingly linked to the European project of a pan-continental European Union, were ubiquitous. The objective was to link what was happening on a local scale to wider historical changes taking place at other scales. In addition, references to other spaces were joined by references to other historical epochs.

Reference to past and present spatial entities

I mentioned above that one aspect of the 'boundless nature' paradigm was the need to take into account not only the scale of nature, but also its time, set in permanence and long historical time. A further temporal dimension is the reference to past historical spatial entities that were seen to be more pertinent than contemporary political jurisdictions. Justification for contemporary spaces (or contemporary space-times) was thus found in reference to shared history, as well as to a shared biophysical environment. In these arguments, past spatial entities were taken to be more relevant or pertinent:

> *History, ancestral channels of communication, everyday languages and practices have always brought the people living in the heart of these protected areas together. These men [sic] of a generous nature are above all men [sic] of the mountains: hard working, brave and obstinate, proud of the cultural and natural heritage that they have inherited. Today, the parks decide to further this common history around a common mountain: a mountain without boundaries*[19]
>
> *(Anon in 'Charte de Jumelage' 1998 : 2) (Parc National du Mercantour/Parco Naturale Alpi Marittime).*

In a similar vein, although phrased in slightly less lyrical language, a bilingual text written by the Parco Naturale Alpi Marittime constructed a curious amalgamation of past times, describing the centuries-long interactions in order to justify the transboundary entity:

> *Men [sic] have inhabited the mountains of the Maritime Alps since prehistoric times, as proven by the thirty-thousand stone carvings in the Vallée des Merveilles and the Vej del Bouc. They constructed paths such as the Salt Route linking up Nice with Cuneo. Contacts between them were facilitated by a common language: Occitan. This strong cultural link allowed the development of similar traditions on either side of the boundary*[20]
>
> *(Anon 1998b) (Parco Naturale Alpi Marittime, Italy).*

> *The two parks don't only have nature in common: they are also united by history, from the ancient times of the stone carvings up to Vittorio Emanuele II and more recent history*[21]
>
> *(Mucciarelli 1998 : 2) (Parc National du Mercantour, France).*

Thus not only was space compressed into one entity, but time underwent the same process of being compressed in order to justify the uninterrupted occupation of the area by one supposedly homogenous population. Homogenous unified space-times were emphasised to the exclusion of other historical periods or understandings. Discourses stressed the historical unity of the area, regardless of the fact that political boundaries were usually the result of political conflict and war. Rather, the biophysical arguments were linked with historical ones in order to project a vision of uniform space:

> *This mountain that nature and history have never ceased to consider without boundaries*[22]
>
> (*Ginésy 1998 : 2*) (*Parc National du Mercantour, France*).

In certain cases, historical arguments were more circumspect, with comments alluding to the fact that history was perhaps not always sunny along the boundary. Both the Franco-German and Franco-Italian boundary areas, for example, were studded with abandoned bunkers, fortifications and fortresses within the protected areas. Along the Polish-Ukrainian boundary, the old USSR fence still remained largely intact. Likewise, Poland was continuing to construct new boundary paraphernalia along its Eastern flank, funded by European Union programmes seeking to reinforce the Eastern boundary of the European Union. While in certain contexts and literature boundary architecture was exploited as sites for tourism and indeed often formed part of outdoor museums and visitor centres, such facts were shunned in transboundary protected area literature. Instead, the emphasis was on the common imprint of history, even if the stories on the ground were different. In the Franco-German case, history was called upon not in order to project a historically unified space, but rather to indicate the shared richness of the two territories. When historical conflicts were mentioned, the healing powers of cooperation were instantly referred to, as was the shared experience of conflict on both sides of the boundary:

> *This landscape is equally rich in history and culture*[23]
>
> (*Anon 1996*) (*Naturpark Pfälzerwald, Germany*).

> *I should like to return to the historical meaning of this area. France and the Pfalz or rather Lorraine and Alsace and the Pfalz had many periods of unity throughout history, but they also have sad and troubled times, specially in the last century, but also in the 19th Century. We often fought each other in wars, but the people in the boundary zones always understood how to build up friendships after conflicts*[24]
>
> (*Conrad 2002 : 8*) (*Naturpark Pfälzerwald, Germany*).

Yet history was not always whitewashed and past conflicts were also alluded to, before the 'healing' powers of a transboundary protected area were mentioned. In the East Carpathians, the bloody history of the area served to justify its unity, regardless of the fact that the effects of war on local populations within each of the three countries were different. Thus history justified the present, suggesting an inexorable move towards integration despite contemporary divisions:

> *This corner of Europe was renowned throughout the continent for both its charm and rich and diverse folklore, as well as for the various dramatic twists and turns to the history of the area's warlike mountain nations that were regularly pitted against one another. Blood was very often shed in these border areas, which were as a result*

abandoned and then recolonized, while churches were built, desecrated or burnt before being worked on again from scratch. It is in this region, so full of distressing experiences from two World Wars leading to devastation and depopulation, that we are today founding our international Biosphere Reserve

(Breymeyer 1999 : 7) (East Carpathians Biosphere Reserve, Poland).

These men already close in the past will rediscover each other in the present in order to grow even closer in the future[25]

(Anon in 'Charte de Jumelage' 1998 : 12) (Parc National du Mercantour, France).

The fear of otherness was replaced by a playful recognition that political boundaries might not reflect present realities:

When we half jokingly say that in this region we are sometimes a little closer to Paris than to Berlin, then that should show you to what extent and how strongly we are bound up together with our French neighbours[26]

(Conrad 2002 : 5) (Naturpark Pfälzerwald, Germany).

This idea of being closer to the neighbouring capital was in itself interesting, although rather than reflecting the proximity of French and Germans in the boundary zone, it could be a reflection of increased internal distinctions between different conceptions of 'German' identity.

These references to a shared historical past and therefore a shared identity however remained ambiguous. If such identities were really shared, then the overt emphasis on reinforcing them would be unnecessary. Yet arguments stressing the reality of shared identity barely resisted moving on to reinforce this reality, seeming to indicate its paradoxical fragility:

Certain projects will be defined in order that the two areas contribute, according to their own specificities, to the bringing together of the inhabitants who share one same transboundary identity[27]

(Anon in 'Charte de Jumelage' 1998 : 14) (Parc National du Mercantour, France).

Yet despite this supposedly unproblematic existence of a transboundary identity, the following section of the Charter quoted above was entitled 'Un objectif: créer une identité transfrontalière'! This issue of shared identity was a leitmotiv on the same level as the idea of boundless nature. Likewise, it was rarely problematised. However, building on Claudio Magris' superb reflections on boundaries and identity, the director of the cultural bureau in Cuneo noted that identity need not be considered rigidly:

I believe we must avoid holding on to a rigid conception of boundaries. There is a perverse form of exclusionary logic that must be avoided and transcended and replaced by a conception of identity (and therefore also of boundary) as a dynamic process, open to all forms of contamination, something that can be lost and found, constantly fluctuating between loss and a return to basics. (. . .) I believe that we must be careful to construct the identity of a geographical region in a historically correct and well documented manner, but subsequently this historical heritage must not be viewed in a fundamentalist manner, but rather on one hand as a contribution to the transformation of culture and on the other as the acknowledgment of a historical heritage that requires preservation[28]

(Cordero 1999 : 49) (Parco Naturale Alpi Marittime, Italy).

Thus transboundary cooperation, rather than simply reproducing an existing rigid spatial identity linked to a fossilised conception of space and time, could take part in the construction of new spaces of belonging. There was a positive embracement of the Other, building on the recognition of existing differences. Such ideals were not only promoted for themselves, but were specifically called upon to reinforce the desired political vision of one transboundary space. Yet it remained controversial in this literature whether such transboundary spaces were constructed or revealed through such experiences:

Visitors passing through the two areas by walking along the same paths, making use of information common to the two parks, will slowly realise that both sites share a common transboundary identity[29]

(Anon 1998 : 2) (Parc National du Mercantour, France).

Conclusions on protected area discourses

In contrast to the largely normative discourse of international publications, the discourse within local protected area publications showed some diversity, offering a selection of visions of boundaries and identity. Nevertheless, as the above quotes illustrated, it was easy to identify a selection of predominant themes emerging in all the sites considered. Thus the pervading myth that political processes (cooperation) were determined by biophysical processes (shared nature) largely took over the debate, no doubt reflecting to a large extent the background of many managers rooted in the natural sciences. Again, this was ontologically suspect, reflecting an impossible merging of realist and anti-realist conceptions of boundaries.

While boundaries were recognised to be more than simply 'bad' things that had to be ignored if nature was to be protected, as in the discourse of international organisations, the idea that shared features provoked shared management was pervasive. In addition to homogenised nature, a crude understanding of history and culture were called on to justify the homogeneity of space. Furthermore, by contextualising transboundary cooperation within a wider discourse of spatial integration, protected area administrations argued for the inevitability of cooperation, associated with the historical move towards a borderless world. However, while all recognised forms of integration to be a good thing, none of the publications considered attempts to define it. The problematic nature of integration only appeared on the ground, in the mouths of people involved in projects and collaboration on a day-to-day basis. It is this discourse that is considered in the next section.

Protected area managers: making sense of (re)territorialisation

Unlike the official discourses on transboundary protected areas that sought to construct homogenous spaces by referring to a selection of well-exercised clichés and myths, individual protected area managers produced more critical reflections. In this section, I present some initial themes linked to (re)territorialisation and cooperation which are further developed in later chapters. The focus is on the creation of imagined communities, leading to an identification of some of the institutional and legal issues that participated in the negotiated understanding of what 'integration' implies and means. This was the first step on the road to cooperation, participating in the definition of 'Self' and 'Other', subsequently suggesting the extent to which they were to be distinguished or integrated.

Imagined communities: emerging spaces of socialisation

The construction of transboundary entities was nothing less than the imagining and subsequent construction of new spaces of socialisation: imagined communities uniting a group and a territory. This pointed to the two stages of the process, in which the imagining preceded the construction. The themes that emerged here largely followed those developed in the protected area literature, hinting at the co-construction of both discourses. In this first section, some of the returning themes of biophysical unity, references to other scales of integration and shared history are mentioned. Some of the statements seem to echo almost perfectly the discourse present in the literature, as this comment made by a protected area public relations attaché implies:

> *Following the meeting in Menton, we first of all put together an atlas (. . .), I can show it to you, it summarises all the scientific data that shows that the two parks are really part of a common natural and cultural entity*[30]
>
> *(Chloé, Parc National du Mercantour, France).*

Thus oral discourses sometimes followed those present in the literature, particularly in interviews which never achieved a satisfactory level of trust. In addition to myths of biophysical integration, references were made to past spatial entities, to times when boundaries lay in other places, again mirroring the written texts. Often, this was linked to the haunting idea of finding some sort of 'original' scale of planning, similar in nature to mythical 'boundless nature'. Allusions to other boundaries and scales were made, such as past Empires and the impact these had on contemporary identity and boundaries:

> *Politically this area has been influenced by the Empires: the Ottoman Empire, the Russian Empire, the Austro-Hungarian Empire . . . We have fourteen ethnic groups in the Delta. This makes the Delta very special*
>
> *(Gheorghe, Danube Delta Biosphere Reserve, Romania).*

In this case, ethnic heterogeneity became a badge of identity and the homogeneity of the spatial entity was seen to come specifically from its internal heterogeneity, as opposed to what was perceived as ethnically 'pure' space surrounding it.

While some of the themes emerging from the interviews echoed the themes developed in the previous section, examples of cooperation and collaborative projects were often referred to much more critically than in the written publications. 'Integration' was not taken-for-granted and instead was identified as a problematic notion. The inherent uncertainties in the process of integration appeared, and indicated its highly uncertain nature in which objectives and end results were constantly transformed along the way.

The creation of an imagined community and the corresponding emergence of new spaces of socialisation underpinned the construction of transboundary entities. Knowledge of the Other was crucial to the process of imagination as both a pre-requisite to projecting possible spatial scenarios and a feature of the subsequent construction. This process of imagination was highly uncertain, emerging as a process of confrontation between Self and Other, leading to an unknown result. In effect, nobody actually knew what a 'transboundary entity' might look like. Referring to existing international programmes such as the biosphere reserve programme did little to remove this uncertainty since there is no formal understanding of what a transboundary biosphere reserve was:

> *But what can we imagine, what do we put behind a transboundary biosphere reserve? What would a transboundary biosphere reserve be? Would it be one team managing all this territory? Is it the two existing teams, with common practices, we have a common office, a common meeting room but we remain two entities because of physical reality, and distance, it's hardly next door. Or else would it be a very extensive interpenetration of activities, I don't know, maybe bordering on having only transboundary activities. Nobody can say, in fact, and it depends on the people you speak to, there is the whole range. But, at some stage, everybody, that is to say both biosphere reserves, will have to agree on what a transboundary biosphere reserve means for them*[31]
>
> *(Hugo, Parc Naturel Régional des Vosges du Nord, France).*

The creation of a new space of socialisation relied on a number of tools, including the use of representation and iconography to project a desired construct (Chapter 8). The uncertainty in what was to be achieved by cooperation was often missed when questions were phrased assuming that 'cooperation' was a taken-for-granted fact. Yet this was rarely the case as conflicting 'imagined communities' and conflicting visions of desired (re)territorialisation remained not only between but also crucially within administrations. Thus the discourses emerging on the ground were much less uniform than those within the more clichéd official publications. Oppositions and contradictions emerged and confronted each other. Resistance was voiced while persuasion and propaganda took place on the scale of individuals.

Initiating cooperation

It was often revealing to discover who was seen as the first proponent of cooperation, as this often conflicting story formed the first stage of a quasi-mythical history, reproduced in a variety of forms. In the Vosges du Nord, the French saw themselves as having put pressure on the Germans to agree to the creation of a transboundary space:

> *And so we were classified as a biosphere reserve, we arranged it that in 92 on the basis of our initiative and pressure, the Naturpark Pfälzerwald was also recognised as a biosphere reserve. Particularly from the start as a common biosphere reserve, in other words pioneering the idea*[32]
>
> *(Théo, Parc Naturel Régional des Vosges du Nord, France).*

In this case, the French director was clear it was in his administration's advantage to bind cooperation activities and projects with a clear spatial entity defined on the ground, something that was not necessarily shared by the German partners. In this case, the situation was negotiated, with the promise of financial gain encouraging the German Naturpark to participate in this form of shared space. At the same time, German managers identified a former German Naturpark President as instrumental in the first stages. Yet it was not necessarily clear from the outset what was actually being sought in such a process. In any case, this was likely to vary also in time, as increased contact between partners changed their respective representations and ambitions – (re)territorialisation was a dynamic and multifaceted process. Chloé, in the Franco-Italian context, identified the different stages of integration clearly, stating that the desired objective was likely to take many years:

> *Globally, the objective is already to make the most of the entire heritage, both the natural and cultural heritage, and to move forward by taking a series of concrete actions on the ground, while at the same time moving towards the creation of a European park on a jurisdictional level. Doing both at once bearing in mind that the evolution towards a European park may take a decade and that it must be visible on the ground*[33]
>
> *(Chloé, Parc National du Mercantour, France).*

This construction of new spaces inevitably implied a form of negotiation between partners, a mutual recognition that we/here and the Other/there held equal standing in the process. This could not necessarily be taken for granted, since all relations were enmeshed in an inescapable web of power relations. Chloé's comment indicated some of the ambiguities of this process of spatialised negotiation, in which both partners sought to 'extend over the Other' their particular social and spatial practices.

> *If one day we had a project, for example, of a reserve that we wished to extend over into Italy, well then we could only act through persuasion. It's the same for them. We don't have any jurisdictional way of doing it. And even when we set up Interreg projects, for*

example, it's true that we are required more and more to chose one park to be responsible for the whole and at the moment we don't know how to do that[34]
(Chloé, Parc National du Mercantour, France).

This quote was more ambiguous than it at first seemed, since Chloé implied that 'one day there may be a project', although at the time this was officially endorsed by the protected area employing her, at least within its published literature. In many ways, this further reflected the diversity of discourses present at any one time within a given situation, where conflicting views coexisted among the involved actors, each struggling to get him/herself heard.

The creation of imagined communities was often explicitly linked to wider discourses on integration at other scales, as identified in the previous section on published discourses. 'European integration', for instance, was frequently referred to in the Franco-German and Franco-Italian contexts. This was not to say that it was unproblematic, nor that it had a universal understanding. References to other scales of integration could be seen in a positive manner, referring to ineluctable shifts, or else on the contrary perceived as an imposed merging that should be resisted.

This must not be forgotten: in transboundary work, even if the boundary no longer exists on paper, even if all this is now part of Europe and all that, these are just words and there is still a political, administrative and economic reality that is radically different and that is even before you start talking about cultural issues[35]
(Hugo, Parc Naturel Régional des Vosges du Nord, France).

Here, Hugo referred to European integration almost as a shibboleth, as an agreed but mistaken discourse that obscured real issues of difference. In the case of the Pfälzerwald, the reference to European integration came almost as an afterthought, as though in deference to what would be expected from the question:

[Juliet] What do you think the basic reason for this cooperation was initially? [Silence] Was it political? Was it to do with nature conservation?
[Maximilian] Expenses. And the European construction
(Maximilian, Naturpark Pfälzerwald, Germany).

The examples above reflected some of the attempts to define a form of imagined community, echoing some of the themes developed in the section on protected area publications. However, unlike the taken-for-granted understanding of integration present in that literature, the highly contextualised understanding of integration only fully emerged in the interviews.

Determining what integration means

While many official speeches called for increased integration, the notion was highly problematic for people actually seeking to achieve this on the ground. Common projects could be one element of cooperation leading to integration.

These projects participated in the process of getting to know the Other, discovering commonalities and differences. As such, they participated in the construction of transboundary entities, yet it was not clear whether such collaboration contributed to spatial integration.

> *Before 1990, we had some contact with the Ukrainians ... It was scientific contacts for scientific cooperation, mainly within Donavsky Plavny. At this time, it was my father who was the director. After this time, it was Ukrainian scientists who came to work, zoologists, herpetologists, botanists. We also had a yearly symposium for sharing results. This is now every two years. The Ukrainians, they like to participate, because they recognise the Romanian expertise. They know that we are more advanced. (...) Still, the Ukrainians have a very high level of scientific knowledge*
> *(Gheorghe, Danube Delta Biosphere Reserve, Romania).*

In areas where boundaries were still very politically charged features and cooperation was regarding with suspicion, such confidence-building projects were very highly regarded. These projects allowed managers to identify partners in the other country, learn about differences and similarities in both work patterns and administrative structure. In many cases, this collaboration on specific projects was considered to be the ultimate objective, as it stopped short of threatening anyone's position, was seen as a possible channel for international funding and achieved identifiable results. As such, it constituted a form of integration, albeit minimal.

> *In the past ten years, it has become easier to have contact with the Slovaks. It was more difficult with the old political system, because even if it was the same political system the barriers were harder. Before, there were very negative effects of having contacts with foreigners. So before we only had contacts on the official level, between directors, but now it is more between lower levels. We have a programme for counting chamois together, for instance*
> *(Jurek, Tatras National Park, Poland).*

Yet paradoxically, in the Polish Tatras situation, the opening up of the country had also brought negative effects for transboundary cooperation. Not only had the Polish national park come up against practical problems in identifying partners in Slovakia, it had also decided to follow a wider understanding of 'cooperation'. This led to a reshuffle of focus away from Slovakia – formerly the only possible partner for 'international' contacts – to other protected areas in more distant locations. Park rangers were more likely to go to the United States or Britain for additional training, rather than Slovakia. 'Opening up' boundaries often had unforeseen results.

Within these projects, there was often a sense of inequality between partners. Managers often found it necessary to justify their relative level of advancement or expertise, stating that the Other was less advanced and therefore more eager to collaborate. While often relatively amusing, such a reflex is part of the process of identifying the Other, implicitly negotiating relative positions for Self and Other within a imagined community. In the case of a tri-lateral entity, such construction of socio-spatial communities was additionally complex, leading to a variety of projects and alliances dependent for example on relative ease of communication and contact:

> *There was also lots of research undertaken because of the biosphere reserve, and we got some money from the Foundation for Eastern Carpathian Biodiversity Conservation. We exchange a lot of experiences with the Polish staff. We have one meeting per year in Poland. We also have one common publication each year. There are also contacts between scientists on both sides. It is much easier for us to have contacts with Poland than with Ukraine*
>
> (*Dominik, Poloniny National Park, Slovakia*).

This construction of Self and Other took on many forms, as each got to know the Other. Recognition of the conditions of work of the neighbours and the additional knowledge this implied, participated in the construction of shared experiences:

> *I admire the level of dedication of the people in Ukraine. Really, they work hard*
>
> (*Nicolae, Danube Delta Biosphere Reserve, Romania*).

The extent to which simple collaborative projects challenged existing practices and representations varied tremendously, leading to a corresponding uncertainty over whether these constituted different levels of integration or simply levels of interaction. In the Alps, transboundary projects that did not upset any existing practices have been ironically called friendly and pain free operations, promoted specifically by those who wished to avoid addressing more politically charged issues. Challenging the status quo was not always in everyone's interest and thus certain actors occupy the land with inoffensive and gratifying projects that promote good neighbourly relations (President of ProMontBlanc, 2002, *pers. comm.*), while subtly holding away other actors who sought to address more emotive or controversial issues. It seemed to imply that projects usually identified as 'confidence-building' for those involved were in fact counterproductive and purposefully designed as smoke screens. This form of cosmetic cooperation was identified as some as the objective of transboundary protected areas. Others, however, remained frustrated by this position and called for more radical definitions of integration:

> *There are no in-depth projects. Real projects that require intensive consultation. As long as it is only about setting up a cycle path from France to Germany, with signposts along it and maybe even a bridge, and even then it is not straightforward! A bridge had been planned between the two countries, I cannot remember exactly where, south of Strasburg, and it's still not built and we have been speaking about it for thirty years. It makes you laugh, but, really, it just goes to show that that's the point we are still at in cooperation, that's to say that in order to keep things positive we don't achieve anything, we don't attack things in depth. Everything is fine because nothing is happening*[36]
>
> (*Hugo, Parc Naturel Régional des Vosges du Nord, France*).

Thus cooperation based only around collaborative projects only went so far, since more than anything else it was designed to replace the need for the creation of a truly transboundary entity. Inevitably, for those attempting to promote a more challenging and far-reaching definition of integration, such 'superficial' cooperation was frustrating. This sense of frustration was shared by people who held a more radical view of what the construction of a transboundary entity ideally entailed. However, such a vision that directly threatened the status quo often created tensions both between and within administrations. It was this appearance of power struggles among actors that uniquely appeared within in-depth interviews, challenging the homogenising published discourses:

> *So in the context of the 1985 convention we set up transboundary paths, putting up information panels on the boundary, we made car parks together, for the 'Erholung' (Recreation) as the Germans say, and they we quickly realised that apart from getting elected people all together once a year for a good meal, and patting each other on the shoulder, we didn't have much to move on with*[37]
>
> *(Théo, Parc Naturel Régional des Vosges du Nord, France).*

The situation was therefore identified as being stuck at a certain stage, while these two managers understood integration to mean more than simply carrying out common activities. Instead, integration for them was seen to be an inescapable process of merging two socio-spatial structures:

> *After all we have a mandate from UNESCO to cooperate because, now if you want, everything has been moving in the direction of making the two partners cooperate inevitably. So now it is no longer possible to go back and we can only go forward, and we realise that in order to go forward and despite the current obstacles, in order to go forward we will have to solve the problems we face in depth. This is something we were not prepared to do, we weren't able to do at the start because we were suffering from the syndrome, as Emmanuel said, of Franco-German friendship. We didn't dare get angry, because, you see ... All that is now over. From the minute we decide to work in depth, we have to work as we do amongst French colleagues, which involves saying so when something is wrong. If there is a problem, we rub up against each other, we discuss things, we debate but above all we don't just remain on the surface of things because if there is to be cooperation then we must get to the bottom of things*[38]
>
> *(Hugo, Parc Naturel Régional des Vosges du Nord, France).*

This Franco-German example was interesting as it reflected the intense power struggles that appeared as soon as conflicting visions of integration were identified. Thus integration, unlike collaboration, was a necessarily contested notion. In this situation, the German director spoke of the different collaborative projects as a succession of independent activities. On the French side, however, the increasing desire to fully integrate the two territories led to increasing frustration. One manager related the different stages of cooperation:

> *And so we had a first Interreg I programme which was a programme of getting things rolling, with activities in line with Franco-German friendship, common maps, we set up*

> *a transboundary walking trail for educational purposes, we had Emmanuel as our common coordinator. We also did some activities linked to tourism: we published a brochure with 66 ideas for excursions within the biosphere reserve. And so on. Nice, but very much to get things going a bit. Things that are easy to do*[39]
>
> *(Théo, Parc Naturel Régional des Vosges du Nord, France).*

Typically, the protected areas followed the confidence-building approach of initially organising a series of uncontroversial and symbolic projects. Yet as time passed and understandings of what the creation of a transboundary biosphere reserve entailed changed, frustration started to build. This conflict appeared at the point when projects started to require closer integration between the two administrative structures, without a shared definition of what this entailed. When left to fester, such tensions inevitably led to resistance by some of those involved.

Thus defining integration was a difficult, contested process. At the very least, it was accepted to be a process of social and spatial change that took place over a substantial period of time, although it remained unclear how long. Determining the speed of change was a difficult issue in itself, as different conceptions of change over time coexisted within and between administrations. This was not a straightforward march, and involved gathering together a variety of individuals and decision-making structures in order to achieve a negotiated result. The complexity of decision-making structures in protected areas meant that information needed to be distributed before such a consensus could be reached:

> *We were waiting for the atlas to really have a solid scientific basis. We have the support of politicians. (. . .) We need agreement from our board of managers in order to apply for biosphere reserve status. (. . .) So now, at the moment, we have the informal support of the president, but it must go through the official process of getting accepted by the board*[40]
>
> *(Chloé, Parc National du Mercantour, France).*

Yet while this sounded straightforward, a different employee linked to the same protected area identified the lack of information circulation as a key point in the lack of progress made towards further integration. He referred to a document comparing the use of graphics within the two protected areas that sought to identify commonalities in order to suggest a common graphical identity for a future transboundary entity:

> *And anyway, this document on the graphical identity of the two parks has only been passed on to Chloé and [another colleague]. When you speak to people about this document, nobody has heard of it. The Italian person also has it . . . (. . .) But I don't think anyone knows about it there either. It went too far, and they don't know how to use it. The Mercantour, in my personal opinion, is not able to diffuse information effectively*[41]
>
> *(Thomas, Parc National du Mercantour, France).*

In other words, the construction of a common graphical identity – one aspect of integration – anticipated the creation on the ground of a shared space. Because this document did not reflect a consensus, it was suppressed, whether voluntarily or involuntarily, as it reflected a change of practices that went further than anticipated. Constructing shared spaces took time and involved an ongoing negotiation of what integration entailed. When this integration was perceived to be contrived or rushed, then counter-productive resistance appeared.

The long quote below further combined some of the issues discussed above, finally concluding that any change must be given sufficient time:

> *Let's say that we don't mind about this but the facts speak differently, the spirit of things and all that, things that we have invested in the transboundary biosphere reserve. Well OK, it's clearly not going to be tomorrow that the two territories will really be a transboundary biosphere reserve in spirit. Everything depends on what you want, what you wish for. At the present time, I think we can say that we expect a lot, we would like to do more, but nevertheless I think it is fair to say that we are not one of the worst European biosphere reserves, because we do still have lots of work carried out in common, lots of things that mean that either with or without the Naturpark as a partner, I mean we do have team meetings, amongst ourselves of course, it happens often, and there is an impetus, we did that first fieldtrip together, and that was the first time a Franco-German group did that, and everyone recognised the need to do more. So some things do go beyond a simple common signpost or a common path. But all this is a long process and I think it would be a mistake to try and do things too quickly, so we must leave time to play its part*[42]
>
> *(Hugo, Parc Naturel Régional des Vosges du Nord, France).*

From this brief discussion of a series of quotes, it was clear that conceptualising transboundary cooperation simply in terms of levels of integration between 'borderlands' – as expressed in some of the traditional approaches mentioned above – was totally inadequate. Spatial integration was more than simple interactions, although such interactions participated in the construction of a transboundary entity as individuals struggled to find meaning and purpose in their work.

Institutionalising cooperation: overcoming legal and institutional boundaries

It was impossible to discuss the creation of transboundary entities without referring to institutional and legal issues. Obviously the simplest expression of this impact of legislation happened in cases where legal frameworks forbade cooperation:

> *The Ukrainians, they would like to cooperate more with us, but the Ukrainian law makes this difficult. Also, the law in Ukraine is based on the old Soviet law, so it does not follow the IUCN protected area categories*
>
> *(Nicolae, Danube Delta Biosphere Reserve, Romania).*

Often, this happened because funding designated for use within a given protected area could not be spent 'abroad', funding travel across an international boundary, for example, or else funding a transboundary project. However, apart from legal restrictions on cooperation, the law had other impacts on the creation of transboundary entities. The lack of common legislation or of an accepted uniform way of doing things was often directly identified by protected area managers as an obstacle to constructing new levels of spatial integration. In this context, integration was considered to be an administrative and legal process which implied a change in decision-making structures and legislation. The lack of any accepted international standard for such areas or even of a recognised framework meant that all actions needed to be negotiated, both on an institutional and on a legal level.

> *And so there will be a time, perhaps in ten or fifteen years, I don't know, where if we really want it to be an united entity, we will have to find appropriate tools. And, for the moment, these tools don't exist, or else we haven't found them and so we really have ... For instance if we give a fine to someone, we cannot give them to someone over in Italy, there are a whole host of legal issues that mean that, well, there is French law and Italian law. It should be possible to move beyond this stage and specifically on the level of the coordinating body, that is to say that for the moment, even if we do lots of things together, there is no way we can force this, and we only do things together because we want to*[43]
>
> *(Chloé, Parc National du Mercantour, France).*

Chloé identified the lack of tools, both legal and practical, as a hindrance to increased integration. These tools included both the simple need to make jurisdictions compatible as well as getting a form of recognition for the unit as a whole. The need for a formal umbrella was what made a certain number of protected areas opt for designation as a transboundary biosphere reserve, since this was seen to offer some sort of soft law international recognition. However, deciding what institutional and jurisdictional shape a transboundary entity was to have was often a long process. In the East Carpathians, for example, the idea dated back to the late Sixties, and subsequently underwent discussion and debate until the choice of setting up a transboundary biosphere reserve was made:

> *It was a difficult process: the first idea was to create an international national park but this was difficult because of the various statuses between the different protected areas in each country. It was difficult to create one area with one name. The biosphere reserve model was the simplest way of having a tri-lateral protected area. Now it would be easier to set up something like this because all three countries have national parks. At the moment, the biosphere reserve is the only possible model if we want to include Ukraine because Ukraine is not part of the PHARE programme or Natura 2000, for example*
>
> *(Dominik, Poloniny National Park, Slovakia).*

This lack of legal recognition for transboundary entities was repeatedly identified as crucial by many managers, compounded in the case of biosphere reserves by a lack of legal recognition within national jurisdictions for the programme itself. In addition to transboundary biosphere reserves and World Heritage sites, one protected area administration explored a further legal framework for establishing a transboundary entity within the European Union: a GIE or 'Groupement d'Intérêt Economique' which would allow different public or private bodies on either side of an international boundary to manage funds together for common efforts. However, while this was being explored it had not yet been applied and its suitability remained under discussion. International programmes awarding official awards or titles were seen to bring about not only a form of soft law status, but perhaps more importantly bring added prestige and the expectation of increased funding possibilities:

> *Now we have reached a stage where there is a need to coordinate, especially as there is an increasing desire to find alternative ways, for instance using the World Heritage Convention, it is something they criticised us for and asked us to present the project together (. . .). I think the World Heritage is something a bit mythical (. . .). We are interested in the glamorous aspects of such a label*[44]
> *(Chloé, Parc National du Mercantour, France).*

In the East Carpathians, the issue of legal recognition was identified by all partners as a very concrete hindrance to carrying out and funding common projects:

> *The draft agreement on transboundary cooperation has been reviewed during the past four years. We are waiting for the signatures from the Prime Ministers. This would mean that all relevant Ministries were formally involved – like the Ministry of the Environment, of Foreign Affairs, of Internal Affairs and so on. Because the biosphere reserve is about more than nature conservation. The biosphere reserve concept is a good one, but it is only carried out voluntarily at the moment by people and institutions*
> *(Andrzej, Bieszczady National Park, Poland).*

Issues of sovereignty inevitably appeared when formalised agreements were sought. In the case of the East Carpathians, there was little debate about what form integration should take and this was assumed to mean the promotion of common projects. Formal Agreements, Memoranda or Twinning agreements often sought the lowest common denominator, remaining as general as possible in order to get official approval:

> *Since the initial agreement was signed by the Ministers of the Environment, no new agreement has been signed by any government, nor by any Ministers of the Environment, to do with cooperation within the Eastern Carpathian Biosphere Reserve. The political situations in each country are fluctuating and have not permitted the various drafts to be signed. In order for these to be signed, they have to remain as general as possible so as not to complicate matters*
> *(Andrzej, Bieszczady National Park, Poland).*

This naturally had an important impact on how transboundary entities were considered and constructed by managers. Likewise, other forms of politically neutral recognition for 'transboundary' sites, such as that offered by the Council of Europe in the European Diploma, reinforced the idea that a transboundary entity existed, when the actual situation on the ground was simply that of adjacent entities. How these recognitions were perceived by managers was thus interesting:

> *The European Diploma is a recognition of the quality of the heritage, its exceptionality, as well as the quality of the management. So let's say that within the scientific community this is well known, although it is a little less known among local people, in fact what happens is that a lot of noise is made when you lose it. (. . .) The application was made together and each park received the same diploma. But the application was common. But there is no legally defined common diploma, so we each got the same, they have to do that*[45]
>
> *(Chloé, Parc National du Mercantour, France).*

In other words, other than recognising excellence on both sides of an international boundary, such quality labels did not constitute the recognition of a shared space. Nevertheless, by creating comparisons, they did participate in widening the scale of management considered. In order to gain such formal recognitions, protected area administrations had to formulate joint applications, although there were often institutional reasons that made such collaboration difficult.

In additional to legal issues, the construction of a transboundary entity implied merging a diversity of administrative and institutional structures, each with a different mandate and territorial extension. While this sounded straightforward in theory, on the ground differences in ways of work, administrative practice, size, accountability, or even levels of funding made this difficult. This meant that the creation of a shared space implied joining together different entities while remaining implicitly founded on a recognised equality between partners. This equality was however often problematic:

> *The Naturpark does not have the same structure as we do, as a territorial collectivity, it's an association, a Verein. They have a board of directors. That's it. So we have turned the situation over and over in our heads looking at every possible and imaginable solution. It remains a complete mystery to us why there is not an acknowledgement . . . It can't work, really, a couple cannot function if it is imbalanced, but here we really do feel that the whole thing is wobbly, both on the human and motivational level, as well as on the level of the respective mandates. (. . .) And it's impossible to work in these conditions, in the spirit of a transboundary biosphere reserve, I mean. It's not possible. If both stay on their own side, then maybe. But that's not our mandate*[46]
>
> *(Hugo, Parc Naturel Régional des Vosges du Nord, France).*

In this case, the construction of a shared transboundary biosphere reserve hit upon the different understandings of what a biosphere reserve should be, partially determined by the differing mandates of the existing protected areas. In other cases, the situation of identifying the Other was not straightforward and uncertainty about who the neighbouring partner administration should be remained:

[Tadeusz] The cooperation is based on personal contacts. There is a bit of a problem at the moment ... Who have you contacted in Slovakia? (...) Now it is complicated because there are two administrations in the Slovak Tatras: the State Forests and the national park administration

(Tadeusz, Tatras National Park, Poland).

Conclusions on the views of protected area managers

Unlike the discourse of local protected area publications, protected area managers did not systematically call on biophysical, historical or cultural images to justify cooperation. Rather, accepting that cooperation was identified as desirable, the more problematic issue of defining what a transboundary entity was dominated. This was illustrated by the multiple and diverse attempts to define integration. Furthermore, it quickly became apparent when listening to individuals that such definitions were rarely shared within one administration, let alone across international boundaries.

In addition to the highly contested issue of defining spatial integration, I have argued that the creation of transboundary entities implies addressing a series of institutional and legal issues. Thus the need to dream up and imagine new spaces of socialisation involved constructing spaces that were socially more complex than adjoining patchwork squares. The uncertainty of the process and the need to continually reassess and redefine the desired result stemmed for managers from the difficulty of defining exactly what a transboundary entity was. Despite an increased use of international models such as transboundary biosphere reserves and World Heritage Sites, this question could only be answered on the ground, in distinct contexts and coordinated by individual managers setting up negotiated processes of (re)territorialisation.

General conclusions

The discussion examined three levels of discourse, illustrating the complexities and contested nature of the process of constructing transboundary entities, as well as the contrasting and multi-scaled diversity of the process. Throughout, I have argued that differing conceptions of boundaries, identity and cooperation led to different reasons for promoting cooperation and thereby founding different understandings of transboundary entities. I repeatedly attempted to identify the ontologically suspect assumption that biophysical and societal boundaries could be assimilated, emphasising the conceptually exhaustive and exclusive distinction between them. The idea of 'natural boundaries' seemed however to be enjoying a renewal, providing a reassuring position for individuals perplexed by myths of borderless worlds as well as offering political legitimacy to natural scientists.

By dwelling on the socio-spatial dimensions of integration and distinction, I attempted to identify elements of (re)territorialisation that emerged during the construction of transboundary entities, describing it as a contested process. The framework for understanding lay in the identification of discourses of othering and the problematic issue of defining social and spatial integration. This implied that rather than being unproblematic 'boundary zones' of localised interaction, transboundary protected areas were contested spaces constructed through relations of power between international, national and local actors. As such, they were prime examples of the emergence of new territorial units illustrating the multi-scaled complexities of regional transformation.

Notes

1 Robert Frost (1914) 'Mending Wall' in 'North of Boston', reprinted in 'The Weekend Book', Nonesuch Press London 1955, p.120.

2 Personal translation from: 'toute société, par un universel d'oppositions, construit un monde de réalités qui passe par un système classificatoire. De ce fait, la construction, par chaque groupe, de son identité entre dans cette logique classificatoire et, en l'état actuel des recherches sur l'ethnicité, il apparaît que l'existence du signifié "Nous/les Autres" est fondamentale à l'organisation du monde de chaque société (...) Ainsi, une délimitation marquée au sol permet de renforcer l'opposition sociale "Nous/les Autres" par une opposition géographique "l'Ici et l'Ailleurs"'.

3 Personal translation from: 'cette zone-frontière a pour fonction de maintenir les différences culturelles de groups en contact. De ce fait, elle est le lieu où se met en œuvre un système organisationnel de relations permettant l'affirmation de deux identités en présence et scindées par la ligne-frontière. Cette zone frontière est, de ce fait, l'auxiliaire naturel de la frontière'.

4 Personal translation from: 'Cette zone cerne la frontière spatiale rigide et est le théâtre de relations d'interdépendance, condition nécessaire à la connaissance réciproque des groupes et au renfort du sentiment identitaire'.

5 Personal translation from: 'zone concrète, un lieu d'alliance et d'innovation sociale qui loin d'être institutionnalisée et instituante, vagabonde à la marge des certitudes de deux mondes en apparence trop clôturés pour se côtoyer'.

6 Personal translation from: 'Instituer un espace à statut particulier au sein d'une société et en accord avec cette même société, suppose d'avoir au préalable identifié la charge symbolique qui est liée à ce territoire. Comme le souligne Augustin Berque, "plus forte sera la charge symbolique, plus intense sera la spaciation". La délimitation du territoire qui va bénéficier de ce statut particulier va permettre le déploiement éco-symbolique et la réactivation des archétypes. (...) En préalable à la création de ce nouvel espace transfrontalier, nous devons nous interroger sur la charge symbolique qui est liée. Peut-être faudrait-il commencer par les deux parcs existants et cerner, notamment, s'il existe une charge symbolique pour d'autres territoires que celui de la "vallée des Merveilles". (...) La multiplication des acteurs concernés par la mise en place d'un espace naturel laisse entrevoir qu'au delà de la simple mise en place d'une structure administrative et de la délimitation d'un territoire, il s'agit en fait de faire dessiner un "espace social" par l'ensemble des groupes en présence'.

7 Personal translation from: 'Comme beaucoup d'hommes de ma génération, parce qu'on nous avait appris que nous étions nés au temps du "monde fini", j'ai rêvé de

retrouver ses paradis perdus et de maintenir, au profit de tous, ces espaces privilégiés où la nature sauvage est encore présente et préservée'.

8 Personal translation from: 'Stambecchi e camosci sono stati i primi portavoce della necessità di collaborazione: non hanno mai preso in considerazione le frontiere e quelli "marcati" in Italia venivano osservati in Francia e viceversa'.

9 Personal translation from: 'Die Natur und die Tiere haben uns den Weg gezeigt und trotz der vielen Schwierigkeiten, die wir zusammen hatten, gibt es doch bemerkenswerte Ergebnisse dieser Zusammenarbeit'.

10 Personal translation from: 'Complémentaires sur le plan géographique, les deux parcs le sont également en matière de faune et de flore. S'ils partagent une exceptionnelle biodiversité avec, de part et d'autre de la frontière, des aigles, des loups, des bouquetins, des chamois ou encore des marmottes, des hermines, chacun d'eux possède des espèces endémiques spécifiques, comme la saxifrage à fleurs nombreuses'.

11 Personal translation from: 'Ce que nous venons de décrire (...) montre que nous sommes une entité naturelle. Aller vers un parc naturel unique n'est qu'une conséquence naturelle de ce que l'on peut observer depuis des années. Voir ce massif comme un entité homogène n'est qu'une conséquence logique'.

12 Personal translation from: 'Le seul examen cartographique montre combien les deux parcs sont géographiquement complémentaires: l'un apporte l'étendue de son territoire et la diversité de ses situations géologiques, climatiques et écologiques, l'autre donne à la réunion de ces deux espaces une épaisseur et une profondeur indispensables à la survie des populations des différentes espèces sauvages que la forme trop allongée du parc français ne pouvait lui donner'.

13 Personal translation from: 'Nun könnte man hinzufügen, kennt die Natur überhaupt Grenzen oder anders ausgedrückt, sind diese Grenzen möglicherweise von Menschen erfunden und künstlich gezogen worden und weist uns die Natur andererseits permanent daraufhin, dass wir diese Grenzen zu überwinden haben und dass wir grenzenlos zusammenleben müssen, wenn wir überleben wollen? Das heißt, wenn die Natur keine Grenzen hat, dann ist sie auch unsere gemeinsame Verantwortung. Keine Verantwortung begrenzt auf Staaten, Gemeinden oder gar familiäre Umstände, sondern eine Verantwortung, die die Menschheit insgesamt trifft'.

14 Personal translation from: 'Weil das eine Symbole dafür ist, dass hier etwas zusammenwächst, was historisch über die Natur, über die Geologie – es sind Buntsandsteingebiete – über den Wald, über die Kulturlandschaften, mit ähnlichen historischen und wirtschaftlichen Entwicklungen, eigentlich schon mal sehr eng zusammen gehört hat. Deswegen hat das heute für uns eine ganz große Bedeutung'.

15 Personal translation from: 'De part et d'autre de la frontière, ligne théorique tracée par des contingences politiques, se sont tissés, au cours des siècles, des liens étroits entre les populations des hautes vallées (liens commerciaux mais aussi culturels). La montagne n'a jamais été un obstacle quand on la franchissait à pied et l'on comprend la similitude de vie existante entre les hommes et ce terroir'.

16 Personal translation from: 'Sie sind das Zusammenwachsen von Europa von ganz großer Bedeutung und haben eine große Verantwortung'.

17 Personal translation from: 'La construction de l'Europe, en matière d'espaces protégés, va bénéficier (...) d'un coup d'accélérateur. Alors que l'intégration politique des États du vieux continent demeure toujours aussi lointaine, deux parcs, l'un italien, l'autre français, s'engagent à construire dans les Alpes du Sud une seule entité commune'.

18 Personal translation from: 'die Schaffung eines grenzüberschreitenden Bioshphärenreservats wird weitere Möglichkeiten bieten, um auf unterschiedlichsten Arbeitsfeldern gemeinsame Projekte durchzuführen'.

19 Personal translation from: 'L'histoire, les voies de communication ancestrales, les langues véhiculaires et les us et coutumes resserrent depuis toujours les liens des habitants partageant le cœur de ces espaces protégés. Ces Hommes au tempérament généreux sont avant tout montagnards: travailleurs, courageux et obstinés, fiers du patrimoine culturel et naturel qui leur a été légué. Aujourd'hui, les parcs décident de poursuivre cette histoire commune autour d'une montagne commune: une 'Montagne sans frontière''.

20 Personal translation from: 'Dès la préhistoire, les hommes ont habités les montagnes des Alpes Maritimes comme en témoignent les trente mille gravures rupestres de la Vallée des Merveilles et de la Vej del Bouc. Ils bâtirent des voies de communication, comme la Route du Sel, reliant Nice à Cuneo par de nombreux cols. Les contacts ont étés facilités par une langue commune: l'occitan. Ce lien culturel fort a permis le développement de traditions similaires de part et d'autre de la frontière'.

21 Personal translation from: 'I due parchi non hanno soltanto natura in comune: sono uniti anche dalla storia, fin dai tempi antichissimi dei graffiti rupestri, fino a Vittorio Emanuele II e alla storia più recente'.

22 Personal translation from: 'cette montagne que la nature et l'histoire n'ont cessé de concevoir sans frontière'.

23 Personal translation from: 'diese Landschaft ist gleichermaßen reich an Geschichte und Kultur'.

24 Personal translation from: 'Ich möchte hier noch einmal auf die Bedeutung eingehen, den der Raum für mich auch geschichtlich hat. Frankreich und die Pfalz oder Lothringen und Elsass und die Pfalz haben viele gemeinsame Abschnitte in der Geschichte aber auch traurige, belastende Zeiten hinter sich, insbesondere im letzten Jahrhundert, aber auch im ausgehenden 19. Jahrhundert. Wir haben uns oft kriegerisch gegenübergestanden aber gerade die Menschen in den Grenzregionen haben es immer verstanden aus ehemaligen Feindschaften auch wieder Freundschaften zu bilden'.

25 Personal translation from: 'Ces hommes déjà proches par le passé se redécouvriront aujourd'hui, pour mieux se rapprocher demain'.

26 Personal translation from: 'Wenn wir manchmal scherzhaft sagen 'wir in dieser Region sind manchmal ein bisschen näher an Paris als an Berlin', dann soll Ihnen das zeigen, wie sehr und wie eng wir verbunden sind mit unsern französischen Nachbarn'.

27 Personal translation from: 'Certains projets seront définis afin que les deux espaces puissent eux aussi contribuer, en fonction de leur spécificité, au rapprochement des habitants partageant une même identité transfrontalière'.

28 Personal translation from: 'io credo che dobbiamo guardarci da un concetto rigido di frontiera e quindi di identità. C'è una logica perversa dell'esclusione che deve essere assolutamente essorcizzata e superata, recuperando viceversa un concetto di identità (e quindi di frontiera) come processo in divenire, aperto ad ogni contaminazione, qualcosa che si perde e si ritrova, sempre in bilico tra spaesamento e ritorno alle origini. (. . .) Credo che si debba essere molto attenti a ricostruire in maniera storicamente attendibile e documentata l'identità di una regione geografica, ma che poi questo patrimonio storico non debba essere giocato in forme integralistiche, ma da una parte come contributo alla trasformazione della

cultura, dall'altra riconoscimento della presenza di un patrimonio storico da tutelare'.

29 Personal translation from: 'Les visiteurs qui traverseraient les deux espaces en cheminant sur les mêmes sentiers et en bénéficiant d'informations communes aux deux parcs, prendront peu à peu conscience que ces deux sites partagent une identité transfrontalière commune'.

30 Personal translation from: 'Sur le colloque de Menton, depuis ce colloque, on a réalisé d'abord un atlas (...), je peux vous le présenter, il fait une synthèse de toutes les données scientifiques qui montrent que les deux parc font vraiment partie d'une entité naturelle et culturelle commune'.

31 Personal translation from: 'Mais, ce qu'on peut imaginer, qu'est-ce qu'on met derrière un réserve de biosphère transfrontalière ? Ce sera quoi, une réserve de biosphère transfrontalière effective ? Est-ce que ce sera une seule équipe qui dirige tout ce territoire ? Est-ce que c'est les deux équipes actuelles, avec un système de fonctionnement type, on a un bureau commun, on a une salle de réunion commune mais on reste quand même deux entités parce qu'il y a une réalité physique, un éloignement, c'est quand même pas la porte à côté. Ou est-ce que ce sera une interpénétration très grande des actions, je ne sais pas, à la limite, que toutes nos actions soient complètement transfrontalières. Personne peut le dire en fait, selon les gens que tu interroges, tu as toute la gamme. Mais, à un moment donné, il faudra bien que tout le monde, les deux réserves de biosphère, se mettent d'accord sur qu'est-ce que c'est, pour elles, la réserve de biosphère transfrontalière'.

32 Personal translation from: 'Et donc on a été classé réserve de biosphère on a fait en sorte que sous notre initiative et notre pression en 92 le Naturpark Pfälzerwald soit également reconnu en réserve de biosphère. Notamment, tout de suite d'emblée en réserve de biosphère commune. En pionniers, quoi'.

33 Personal translation from: 'Globalement, l'objectif est déjà de vraiment faire une meilleure valorisation de l'ensemble du patrimoine, à la fois du patrimoine national et du patrimoine culturel, donc d'avancer par des actions concrètes sur le terrain et, parallèlement, d'aller vers un parc européen à travers une évolution plutôt juridique. De mener de front les deux en sachant que l'évolution vers un parc européen peut prendre une dizaine d'années et que cela doit devenir concrètement visible déjà sur le terrain'.

34 Personal translation from: 'Si un jour on avait un projet, par exemple, d'une réserve qu'on souhaiterait vraiment étendre sur l'Italie, on ne peut agir que par persuasion et eux aussi. On n'a pas d'autres moyens juridiques de le faire. Et même quand on fait, par exemple, des dossiers Interreg, c'est vrai que de plus en plus on nous demande que ça soit l'un des deux parcs qui soit responsable pour les deux parcs, et ça pour l'instant on sait pas très bien comment s'y prendre'.

35 Personal translation from: 'Il ne faut quand même pas négliger ça : dans le transfrontalier, il y a quand même, même si la frontière n'existe plus sur le papier, même si tout ça, c'est l'Europe et autre, ce sont des mots, il y a quand même une réalité politique, administrative et économique radicalement différente, sans parler du culturel'.

36 Personal translation from: 'Il n'y a aucun projet en profondeur. Des vrais projets qui demandent une forte concertation. Tant qu'il s'agit de mettre un piste cyclable qui part de France et d'Allemagne, des panneaux tout le long et, éventuellement, un pont, et encore! Il y avait un pont qui était prévu entre les deux pays, je ne sais plus à quel endroit, au sud de Strasbourg, et il n'est toujours pas construit et ça fait 30 ans qu'on en parle. Ça fait rire, mais, en fait, ça prouve bien qu'on en est encore là de la coopération, c'est-à-dire, on essaie, pour que tout aille bien, que les choses

n'aillent pas finalement, on n'est pas en profondeur. Tout va bien parce qu'il ne se passe rien!'.

37 Personal translation from: 'Donc on a effectivement dans le cadre de cette convention de 85 on a mis en place des itinéraires transfrontaliers, de faire en sorte que sur la frontière il y ait des panneaux d'information, on a fait des parkings ensemble, pour la "Erholung" comme disent les allemands, et puis on s'est très vite rendus compte que à part se retrouver une fois par an entre élus pour faire une bonne bouffe, et se taper sur l'épaule, on n'avait pas grand chose pour avancer'.

38 Personal translation from: 'On est quand même mandaté par l'UNESCO puisque maintenant, si tu veux, tout est allé dans le sens de, inévitablement, mettre les deux partenaires dans l'obligation de coopérer. Donc maintenant, on ne peut plus faire machine arrière et on ne peut que avancer, et on se rend compte, malgré les blocages actuels, que, pour avancer, il va falloir régler les problèmes en profondeur, choses auxquelles on n'était pas prêt, on n'était pas en mesure de faire au tout début, parce qu'on était encore sous le syndrome, comme l'a dit Emmanuel, de l'amitié franco-allemande, on n'ose pas se fâcher parce que, tu vois ... Tout ça est fini maintenant. A partir du moment où on travaille en profondeur, on fait comme entre collègues français, franco-français, si ça va pas, on le dit. S'il y a un problème, on se frotte, on discute, on débat mais on reste surtout pas en superficie comme ça puisque, justement, pour qu'il y ait coopération, il faut que ça aille en profondeur'.

39 Personal translation from: 'Et donc on a eu un premier programme Interreg I, qui était un programme de mise en jambes, avec notamment des actions du type amitié franco-allemande, avec des cartes communes, nous avons faits un itinéraire passe-frontière, à but pédagogique, on a eu un coordinateur commun: Emmanuel, un français pour les deux. On a eu aussi une action touristique: on a sorti une brochure de 66 idées de sorties pour découvrir la réserve de biosphère. Etc, quoi. Bien, mais de mise en jambes, quoi. Des choses un peu faciles à faire, quoi'.

40 Personal translation from: 'On attendait l'atlas pour vraiment avoir une base scientifique solide, on a l'accord des politiques. (...) Il faut que notre conseil d'administration soit d'accord pour qu'on présente notre candidature comme réserve de biosphère. (...) Donc, là pour l'instant, on a un accord de principe du président, il faut que ça passe officiellement au conseil d'administration'.

41 Personal translation from: 'Et d'ailleurs, ce document (sur l'identité graphique des deux parcs) n'est divulgué qu'au niveau de CM et de LO. Quand tu leur parles du document, personne n'est au courant. Et l'Italien aussi, il l'a ... (...) A mon avis, ils n'en savent rien non plus. C'est allé trop loin, ils ne savent pas utiliser ... Le Mercantour, à l'heure actuelle, à mon avis et cela n'engage que moi, n'arrive pas à diffuser ces informations'.

42 Personal translation from: 'Disons que, nous, ça ne nous gêne pas mais il y a les faits, les actes enfin, l'esprit, tout ça, on nous a mis dans la réserve de biosphère transfrontalière, OK, mais ce n'est pas encore demain que les deux territoires seront réellement dans l'esprit de réserve de biosphère transfrontalière. Tout dépend ce que l'on demande, quel est ton degré d'exigence. A l'heure actuelle, on peut dire qu'on est peut-être quand même, malgré tout ce que nous, en gens exigeant, on voudrait, je te l'ai déjà dit, faire plus, en profondeur, on doit peut-être quand même pas être l'une des plus mauvaises réserves de biosphère européennes parce qu'on a quand même beaucoup de travail en commun, beaucoup de choses qui se font, avec ou sans d'ailleurs les Naturpark comme partenaire, mais on a des réunions d'équipe quand même, entre nous évidemment, c'est fréquent, il y a quand même une dynamique, on a fait ce premier voyage d'étude, en tant que franco-allemand c'était une première et c'était bien, et tout le monde d'ailleurs a souligné la nécessité d'en

refaire d'autres. Il y donc quand même des choses qui vont au-delà d'une plaquette ou d'un sentier, qui est de l'ordre du symbolique mais qui, finalement, fait que les hommes travaillent pas ensemble. Mais tout ça est un processus long et je crois qu'il ne faut pas vouloir effectivement aller trop vite et laisser le temps au temps'.

43 Personal translation from: 'Et il y a un moment donné où, mais ça sera peut-être dans 10 ans, dans 15 ans je ne sais pas, où si on veut vraiment que ça soit un organisme qui ait une unité, il va bien falloir trouver des outils. Et, pour l'instant, ces outils n'existent pas vraiment, ou alors on n'a pas trouvé et on a à faire vraiment, au niveau, par exemple, quand on met des contraventions, comment on peut les mettre en Italie, c'est pas évident, il y a plein de données juridiques qui font que, bon, il y a la loi française et la loi italienne. Il faudrait arriver à dépasser ce stade et notamment aussi au niveau de l'institution qui pilote, c'est-à-dire que, pour l'instant, quand même, on est, même si on fait beaucoup de choses ensemble, il n'y a aucun pouvoir de l'imposer, c'est parce que l'on veut bien les faire ensemble'.

44 Personal translation from: 'On arrive quand même à un niveau maintenant où il faut coordonner, d'autant plus qu'il y a une volonté quand même de plus en plus forte d'essayer de trouver quelque chose, notamment le patrimoine mondial de l'UNESCO, c'est quelque chose qu'ils nous avaient reproché et ils nous ont demandé de présenter le dossier ensemble. (...) Je pense parce que le Patrimoine Mondial c'est quelque chose d'un peu mythique (...) C'est l'aspect glorieux du label ...'.

45 Personal translation from: '(Le Diplôme Européen) c'est une reconnaissance de la qualité du patrimoine, son côté exceptionnel, et de la qualité de la gestion. Disons que, au niveau des scientifiques et tout ça, c'est très connu, au niveau de la population un peu moins, en fait, on fait beaucoup de bruit quand on le perd. (...) La candidature a été faite ensemble et donc chacun des parcs a eu le même diplôme. Mais la candidature est commune. Et là, au niveau juridique, il n'existe pas de diplôme conjoint, donc on a eu le même, ils sont obligés'.

46 Personal translation from: 'Le Naturpark n'est pas une structure comme nous, collectivité territoriale, c'est une association, un Verein. Ils ont un conseil d'administration. Voilà. On a passé x fois dans notre tête toutes les solutions possibles et imaginables. Ça reste pour nous un mystère obscur de savoir pourquoi il n'y a pas une espèce de prise de ... Ca ne peut pas marcher, enfin, un couple peut marcher s'il est un peu équilibré, mais là on a quand même l'impression que l'on est bancal, tant au niveau des moyens humains que de la motivation et du mandat. (...) Et là, on ne peut pas fonctionner comme ça, dans l'esprit d'une réserve de biosphère transfrontalière, on s'entend bien. Ce n'est pas possible. Si chacun reste de son côté, pourquoi pas. Mais ce n'est pas ça le mandat'.

Chapter 7

Cooperation: Understanding Acceptance and Resistance

You that I stretch my hand to, can you stay
Unmoving on the further bank? The wine
Runs red with blood that now is past recall.
Drink up the gap between us, then, and live
(Paul Griffin)[1]

Defining cooperation in transboundary protected areas

When initially looking for a framework that might enlighten me on what taking place in the field, I sought out 'theories' of cooperation. I was surprised to discover that the issue was more problematic that I had initially imagined. In discussing transboundary cooperation in Cyprus, Hocknell also mentions that few comprehensive theories of cooperation exist (Hocknell 2000 : 35). In much of the literature, the definition of cooperation is vague or conveniently replaced by degrees of interaction between actors or areas, often without specifically distinguishing between the two. Thus spatial metaphors of 'integration' replace conceptualisation of the process. Cooperation remains conveniently taken-for-granted. In the first interviews I carried out, I unconsciously perpetuated this by asking people 'how they were cooperating'. This obviously did not allow me initially to gain particular insight into what cooperation might actually mean on the ground. Interviewing is a learning process, and subsequent interviews yielded much richer material.

As I read through my interviews, it became apparent that there were elements of 'working together in separate countries' that were quite distinct from the explicit construction of a shared spatial entity. Thus in order to avoid equating the two I have deliberately attempted in this chapter to isolate the non-spatial dimensions of the process, all the while recognising the interaction between working together and constructing new spaces. This leads me here to focus on the joint processes of acceptance and resistance to the ideas and consequences of cooperation in the context of transboundary protected areas. As Ò Tuathail has noted, the main struggle concerns meaning: 'the struggle over geography is also a conflict between competing images and imaginings, a contest of power and resistance that involves not only struggles to represent the materiality of physical geographic objects and boundaries but also the equally powerful and, in a different manner, the equally material force of discursive borders between an idealized Self and a demonised Other, between "us" and "them"' (Ò Tuathail 1996b : 15).

I argue here that rather than relying on abstract ideas of cooperation, situations can be understood as ongoing processes of identity construction in which individuals give meaning to discursive boundaries. Throughout the discussion, I attempt to identify what definitions of cooperation inform the choices and actions of protected area managers. In order to understand the negotiated temporal and spatial dimensions, I explore assumptions regarding implicit and explicit power relations between actors. This leads me to discuss issues of acceptance and resistance, integration and distinction, on a social and spatial level. I end by suggesting that because of the inherent weaknesses of the existing literature in the field, other more lateral analyses are needed, stepping outside the usual frameworks and including an improved definition of cooperation as a negotiated process. Rather than being an unproblematic process that leads to higher spatial integration, it is an unscripted process which if carried out comprehensively leads to unsuspected results.

Examining cooperation: three trends

The existing literature on cooperation can be broadly divided into three trends: an examination of the process on the scale of the individual; a reduction of the process to degrees of interaction; and finally a drafting of theories of international cooperation. I offer a brief review of each, laying emphasis on some of the elements that might apply to the study of interactions in transboundary protected areas. However, as the review indicates, none of these approaches provides a comprehensive framework within which to understand the issues emerging from the fieldwork. In the second section of this chapter, I suggest some ways in which cooperation should be approached in the future in order to overcome the existing shortcomings of existing approaches.

The first approach to cooperation dealing with processes on an individual scale is best represented by *game theory*, describing the motivations of individuals when faced with the choice of using cooperation as a strategy for individual gain. Game theory is usually associated with the 'prisoner's dilemma': a metaphor and illustration of cooperation in controlled circumstances (Axelrod in Heylighen 1992). Implicit in this approach is the notion that there can be a strong incentive to cooperate when individual actors are too weak to accomplish a given task alone. It is difficult to see how this might be translated into a frame for practical analysis. At most, it indicates that 'cooperation' (whatever that might be in real life, outside the prisoner's cell) is a behavioural strategy that can be adopted to further personal interest in a given situation. This is a depressingly narrow vision of a complex process and is not much use for practical analysis.

The second trend within the literature avoids conceptualising cooperation by replacing it with a description of degrees or taken-for-granted stages (Table 7.1). These are invariably presented incrementally, in a form of mythified progress towards an absolute, yet largely unattainable, goal. Irrespective of scale or actors considered, authors identify categories, with or without specific

Table 7.1 Comparison in scales of cooperation within spatial entities

States (Taylor 1990)	*Borderlands* (Martinez 1994)	*Protected areas* (Zbicz 1999d)
	Alienation	No cooperation
Coordination	Co-existence	Communication
Cooperation	Interdependence	Consultation
Harmonization	Integration	Collaboration
Association		Coordination of planning
Parallel national action		Full cooperation
Supra-nationalism		

spatial dimensions. Thus Taylor's stages of cooperation between states (Taylor 1990) is largely a-spatial, while Martinez's descriptions of cooperation between adjacent borderlands specifically implies degrees of spatial integration (Martinez 1994), as discussed in the chapter on the creation of transboundary entities (Chapter 6). Zbicz's description of cooperation between protected areas, like Taylor's, avoids all reference to space despite dealing with intrinsically spatial objects (Zbicz 1999d).

Despite assuming that situations can fluctuate, none of these typologies explains what happens when this comes about, other than saying that 'cooperation' increases or decreases. All actors on each side of the boundary are taken to behave as uniform or homogenous entities: quasi-actors or subjects behaving as one. These tautological approaches explicitly promote the idea that cooperation is something linear, respectively with 'supranationalism', 'integration', or 'full cooperation' as the end, and implicitly desired, result.

In many of the studies using these approaches, situations are ranked. This led Zbicz, for example, to publish in a thesis carried out on the basis of postal surveys of 147 adjoining protected areas, that 17.7% were not cooperating, 38.7% were cooperating at the first level, while full cooperation had been achieved by 7.5% (Zbicz 1999d : 269). Quite what this might mean was unclear. Although she explained the research process fully, the assumptions on which such a methodology rested were immense. In addition to all the pitfalls of carrying out cross-cultural surveys at a distance, one basic flaw was the assumption that 'transboundary dyads', combinations of adjacent protected areas within separate countries, behaved like as many individual actors. A protected area administration was considered to be one actor or subject, just like states were within most of political science. Yet it would take no longer than one afternoon of interviewing to understand that this was far from the case on the ground.

The third trend within the literature is written about by scholars with political science and international relations backgrounds. It is by far the most pervasive. Here, 'international cooperation' is seen as a component of traditional international relations, focussing on the policies adopted by individual states in relation to others. Their perspective tends to be on achieving peace – or absence of conflict – rather than on cooperation as a means of solving extraneous problems. Various traditions exist, including political realism and neorealism, as well as integration theories which seek more directly to refashion the state system, within currents such as functionalism, neofunctionalism and regionalism. These are briefly reviewed below.

In *Political realism*, 'cooperation, in whatever form, is likely to be *ad hoc* and determined largely by the interests of the most powerful state actor or hegemon. Cooperation, in other words, is both a function of, and conducted within, the parameters of power politics' (Hocknell 2000 : 37). Realists hold that cooperation is no more than 'power politics in disguise' (Groom 1990 : 9), and thus 'that conflict is the norm and cooperation is rare or even non-existent' (Zbicz 1999 : 42). Following this, several *neorealists* have further developed these ideas, with qualifications (Keohane 1984, Haas 1990 quoted in Hocknell 2000 : 38). Even the more 'liberal' neorealists, however, continue to focus on political power, suggesting that, short of changing underlying Darwinian principles of the struggle for life and the survival of the fittest, effective and extensive cooperation is virtually impossible. The scale in this case is overwhelmingly the state, and does not offer any practical analytical framework for my work, other than to perhaps illuminate some of the logics leading to institutional resistance to the idea of cooperating in a transboundary situation.

In contrast, *functionalism*[2] emerged in the 1940s within the wider body of 'integration theories' which seek to refashion the state system. The focus is on the functions necessary to solve specific problems, avoiding divisive ideologies and leading states to become increasingly enmeshed in an interdependent network of international agencies penetrating deep into economic life. Its essential starting point is to concentrate on particular tasks, discovering an irreducible set of relations between things which are distinguishable from relations suggested by a constitution or a dogma, and which, if left to themselves, suggest an 'ideal geographical extent in which the problem could be tackled, and the most appropriate administrative arrangements' (Taylor 1990 : 126). International organisations play an important role in this approach, and these are seen to help shape a socio-psychological community, which in turn reflects a cooperative ethos across national boundaries, also lessening potentially fanatical attachments to the state. What can be gleaned from this theory is the explicit idea that certain geographical entities might assist a particular problem, assisted by appropriate non-governmental or governmental structures. It does not, however, assist in understanding how individuals deal with cooperation on an everyday level. Emerging in response to certain shortcomings of functionalism, *neofunctionalism* promotes central institutions which are supposed to play a creative role in achieving the overall objective of peaceful interaction between states. Regional integration is seen as the key to stability, with the formal, legal and coercive powers of the state relegated to second place, a substantial change of scale from the previous theories. Critics of neofunctionalism note that the importance of the formal powers of central government in the political process is glossed over, while too much emphasis is laid on the importance of group activity and informal power relationships (for further discussion see Hocknell 2000 : 45–53).

Regionalism is a further example of integration theories, sometimes taken to be a sub-current of neofunctionalism. A recurrent theme of regionalism is the definition of 'that particular scale of geographical area which is best fitted to the performance of tasks judged crucial for the welfare of individuals, or for the advantage of governments' (Taylor 1990 : 151). While functionalism laid emphasis on choosing the particular structure appropriate to a given situation, regionalism further emphasises defining a particular geographical area in response to a need. This may extend beyond the boundaries of existing states, since the strategy is to make popular loyalties focus on the institutions, symbols or even 'what have been called the icons of the larger area' (Taylor 1990 : 151). As in functionalism, the idea that problems can be contained or isolated in some way is fundamental, since their solution lies in the definition of a geographical area in which this is to be carried out. However, while this idea is in itself interesting, there is a difficulty in agreeing on a coherent definition of 'region', problematically defined as a homogeneous spatial entity with regard to several attributes (Russett 1967 in Taylor 1990 : 152). Regionalism became unfashionable as early as the 1970s, partly because it quickly became apparent that the increased importance of regional systems was not leading to the emergence of new political actors, but rather that the existing character of the international system persisted unchanged despite increased regional cooperation. Nevertheless, regionalism introduced a territorial dimension to the largely aterritorial and unidimensional view of the functionalists. This has led to substantial theoretical debates about the appropriate size of these units, and their corresponding boundaries, more or less sharply defined.

Other fields within international relations have further concerned themselves with transboundary environmental issues, including the economically-oriented *'new institutionalism'* – a branch of political economics – or regime theory, an offshoot of functionalism and interdependence theory, dealing with issues of governance rather than government. The assumption in regime theory is that although states may never yield their independence completely to a form of 'world government', they may be willing to 'yield certain aspects of sovereignty in specific issue areas, in order to reap joint gains from cooperation' (Zbicz 1999d : 54). The success of regime theory has led to its multiplication in different directions, e.g. into 'power-based realism', 'interest-based neoliberalism' and 'knowledge-based cognitivism'. Further reviews of these schools can be found in Groom and Taylor (1990), Zbicz (1999d) and Hocknell (2000), with the latter two more specifically discussing their relevance to transboundary situations.

Conclusions

This survey has briefly reviewed how cooperation has been discussed in three main trends. In certain cases, the definition of a most appropriate scale for cooperation has been a concern, something that has indirect relevance to protected areas. Yet this only relates to determining the best size of a spatial entity likely to assist cooperation – ranging from discussions of NATO, the European Union or Euroregions, and remaining largely speculative. There is no specific conceptualisation of the effect of scale on human relations or interactions, and cooperation itself remains largely undefined.

One belief common to several of these approaches is that informal contacts and activities in adjacent areas carried out jointly, both at governmental and non-governmental levels, may lead to more formal political integration on a state level. Thus 'informal cooperation between adjoining protected areas may possibly lead to higher level formal cooperation and improved relations between neighbouring countries, and even political integration (. . .). The belief that this is true has caused central governments eager to foster regional integration to direct local authorities and managers of adjoining areas to cooperate' (Zbicz 1999d : 46). This is particularly true of the approaches that rank levels of interaction. Yet this increased political integration may be far from the daily objectives of protected area managers, and remains largely in the background. Since such thinking serves to underpin international programmes such as Interreg or Phare that directly fund transboundary projects, it is not irrelevant. However, it does little to explain processes taking place at the scale of individuals and administrations involved in concrete activities.

Is cooperation about places, or people and institutions?

Establishing 'cooperation' between adjacent protected areas is about changing people's habits. Previous chapters explored the problematic nature of constructing transboundary spaces, affirming the central role identity construction and othering play in the process. Uncertainty and risk are inherently linked to change, yet individuals have started 'cooperating' across international boundaries, exploring new ideas and seeking out common projects. This section explores the rationales that motivate individuals to explore new avenues and attempt to change professional practices, discussing how these tie in with the construction of transboundary entities. Unlike previous chapters, the focus here is not so much on the spatial dimensions of territorial change, but rather on the personal and institutional complications of planning across a larger geographic entity.

If 'cooperation' is about individuals, then it is illuminating to take a look at the photos in the literature portraying people. Some photos have achieved iconic status, such as the handshake between an American and a Canadian ranger in Glacier/Waterton that appeared in both the 1996 IUCN publication (Hamilton *et al.* 1996 : 29) and the subsequent 2001 IUCN Best Practice guidelines (Sandwith *et al.* 2001, cover page). All would agree that 'cooperation' is more than simply shaking hands. The iconic success of this photo is indicative of a wider difficulty: it is not easy to actually portray what cooperation actually is. Would a photo of a round table meeting be more appropriate? How have the people involved come to define it, beyond the easy clichés?

The literature on transboundary protected areas produced by international organisations such as IUCN and UNESCO is surprisingly silent on defining cooperation. It is taken for granted that transboundary protected areas are based on cooperation, but this is not defined. The most recent IUCN document on the issue only defines 'co-operative management' in this context, stating that 'co-operation between the two or more individual protected areas is a prerequisite for recognition as a TBPA (...). As a rule of thumb, the level of co-operation should reach at least Level 1 (as set out in Box 3.9) in order to be recognised as a TBPA' (Sandwith *et al.* 2001 : 3). The box in question reproduces Zbicz's categories (Table 7.1). Cooperation, therefore, is reduced in its essence to simple communication. In another document, subsequently reformulated as IUCN policy (Hamilton *et al.* 1996 : 2), Hamilton states that transboundary cooperation 'can be of many kinds and degrees. It can range from park managers feeling comfortable enough with each other to pick up the telephone and talk about a problem or opportunity, to a formal international treaty that endorses cooperation between agencies administering the protected areas' (Hamilton 1998 : 27). In other words: cooperation is defined by cooperation, another wonderfully self-referential definition.

In order to question this tautological definition, this section starts with the emergence of cooperation in the case study areas, examining the conceptual paradox of identifying one individual, administration or institution that sparked off cooperation with adjacent protected areas. Issues of acceptance and resistance within protected area administrations are argued to be linked to how, when and by whom the idea of 'transboundary cooperation' first appeared, and more importantly what underlying definition of cooperation this entailed. The discussion then examines the problematic role assigned to coordinators of cooperation, discussing the various institutional structures that have been used to foster increased contact between administrations. The difficulties in identifying the Other with whom to cooperate, both on an institutional and cultural level are then addressed, hinting that such an issue is far less straightforward than is generally assumed. The extracts presented below stem from all the case study areas and present a broad picture. However, for methodological reasons, direct quotes from the Vosges du Nord/ Pfälzerwald and the Alpi Marittime/Mercantour dominate.

'Initiating' cooperation

Mirroring the confusion within the literature, there was no clear wide-spread definition among protected area managers of what cooperation actually was. Because of the wide range of case study sites, with different political contexts, cooperation ranged from simple exchanges of goodwill and initial attempts to exchange information, to intense shared projects. Yet cooperation was more than simply varying degrees of the same process, but rather covering extraordinarily diverse spatial and social practices. One manager attempted to define what cooperation was and wasn't for him, noting that it was more than just carrying out easy, non-threatening activities, but rather implied moving on from persistent myths of 'friendship across boundaries':

For me, Franco-German friendship isn't transboundary cooperation. When I was speaking yesterday about Franco-German friendship, it may be more relevant to generations older than ours (...) it's the Stammtisch and big meals, fieldtrips, ceremonies and things, thumping each other on the back and it's very good, and then we do things that fundamentally don't raise any issues. We only ask each other questions that don't cause anger. And especially, we don't ask questions that do create conflict. In order not to spoil Franco-German friendship. A sort of myth saying 'Beware, Franco-German friendship is precious' (...). But the problem is that it hides a certain number of other questions – and I'm not kidding, Franco-German friendship means we are still mates, the war and all that, it's over, it's great, but the landscape, the planning, the road you want to impose on us, that's all shit (...). When you steal our ideas, that's shit! And the French find it hard to say that. And that is transboundary cooperation[3]

(Théo, Parc Naturel Régional des Vosges du Nord).

It was difficult in each case study to determine how, when or by whom it started. Examining descriptions of the first stages of the process nevertheless shed light on what it meant to individuals. Stories from one site didn't always overlap, but diverged and conflicted. It was problematic to identify one deciding factor, one key spark that led to cooperation. Unlikely founding myths appeared and became widespread. A fundamental paradox appeared when managers were asked to describe how cooperation first started: the frequent combination of descriptions of pressure and negotiation, suggesting and creating inequality between actors. Negotiation or diplomacy designed to convince people that cooperation was necessary was described or implied, hinting that instigating cooperation was rarely an unproblematic process of rational decision between equal partners. The impetus for initiating cooperation came from various directions. In all the stories, the deciding feature was identified as being a professional group (such as scientists), an individual, or an institution (MAB Committee, UNESCO, IUCN, and so on). In the following paragraphs, I address each of these.

Individual scientists were often described as key players in the early stages of establishing cooperation, organising shared scientific meetings to exchange information and compare results, sometimes publishing these together in joint publications. Several managers, themselves often trained scientists, mentioned that due to the nature of scientific work, scientists were more used to international contacts. This was seen to imply that they were more open to the idea of transboundary contacts, working as efficient catalysts. In the Polish and Slovak Tatras, joint meetings between scientists stretched back to 1957. In 1987, these contacts were coordinated and organised by the two Academy of Sciences, institutions that remained important players in many former Socialist countries. Before the political changes of the 1990s, friendly exchanges between the two countries were also undertaken between 'syndicated workers', involving fierce skiing championships that were invariably won by the Slovaks, according to Krzysztof, much to the despair of the Polish managers!

The first idea for international nature protection came up in 1967–68 during the Prague Spring when everyone was open for free ideas. There were meetings between the three

countries, but after 1968, the Russian army came into Prague and the idea was dropped. The idea came back in 1989 when there was a meeting of the three sides
(Dominik, Poloniny National Park, Slovakia).

The East Carpathians similarly developed exchanges between scientists, with one meeting per year held in Poland leading to a joint publication. In the Danube Delta, meetings previously held every year were organised every two years from 2000 due to lack of funds. These exchanges were often considered very successful and popular, despite occasional difficulties in comparing results stemming from different research methodologies. Such meetings undeniably participated in confidence-building and increased knowledge of the Other, creating non-threatening venues for exchanges between people sharing similar professional expertise and status. It was clear to managers that this was cooperation. If frustration was expressed, it was usually linked to the inability to organise permanent joint scientific committees, despite various attempts (East Carpathians, Tatras, Danube Delta, Vosges du Nord/Pfälzerwald). This institutionalisation of contacts was surprisingly problematic, and indicated that cooperation might be about more than simple exchanges.

In addition to scientists, the first steps of transboundary cooperation were frequently associated with individuals who identified a shared interest in the wider transboundary area. This was often linked to wider issues of institutional 'integration', personal or family history or else a belief in the inescapable reality of a shared nature and/or shared destiny:

We have to be coherent. Either you believe that there is here a common natural space in which there is a community of interest, it's like the European Union: either we believe there is a community of interest, or we don't. If we do believe in it, in this community of interest, then we must act so that policies and actions are progressively coordinated and led together. So I do believe in this community of interest. There may be others who for various reasons, believe in it less, you know[4]
(Théo, Parc Naturel Régional des Vosges du Nord, France).

This manager described cooperation as being the logical step following the identification of a shared destiny within a shared space, all the while acknowledging the problematic nature of assuming this feeling to be shared by all. Opposed to this identification of wider continental issues, others across the boundary suggested more concrete reasons linked to a particular individual:

Probably, we should also mention ... Well, my impression was that also one of the reasons that the cooperation has been started, apart from what Mr 'Maximilian' just mentioned, was that our former President, he had a very strong personal commitment to French-German cooperation. Because during the Second World War he was in exile in France and he speaks French very well. And so that is one of the reasons I think that cooperation has been established, apart from the points that were just mentioned
(Daniel, Naturpark Pfälzerwald, Germany).

It was interesting to note that different stories abounded in any one site, with individuals putting forward their own version of the tale, situating themselves and their actions within it. In the East Carpathians, in parallel to citing institutional or scientific arguments, many managers went to great pains to explain how their personal and family histories were physically inscribed in the wider transboundary area. One Polish manager, for example, went through the list of his Slovak and Ukrainian colleagues explaining why each had personal reasons for wanting to cooperate. According to him, each was personally motivated to attend rotating meetings in order to visit, say, a grandmother, a cousin or simply a village that had been occupied by past members of his/her family. Another, from Poland, met his Ukrainian wife at one of these meetings, giving him a further engagement with transboundary matters. Thus personal belief in the appropriateness of cooperation was linked to a belief in local rootedness, both for Self and Others stretching across existing political designations. In this particular site in the Carpathians, the personal engagement of managers with the wider transboundary site was suggested as a determining reason for transboundary contacts. Personal curiosity and the possibility of using transboundary travel for purchasing cheaper goods also cropped up in such discussions, and meetings were often seen as welcome opportunities for shopping trips. In other sites, however, individuals went to great lengths to specifically distinguish themselves from their foreign colleagues, distancing themselves from their neighbours and citing curiosity about difference as an impetus for exchanges (Danube Delta). This was particularly true in Central Europe before the 1990s, when other foreign travel was restricted (Polish and Slovak Tatras).

These tales of personal engagement reinforced the apparent equality between the three sites, and the shared interest in 'cooperating'. However, such equality and shared motivation were not necessarily identified by all. In most cases, individual stories reinforced the inherent inequality in the first stages of cooperation, stressing the need to convince others, both within their own administration and within adjacent protected areas. Attempts to convince partners of the need to cooperate abounded. The Franco-German Vosges du Nord/Pfälzerwald was a good example of a situation in which individuals gave conflicting versions of a story. This illustrated the discursive nature of information produced during interviews, and hinted at the difficulty of finding an 'objective truth' when identifying the origin of cooperation. For example, one manager in the Vosges du Nord recounted the decision to apply the biosphere reserve model to the existing protected area, initially within the French park but with the idea of extending it across the boundary. He raised the issue of initial acceptance or resistance to the idea:

> *We here, well the structure here wanted this label [designation as a biosphere reserve], it didn't spit on it, nobody came and imposed it. It asked for it. It's us, within the park, who put together the proposal but they could have refused by saying 'what on earth is this?'. So, it's different if you are hoping for something, you are going to make it live, rather than if you don't have any high hopes and it's just something you got at a certain point. That's important. Now, it may be that within our team, not everything was*

properly understood as such but, in any case, here, on the level of the technical team, people can see that these are not only laurels handed out so that we can put our feet up, it's really an encouragement and work to be done. Nothing is taken for granted. The difficulty in carrying out a biosphere reserve is the same as for all other things where you have no other power than persuasion[5]

(Hugo, Parc Naturel Régional des Vosges du Nord, France).

The use of words was interesting here as he explicitly situated himself spatially ('we here') and institutionally ('the structure'), in opposition to an Other (implicitly 'them there') which he described subsequently. In this quote, the fact that his administration actively sought something was implicitly offered in opposition to the fact that the other protected area did not. Hugo went on to compare this situation to that of the neighbouring Naturpark in Germany, noting that they were in the end 'convinced' of the necessity of seeking biosphere reserve status in order to have a framework to cooperate with the French. According to him, this was largely for financial reasons. In order to explore this situation further, I asked the same question to another manager in the Naturpark:

[Juliet] What do you think the basic reason for this cooperation was initially? (Silence) Was it political? Was it to do with nature conservation?
[Maximilian] Expenses. And the European construction

(Maximilian, Naturpark Pfälzerwald, Germany).

This pragmatic response by a German manager seemed to fit in with the French manager's version of events, despite the reference to the need to construct Europe. This was added almost as an afterthought, completing his pragmatic financial calculation by a reference to wider issues of spatial integration. When I related this to Daniel, another German manager, he suggested a more idealistic version of events, contradicting what Maximilian had said:

[Juliet]: Because when I met him [Maximilian] last time I did not have the impression that he had tremendous feeling for this cooperation. It was more a pragmatic way of getting extra money.
[Daniel]: I would say you are wrong. It changes. There are days when he is really committed and he says well Mr 'Daniel' let's go to La Petite Pierre [the headquarters of the French park] so ... He took an initiative this Spring and we went there three or four times together ... Because things were not moving in terms of Interreg III, because the French colleagues ... Théo had some difficulties with his colleagues as far as motivating them was concerned because they said well, we have an overload of work, we don't want any additional transboundary projects. So all the propositions came from the German side

(Daniel, Naturpark Pfälzerwald, Germany).

This was a very different version of events from that told by the French manager. On the contrary, Hugo's version of the story indicated a need to convince their German partners of the need to cooperate. He paradoxically expressed the dominant partners' need to 'impose' cooperation on a more reluctant side, invoking the supposed advantages or gains to be obtained by the cooperation in the framework of a transboundary biosphere reserve.

In a subsequent interview, Hugo noted that for the next stage of Interreg III projects, all the proposals came from the French ... Beyond the need to identify 'the truth' of the situation, the attempt to ascribe impetus and initiative to one side or the other was revealing. Daniel's version portrayed a dynamic German side, held back by French immobility. Simultaneously, Hugo was blaming the lack of progress on German lack of enthusiasm. If anything was shared here it was the recognition that initiative and launching new projects was a positive thing to be valued, regardless of who was actually doing it. It was less important to actually determine the details: what mattered was projecting the impression of dynamism of 'us here' being held up by 'them there'. The managers' need to articulate positions valuing the contributions of their own side could be understood almost as a pre-emptive defence against possible blame.

'Imposing' cooperation

These examples posed the question of the 'imposition' of cooperation either by an outside institution or by one protected area on another. Was it in fact paradoxically possible to 'impose' cooperation successfully on reluctant people or on a reluctant administration, rather like a theoretical *game* in which circumstances encouraged collaborative strategies? Could resistance be overcome by negotiation or persuasion using convincing arguments, or was it something that emerged almost 'naturally', as was suggested earlier in some of the biophysical arguments given by managers stating that shared problems inevitably led to a shared response (Chapter 6)? Much of the literature promoting transboundary cooperation assumed free will or free choice for all actors. If people were really free to make independent choices in all circumstances, then it would of course be impossible to force people to work together against the grain. Yet it appeared to be taking place, partly because of financial implications. Explicit situations emerged in which individuals were not in a position to choose freely. Two distinct scenarios in which cooperation was imposed could be identified: the first involved external pressures from institutions or funding agencies, the second involved local pressure and lobbying from adjacent protected areas. These two scenarios are discussed in the following paragraphs.

The first scenario involved circumstances that concurred to make cooperation an unwelcome necessity that had to be enforced:

> *The difficulty is normally I would say ... normally I would say I don't waste any time with people who do not want to cooperate at all. Because it is not worth it to waste time and energy to convince these people if you already tried and you failed. Normally I would do it this time, and I find the people who really want to cooperate and I work with them but, if you are in a fixed framework of EU funding, and you have the obligation to carry out that project until the end with these people, then what do you do? You know*

you have to work with these people and you know they don't want to cooperate. But we have to carry out that project

(Daniel, Naturpark Pfälzerwald, Germany).

Thus financial constraints lead managers to find themselves in Catch 22 situations. Without cooperation, administrations could not access additional funding at a time when lack of funding for core activities was considered a crucial issue. It was clear that without financial incentives, there would have been no cooperation. Indeed at the end of funding periods in similar situations, activities and contacts ground to a halt (Vosges du Nord/Pfälzerwald; Danube Delta; East Carpathians). This meant that it was in fact possible for an outside institution to impose cooperation for a limited period. These institutions were funding agencies such as the European Union's Interreg programmes, the Global Environment Facility (GEF, funded by the World Bank) or UNESCO's MAB and World Heritage Programmes. They were able to yield sufficient financial power or clout to challenge existing relations and practices between adjacent protected areas.

In the Danube Delta and the East Carpathians, large amounts of money were spent within GEF projects lasting three or four years. These were designed specifically to 'build capacity' and encourage transboundary cooperation by offering direct financial incentives. While the money lasted, common activities did successfully take place. However, the idea faltered in its long term consequences. These projects were implicitly designed to demonstrate the overwhelming benefits of cooperation, leading to permanent changes in work practices. That this did not happen illustrated the difficulties of attempting to impose changes in work patterns. Although an initial large input of money over a short period of time could not challenge resistance overnight, it did show administrations that cooperation was an internationally popular strategy for which funding might be secured from other sources in future. However, international funding programmes were not unproblematic sources of money: while the popular Interreg programmes led to transboundary projects, critiques noted that they were not always actually facilitated by them. This led some to argue that 'the vast majority of border regions initiatives came into existence because of EU and intergovernmental funding but were then hampered by the very political and administrative system which encourages them in the first place' (O'Dowd *et al.* 1995 in Van Houtum 2000 : 66).

In addition to initiatives from large bodies such as those mentioned above, other smaller bodies also played a part in promoting transboundary cooperation, not being able to actually impose it with financial incentives. National MAB Committees, for example, sometimes acted as catalysts or at least distributors of information in Germany, Poland, Slovakia, Romania and Ukraine. In Italy, the MAB Committee was not operating at the time of the fieldwork, and in France was largely replaced by the dynamic coordinator of the French biosphere reserve network. Their ability to perform this role of catalyst effectively depended on their role, prestige and political position within countries. In the case of Poland, the MAB Committee took on the issue of transboundary biosphere reserves and actively promoted the idea on an international level. In this case, however, it was not so much cooperation that was lobbied for, but rather the creation of transboundary entities. The ability of such 'soft' bodies to impose or encourage cooperation was however limited when problems or setbacks appeared:

> *The Poles and the Polish MaB Coordinating Council have been the main driving forces but now the process is going in circles*
>
> *(Andrzej, Bieszcsady National Park, Poland).*

Other attempts to impose or lobby for cooperation included similar lower-key yet potentially more successful long-term strategies. These did not involve outside pressures but rather concerned gentle persuasion or active lobbying of one protected area administration by another. This muscular negotiation for the creation of a common entity involved attempts to institutionalise cooperation under a shared institutional structure. In the Alpi Marittime, the Italians were keener to formalise transboundary contacts within a transboundary biosphere reserve than the French. This led to often intense lobbying by the Italians, anxious to promote their ideals:

> *For example the discourse around the biosphere reserve, we, who are rather more attached to it, have written a letter to the President of the Mercantour park and they have not yet responded, that is to say we wrote a letter in which we said: we have already prepared the project for the Italian part, we wanted to present it this year (to UNESCO) but seeing that we won't be able to, therefore we will have to finish another year, but we asked them what is their opinion, because we'd like the reserve to be transboundary, seeing that there seemed to be support for the idea when we organised the symposium and we wanted to know on a practical level what his intentions were, what he intended to do, whether we could already write in the application that this was the first half of a second half that would be submitted later on. This was the least we could do, and he hasn't responded yet*[6]
>
> *(Alessia, Parco Naturale Alpi Marittime, Italy).*

This was an interesting example illustrating the practical difficulties of such negotiation, in which steps have to be carefully calculated in order to promote an ideal yet avoid offending anyone by presuming too much. Here, such contacts took place directly between directors. In other cases, the negotiation and subsequent management were handed over to 'coordinators', institutionalising the role of go-between. This particular strategy is examined in the following section.

Institutionalising cooperation: the role of 'coordinators'[7]

In one site, specific coordinators were employed at different stages to facilitate cooperation. The first time I met one individual working as a transboundary coordinator his job was on the line and he needed to justify both to his boss, his colleagues and to the neighbouring protected area that his role was useful. The first interview was conducted in a group, following a decision of the director. The context was formal and questions were answered succinctly:

> *[Juliet] What is your day-to-day work as coordinator of the cooperation?*
> *[Manager 1] Well it depends, on one hand there is project management, projects that I am in charge of, on the other hand there is coordination between the two managing bodies, close interaction with the main actors, and there is facilitation of working groups, meetings, sometimes it is translation work, sometimes it is moderating or facilitating meetings, bringing people together who want to cooperate or initiating new cooperation. Quite a wide range of tasks*
>
> *(A manager).*

This indicated the diversity of work carried out by coordinators acting as key actors and facilitators of transboundary cooperation. The diversity of the tasks also meant that it was not always clear to managers exactly what such coordinators did, leading some to suspect that they got away with doing very little compared to those carrying out concrete projects with immediately tangible results. Despite such suspicion, the idea of having one or several coordinators was widespread. At the time of the fieldwork, a wide diversity of situations existed. In one site, contacts were coordinated by a senior manager in one country and the Director in the other. In another, a coordinator was in the process of being employed and was dismissed by the time this book was drafted ... In the two other sites, the work was carried out by the Directors. In another, a senior manager coordinated transboundary and international contacts, while confusion reigned in the other following the split of the former administation. In another, following a succession of scenarios, only one coordinator remained when formerly there had been one in each country. In another, contacts were previously coordinated by the Director of one side and a combination of senior managers on the other:

> *For the moment, we have meetings roughly every two months. At the level of the [park], it's a bit me who is in charge of coordinating. [One colleague] took care of all the jurisdictional aspects, [another] coordinates the political dimension and so we have quite regular meetings with the director of the [neighbouring park] and for the moment ...*
>
> *(A manager).*

In the absence of a formalised structure with one person specifically designated for coordination, responsibility rested on other members of the administrations. With no specific job description, this supposed continuity within mandates and a continuing personal dedication. However, different administrations in separate countries had varying speeds of turnover of staff, as the director once wrote: 'During the time I have been director of the Parco Naturale dell'Argentera, now Alpi Marittime, the Mercantour has had five different directors and the managers of the different departments (scientific, communication) also changed several times. It's important to underline this because collaboration between institutions is before all else collaboration between people: parks are made up of the people who work in them'[8] (Rossi 1998 : 7). Because of this difficulty in ensuring continuity, the idea that one person should take on the specific tasks of cooperation was widespread. In one site, where such a role had never been explicitly ascribed to one person, the task was identified as being exclusively administrative:

> *We have work that is getting harder to manage because it concerns communication, land planning, scientific work. There is a need for someone to coordinate all that. There is a real need for someone who can manage a whole list of (. . .) projects in order to focus on the coming years in the different fields, (. . .). But the first project is to have someone as a coordinator*
>
> (*A manager*).

There was a definite circulation of information concerning ideals and institutional models between the different sites around Europe, fostered during international meetings and through publications drafted by individuals employed as coordinators. This exchange of information regarding institutional solutions for coordinating cooperation was often directly referred to:

> *And then another complicated thing that we have to define a bit is the management structure if the reserve is transboundary. And indeed one of the [programmes] that we would like to do is precisely concerned with the creation of a common management body. That means we want to present this project so that they fund the creation of a common management body. Indeed with people who work within it, a common office, maybe not so many, two or three. But . . . Someone responsible, in other words. A bit like what the Polish, the Czechs have, no? Etcetera. Or else something like the . . . What's his name?*
>
> (*A manager*).

This attempt to emulate other sites followed explicit attempts by international organisations and non-governmental organisations to promote 'best practice', aided by a team of 'experts'. One individual, for instance, was a coordinator in one site, a consultant for EUROPARC and a member of IUCN's task force on transboundary protected areas. Likewise, another coordinated cooperation, served on the governing board of EUROPARC and was a member of IUCN's task force. Still another manager also appeared both in EUROPARC and UNESCO's expert groups. Thus 'experts' were recognised, and promoted their own experience, benefiting professionally from this role and creating a core group of individuals who cropped up in all relevant international meetings. The need for coordinators was unsurprisingly systematically promoted by these individuals. On the ground, however, support for coordinators was variable, linked to funding circumstances and individual positions. One manager relates his version of the changes that took place since he first started work, indicating his central role in maintaining continuity:

> *So when I came the idea was to have two coordinators: [one from each country], and it was one of our major tasks to elaborate the (. . .) programme, to prepare [it], which then I had to do alone because [my counterpart] in between didn't have a job anymore. (. . .) They would not have had enough money to employ two coordinators so they said*

> *if you get fifty per cent from [a funding body] we can do it. (...) And then [he] left in 99. And then I was alone again. And still I am*
>
> *(A manager).*

In contrast to this, a colleague in the neighbouring country suggested that the balance of work between the two coordinators was unequal, leading to increased frustration on his side:

> *Two coordinators, at a certain point, it's really quite a weight. (...) Salaries and additional charges. So then, well, we judged on results: it's clear I may not be objective in saying this, but it's [our employee] who had to do almost all the work. There were a whole lot of forms to fill in: that was for [him]. [He] was here all the time, and [he] was fed up of always having to work in a vacuum. He was fed up with having to make people do things that they didn't want to, and so I don't want to put all the blame on [the other coordinator's] shoulders, but at one point I think [he] gave up, and we saw clearly when [our employee] left, nothing happened at all. On the other hand when there was only a coordinator on [our] side, he would travel to [the neighbouring country] a lot, he wasn't only in his office here. We didn't see that when [he] left. [The other coordinator] didn't come to [our country] any more frequently, on the contrary. (...) So this doesn't allow us to deal with problems, with rumours ...*
>
> *(A manager).*

This frustration and the feeling that coordinators could choose to have an easy life and could afford not to achieve much was compounded by their often vague job descriptions. 'Coordination', in practice, could mean a whole host of things, some of which were bound to be intangible. For managers used to projects in which concrete results were valued, such interpretive work seemed superfluous at best, when it was not considered simply naïve and unrealistic:

> *She doesn't realise but they are all people living in a world where it's easy to say 'let's do this, or that ... '. These are all very much administrative projects and [she] is only administrative, she only drafts projects, she doesn't do anything on the ground, so she meets all sorts of people, she writes reports, things like that, meetings, but [she] is someone who only deals with public relations, everything concerning meetings and so on, reunions, discussions, because, well, that's her job. But afterwards, on the ground, nothing is happening at all. Nothing*
>
> *(A manager).*

Coordinators as cultural facilitators

One of the more interesting aspects of coordinators as go-betweens was that beyond being simply administrators, they took on the role of cultural facilitators. Since many of the problems encountered were intrinsically linked to cultural differences in work patterns and communication, coordinators were identified by some managers as having a key role to play in bridging these gaps. However, this aspect only appeared explicitly in two cases.

> *I think there are more advantages of being two [coordinators]. There are advantages and disadvantages. The big advantage is that we could exchange a lot about (...) culture, whatever it means. With all the differences, the cross-cultural problems. He could explain why [Country A] people may react like this. And together we could develop strategies of how to set up meetings and working groups etcetera. We could share a lot of ideas and we could also push together for the creation of a transboundary biosphere reserve, each one on his side, and it was the link. He and I were the link because there is not so much communication between the directors and you need someone to fill the gap and we did*
>
> *(A manager).*

This position was nevertheless problematic. It required a high level of personal reflexivity, as well as the capacity to stand outside of accepted cultural practices, taking a critical stance on practices perceived as 'normal'. This ambiguous position of go-between held difficulties for both parties, challenging the non-problematic distinction of Self and Other. Within such a clear-cut dichotomy, it was not always entirely clear what the role of such a facilitator should be:

> *[He] was perceived to be the spy working for the [people of country A]. That's it. A sort of traitor, who passes on information, who informs the [people in country A] about [the people in country B]. That's the whole principle of these coordinators. It is to say that us, (...) we have a certain approach, and we are looking for partners on the [other] side working on orchards, for example, and so instead of sending out a little [person from our country] who would phone left and right to the [farmer's associations] and all that, and who'd be sent packing, we might as well take a legitimate [person from that country], and ask him to carry out this information and intelligence work. It's really like a military liaison, you see. But that's it, otherwise we lose a vast amount of time. And to have a [local person] tell you 'watch out, [the people in that country] think in this way. They work this way. And at this speed'*
>
> *(A manager).*

This position of go-between was not always understood. The individual managers involved in this situation clearly held very different opinions, with some supporting the idea entirely and others inherently mistrustful, rejecting the need for such mediation. Managers negotiated and presented their position and its legitimacy within contexts of uncertainty and mistrust between the two administrations, projecting their own stance to the researcher asking questions. Beyond personal dislike, many of the descriptions of the coordinators hinted at the difficult role of mediator between two bodies operating in different countries, with different attitudes to decision-making:

> *He doesn't play the game of frankness and honesty. Me, I'm starting to get tired. Because he doesn't tell the truth. He's not a frank person. And he has an attitude of always following protocol, a respect for formalism*
>
> *(A manager).*

The need for such a person was assessed periodically in one site, in line with funding deadlines. The question then was not only what the coordinator's role should be, but also whether such a person was in fact useful. Part of the issue was of course defining who was qualified to take such a decision. In small protected area administrations, beyond managerial strategies, choices of maintaining or cutting professional positions had direct human impacts. In marginal, remote areas, there was often a dearth of qualified jobs for professionals: arguing that coordinators were necessary thus directly served to create and maintain such jobs. It may be judged unduly cynical to advance that coordinators were involved in a self-perpetuating industry, justifying their own existence in multiple fora. However, this position was supported by some of the declarations of managers in three of the transboundary sites. One person put it in no uncertain terms, describing a person not yet employed by the protected area's administration but who was seeking a job as a coordinator:

> *She is pro-European but she had an advantage and that's that she has the two nationalities, (. . .). Well, she hasn't been [a citizen of country A] for long, I think it's been for about a year. But she took this (. . .) nationality in order to be able to work with the park [in country A]. The whole thing is financial. (. . .) So, she can play both fields. It works up to a point but it's true that she is getting a rock-solid contract, in which she makes piles of money . . .*
>
> *(A manager).*

As in the case of the coordinator quoted earlier, the figure of the 'traitor' was not far away in the description. The choice of words emphasized both the cunning and dishonesty of attempting to adopt a dual identity, fitting into the marginal spaces between the two countries. This was described here as no more than a cynical ploy designed to gain additional money. As earlier quotes have indicated, however, gaining legitimacy as a go-between was always problematic. The objective here was to avoid creating institutional imbalances, something that was recognised as difficult in situations with only one coordinator. In one case, although it was laid out who was meant to be dealing with whom, things were not straightforward:

> *So . . . well . . . another problem of my role is that I am not the managing director, I am the coordinator, but to some extent I have to do things which normally would be done by the managing director and I have directly . . . most of the decision-making is between director [of the neighbouring country] and myself. And then I have to negotiate that with my managing director, which doesn't make it easy. It normally should be the job of the [other] director and [mine], but . . . (. . .) I do not have the mandate to make decisions. So I always have to counterbalance things which takes a lot of energy sometimes*
>
> *(A manager).*

Here, although the Other was clearly identified, he did not hold similar decision-making power and accountability, making exchanges asymmetrical. This was a similar situation to that previously existing in the same site when the only coordinator was from one country.

Inevitably, when situations degenerated, mutual blame of the Other 'not being committed' enough tended to fly, usually followed by declarations of personal compromises made to try to save the situation. Beyond personal recriminations, this did indicate the institutional challenges of organising common work. Here, cooperation was understood to be about more than simply exchanging information and challenged existing work practices in a substantial manner. Individual authority was tested by the change in scale; cooperation became more than simply coordinating common projects. Rather, it implied a radical reorganisation of authority and decision-making processes within existing administrations:

> *But now, with [another international] programme coming up, with all these projects, we get a big problem. (. . .) So I ask myself what happens. It might happen that they want me to do all this but I can't. And I don't want. It might also happen that they see the necessity to employ someone else. It should not be too difficult, normally they have quite good funding. Much better funding than we have. (. . .) There are some colleagues [from the neighbouring country], they say we don't need a coordinator. [One person], for instance, he says 'I don't need a coordinator'. Which I do not agree on in many cases there were you always needed a coordinator, but okay, he says I don't need a coordinator. There are others they say we don't have time for transboundary projects so we don't need a coordinator. There is [still another] who says he doesn't have money to employ a coordinator. I don't know if there are any more reasons, probably you know more than I do. I don't*
>
> *(A manager).*

The question was obviously not only financial, although this was often cited as a sufficient reason not to explore alternative institutional structures. Mutual blame of 'lack of commitment' flew from both sides, with mutual accusations following periods of stagnation. The problems described were often inherently cultural, linked to different habits of work and decision-making. Both sides identified lack of circulation of information as a problem, although all had conflicting takes on the actual reason. Many of these problems were seen to be linked to institutional and administrative differences. The implication of these for cooperation is explored in the following section.

Who is the Other? Institutional and cultural issues

Identifying the Other

If cooperation was about negotiated identity construction in which Self and Other were distinguished before being put into contact, then identifying the Other grounded the process. In transboundary contexts, however, clearly identifying an Other with which to engage was not as straightforward as implied in the literature. The problem individuals faced in identifying counterparts took place on at least two levels: first of all identifying an appropriate partner institution with whom to engage, and subsequently finding an individual person with a similar job description. The search for an equal counterpart, for a sort of mirror reflection on the other side of the boundary, was often the search for a chimera.

The reification of the Other as a single entity, either described by nationality ('the Germans'), by location ('the French side') or by institution ('the park') systematically took place, particularly when managers described conflict situations. Discourses also described 'the coordinator' as separate from other employees assimilated into one homogenous group, emphasizing the role of go-between and ambassador, while belonging to neither side. The Other was rarely some cloned Self simply located on the 'other side'. Rather, the Other was fundamentally different and had to be engaged with as such in order for the full richness of the encounter to develop. But before any engagement could take place, this Other had to be identified:

> *When I look through all the files, at least in my own field of nature protection, first of all I have a big problem and that is that I don't have an equivalent in the German team. The only person I cooperate with is Lukas, a forester handed over to the Naturpark to work on the lynx project, and he is the only one. He is very competent and it's working very well. We did a fieldtrip to the Harz together, we do our job. But it's limited. I don't only have the lynx project. So there is no response. So every time I have to go and look within the different levels: Mayence, Oppenheim, Neustadt, the Landkreisverwaltung-thing. I'm starting to build up my network*[9]
>
> *(Hugo, Parc Naturel Régional des Vosges du Nord, France).*

In this case, there simply was no obvious partner for this manager and so rather than remain resigned to the fact, he decided to actively seek out various individuals with responsibilities he could piece together in order to create a composite Other. This creative construction of an Other was however time consuming and supposed a high level of personal dedication. In three sites, contacts between adjacent protected areas were largely restricted to contacts between directors and senior managers. In another, there was some evidence of a shift from formal contacts between directors to issue-specific contacts between technicians working in the field, counting chamois or making surveys. Restricting contacts to 'directors' was however the easiest model that first appeared, in which it seemed obvious that the Other was an equal:

> *At the moment, there is no specific cooperation between technical teams. The cooperation is mostly between the directors. We have one scientific conference every year, for scientists to exchange information*
>
> *(Andrzej, Bieszczady National Park, Poland).*

However, in more in-depth discussions it emerged that if all were equal, some were 'more equal than others', to quote a much-used phrase. Being a 'director' of a protected area did not mean the same in each country as levels of authority, decision-making and accountability varied tremendously. In the East Carpathians, the Polish director had virtually full decision-making power over the state-owned land, while in Slovakia, at the other extreme, this role was largely consultative, despite all three sites being designated 'national parks'. This was obviously compounded when protected areas were of different IUCN categories, but this did not seem more of a determining factor in creating inequality than differences in institutional structure and national legislative practices. In situations where it was clear that the Other did not hold an equivalent position, contacts could work out successfully if this was sufficiently recognised.

Basing cooperation on personal relationships

A reliance on personal relationships was extremely widespread, both in the cases where cooperation was still tentative (Danube Delta; East Carpathians) and in cases where the process was well established (Alpi Marittimi; Vosges du Nord/Pfälzerwald). That cooperation rested on personal contacts was repeated endlessly, illustrating the lack of any more formalised engagement:

> *We wanted to organise more exchanges but we do not have any money to do this because we cannot use Romanian government money for this. So now there are some contacts between scientists but there is no systematic exchange. But personal relationships are good*
>
> *(Nicolae, Danube Delta, Romania).*

On a personal level, these contacts occasionally gave people great satisfaction and turned into real friendships, actively participating in fostering good neighbourly relations and building trust. Nevertheless, relying exclusively on personal contacts put a lot of pressure on maintaining good relations and avoiding problematic issues. When and if things went wrong and confidence was lost, the whole process risked collapse. Thus rather than being a positive and fulfilling process, the reliance on personal relations backfired. Conflicts became personalised and confidence was lost. In one site, an attempt was made to formalise personal contacts by creating pairs of people, helping to identify clearly for each person a neighbouring equivalent:

> *I said that in this context we could probably try to set up (. . .) project teams, and we should find a way that one of these two is the major responsible because we need one person who is really in charge of presenting reports and things at the right time (. . .) then we decided who is going to be the main responsible, so we had a good mixture on both side and it worked out quite well (. . .) then after some time these couples had troubles in terms of personality . . . In some projects we had difficulties because they didn't meet regularly, they didn't want to communicate (. . .) I tell you, if they don't like each other, it will never work. Even if you . . . Well, it depends*
>
> *(A manager).*

In one acute example, communication within one of these pairs broke down completely, leading one to refuse all further common work with the other. Admitting this was a big step, and was seen to be breaking a taboo. In all the other sites, nobody admitted to not liking individuals in the neighbouring country. This may well have been because this was not the case, and individuals did get on well or maintained professional conduct that did not allow for personal feelings to become involved; or it may have been because the levels of interaction between partners had not yet reached the point where discussion, conflict and negotiation strained personal relations. The results from the one site indicated that personal likes and dislikes were in fact an issue and that relying exclusively on personal contacts was a risky strategy in the long term. In the following two quotes, two managers from adjacent countries express similar views about the collapse of personal relationships:

> *There is a very strong conflict between people, now, (. . .). And in this case [my colleague] made a massive effort, he took all the initiatives and all that, and (. . .) he said 'ah no, I'm fed up with [*abusive nickname*], as he calls him. And so now nothing more is being done*
>
> *(A manager).*

> *Personalities is also a very big issue. It is very much underestimated. It is a big issue. I can tell you. If people don't like each other, nothing works. We have cases where (. . .) colleagues [from the two countries] don't like each other. No way. No way*
>
> *(A manager).*

In this site, the initial 'designated' Other was clearly identified, only to be subsequently recognised as so fundamentally different as to make cooperation impossible. This led to the conclusion by one side that the initial partner administrations were not altogether the most appropriate for sustained cooperation. A decision was then taken to widen cooperation, jumpstarted by the establishment of specific working groups. This negotiated process and increased knowledge of the Other took time, stretching over several years. During this time, initial enthusiasm was followed by a stark acknowledgment that things were not working out as planned, and that existing partners were inadequate.

> *But let's be clear: I remain an ardent proponent of the cooperation. It's not because I hold a critical view of the institutional structure that according to me doesn't work on [their] side, while it is going quite well on [our] side (. . .). But I continue, I continue, but without relying and passing through the [existing administration]*
>
> *(A manager).*

Identifying an Other with whom to cooperate was thus a problematic process. Two protected area administrations, rather than behaving as individual subjects, were clearly shown to be composed of individuals seeking to make sense of a situation on which they only had partial information, attempting to identify someone on the 'other side' with whom to cooperate. Identifying differences and understanding how these were articulated led some managers to reach a clearer understanding of Self, upholding the idea that identity was something that was negotiated discursively. One manager described how titles implied different job descriptions and levels of authority:

> *When you are speaking to a president, to a Geschäftsführer, in any case a Geschäftsführer is different from a park director ... (. . .) It's true that with the Germans we discovered all that. That there is a different role for technicians and elected people. Elected people delegate much more. While in France, elected people, sometimes they wear the clothes of the technician, in a way. There is this whole game that they*

> *play. And so between the idea of a Geschäftsführer (...) and the vice-director ... It's true that me, when I introduce myself in German, I say 'Leiter'. I don't say Geschäftsführer, because that notion, it's really ... It's almost the accountant, really. It's the head of personnel, or else, someone who is the director but who is only there in order for everything to work well. But I don't have that role. While I have a general secretary who is in charge of logistics, and I direct and I also play a technical role, on particular issues and projects*[10]
>
> *(Théo, Parc Naturel Régional des Vosges du Nord, France).*

The process of identifying the Other was therefore inherently linked to understanding the system within which this Other functioned. Getting to know and understand the institutional differences and patterns of work in the neighbouring country were inseparable from getting to know the individuals. Again, this process of identifying institutional processes was problematic, as differences needed addressing if cooperation was to be formalised in some meaningful way.

The institutionalised Other

Problems in establishing and maintaining cooperation were often identified as stemming from institutional differences, sometimes taken to be irreconcilable. Inequality in institutional form or framework was repeatedly mentioned to be a fundamental issue withholding real exchanges. It might have been because this was deemed an uncontroversial fact that depersonalised resistance. It was easier to say that cooperation was impossible because of differences in institutions, rather than saying that the neighbours were too different or that fear of changes in work patterns or loss of authority created resistance. Whatever the underlying reasons, institutional issues were considered crucial and much energy was put into imagining alternative scenarios: coordinative, consultative bodies and committees that could bridge the boundary. In the East Carpathians, no less than four separate bodies were imagined over less than five years: a Coordinating Council, followed by a Consultative Council, a Scientific Council and finally a tri-lateral Foundation. Each was created with a specific purpose in mind and all but the last ended up being deemed inefficient and dropped. This reflected the practical difficulties of establishing shared institutional structures, including difficulties in funding them and determining their mandate.

Differences in size between administrations were repeatedly identified as a problem, often tied to differences in specific role and mandate (Danube Delta; East Carpathian; Vosges du Nord/Pfälzerwald). In the most acute case, the Danube Delta was managed on the Ukrainian side by a team of 36 people, including 4 researchers, while on the Romanian side there were two distinct bodies, the biosphere reserve Authority employing over 100 people, and a research institute employing 112 people. In such situations, establishing exchanges between two administrations was far from easy. This difference in size, which sometimes also reflected a difference in territorial extension, was systematically accompanied by differences in mandate:

> *We, over time, have managed to really be an animator of the territory. They are not. The result is that on nature protection, education, tourism or even the forest, the Naturpark each time does not have the information, or else is not at all involved because people don't see why they should inform it. The result is that if we ask to see the Naturpark to discuss something, they wonder why we want to see them because they are not informed. (...) I went and did meetings in German, in Germany with Germans, informing my colleagues (from the Naturpark) about this ... but then they didn't come because either they didn't have the information, because they had too much work, because they are only a small team, it's true there are fifteen of us ... so, you see, we went and looked for a different partner*[11]
>
> *(Hugo, Parc Naturel Régional des Vosges du Nord, France).*

Here, the manager identified being an 'animator of territory', a uniquely French term, as the real mandate of a biosphere reserve, equating it to that of the French model of regional natural parks. This was substantially different from the practices of a Naturpark. The inability to engage with each other within a shared mandate led the French park to implicitly reject the Naturpark as a legitimate partner. The Other was too different, too alien and could therefore not be engaged with as an equal. Chapter 5 described the spatial consequences of the creation of coextensive spatial entities when an existing protected area was designated as a biosphere reserve. In addition to the problematic spatial consequences of this double identity, institutional issues were created, even in cases where the territory was managed comprehensively by one administrative body. The need to change existing practices to coincide with the designation as a biosphere reserve reflected different understandings of the concept. One of the features of the model was that it could be adapted to local circumstances, yet this flexibility created specific difficulties in transboundary situations when interpretations differed. These differing interpretations appeared in all the transboundary cases examined. These were not only linked to appropriate institutional structures but also to the underlying objectives of such a model.

These interpretations had direct impacts on the individuals concerned who were faced with a choice of conflicting mandates. In complex situations of overlapping institutional identities, one possible strategy for individuals was to identify more clearly with one designation, relating and allying to one specific mandate. This led managers to be variously committed to transboundary cooperation. This was expressed by a French manager who described the splits within the German Naturpark:

> *Because after all, the fact that the Naturpark Pfälzerwald remains, in fact, the body responsible for the biosphere reserve, it is officially identified as such, it is ... it doesn't have the legitimacy that another structure could have. They remain in the famous German Kistenkultur, that is to say that it remains ... they are not able to get out of their Naturpark box. (...) So there there is a real problem, which means that in the current team of the Verein, the four people, some people have very clearly remained 'Naturpark', and then people like Daniel are 'biosphere reserve', so there are more and more internal problems*[12]
>
> *(Théo, Parc Naturel Régional des Vosges du Nord, France).*

This lack of legitimacy of institutional structures had a very strong impact on cooperation. Certain individuals in Germany did not recognise that the biosphere reserve was a legitimate body, but instead took it to be a simple designation. Furthermore, the French managers, by not recognising the legitimacy of the institutional structure in Germany, effectively lost their original partner. This difficult situation was further compounded by culturally-different needs for legitimacy. A French manager suggested that Germans were particularly attached to institutions that provided official sanction for their actions, commenting on the existing scenario across the border:

> *I am intimately convinced that the Naturpark is not the right institutional structure for the biosphere reserve (...). So there is the need for another structure, recognised as such, with strong status, because Germans like to have strong status to rest on and that provides them with legitimacy. (...) Two three people: you don't need a huge team*[13]
> *(Hugo, Parc Naturel Régional des Vosges du Nord, France).*

This example was interesting in that within transboundary contacts, alternative institutional structures for biosphere reserves were confronted. Institutional structures and mandates were seen to define legitimacy, determining whether an administrative body was an 'appropriate' Other with which to engage. Cooperation led to sufficient mutual knowledge to permit judgements on the Other, although here the French conclusions were largely negative and self-justifying. Although the conclusion in this case was that common work could not continue with the Naturpark as sole partner, it did lead the French to actively seek out others. At the time of the fieldwork, while no new administrative structures had been formally established on the German side, there was strong indication that this would happen in the near future (Issues of institutional legitimacy are further discussed in Chapter 9).

Conclusions

This chapter began with an examination of the three main trends of literature dealing with cooperation. Despite being composed of three substantially different orientations, I suggested that this literature did not address the complexity of cooperation as a social and spatial process, other than taking it to be a tautological course of action that defined itself. Game theory, first of all, addressed the problem on the scale of the individual but did not offer a framework that could be extrapolated outside precisely controlled conditions in which collaboration was a pragmatic strategy. Secondly, approaches describing stages of cooperation took the process for granted, replacing it with spatial metaphors of integration, and positing a cumulative move towards greater integration. Thirdly, the literature on international cooperation, ranging across political realism, functionalism, neo-functionalism, regionalism and new institutionalism did no more than indicate that the analysis of cooperation could take place on a variety of scales while avoiding the central issue of actually defining it.

Unlike previous chapters, the argument centred on the people and the institutions involved in the process of changing work practices, seeking to understand how these were bound up with the construction of spatial entities. This chapter, unlike others, was therefore not specifically about space or spatial interactions but sought to understand the social and institutional dimensions of introducing cooperation.

Protected area managers contributed diverse and sometimes contradictory definitions of cooperation which informed both their individual actions and their personal interpretations of situations. This indicated that it was not simply an unproblematic process stemming from a rational decision made between equal partners for expanding existing work across a larger area. Instead, contrasting definitions of cooperation co-existed within and between administrations. These ranged from considering friendly contacts and common meals to be the desired result of cooperation, to attempts to create institutionally complex shared bodies employing specific coordinators. Stories and narratives describing cooperation in specific sites were contrasted, allowing me to suggest that protected area managers construct interpretations of situations and situate themselves within them as part of an ongoing process of identity construction. This necessarily involved spatialised references to Us/Here and Them/There, linked to the reification of the Other described by nationality, location or institution. Methodologically, the multiple stories further illustrated the discursive and conflicting nature of information collected during interviews and the corresponding need to move beyond finding 'the truth' when analysing social processes.

In order to discuss some of the difficulties of establishing cooperation on a practical level, I paid specific attention to the role of coordinators who acted as mediators, bridges, go-betweens or 'spies' between administrations. The difficulties such individuals felt when adopting 'in-between' identities, attempting to span two different systems while fully belonging to neither, further strengthened the need to consider cooperation as a negotiated practice. By challenging the unproblematic distinction between Self and Other, Here and There, coordinators were potentially in a position to assist others in moving beyond such dichotomies. Yet the practical difficulties they encountered in justifying and carrying out their role indicated that bringing about such changes was far from straightforward. The different interpretations of cooperation present among the managers were compounded by the intrinsic inequalities between institutional bodies that demonstrated the problematic dimensions of considering cooperation to be a freely-chosen strategy. In order to illustrate this, I drew out cases where cooperation was 'imposed' from outside, demystifying the assumption that cooperation happens between equal partners. I identified that this was linked both to international and local pressures and involved a variety of different institutional actors including international organisations, international funding agencies and adjacent protected areas.

This discussion of cooperation in protected areas has indicated the need for an improved definition of cooperation. This must be seen to be a negotiated process that involves multiple individual and institutional actors, and rests on the combination of *social* processes of acceptance and resistance to institutional changes, as well as *socio-spatial* processes of integration and distinction. Paradoxically, such a definition must move beyond the assumption that cooperation erases boundaries. If cooperation rests on the assumption that there is some interaction between at least two partners, then these must be spatially and institutionally differentiated. Yet without a (physical or conceptual) boundary there is no such Other and therefore no possible cooperation between distinct partners. Equating cooperation with the creation of transboundary spatial entities is conceptually problematic and therefore literally and figuratively misplaced. It may be that other academic traditions outside the traditional domain of geography or political science have dealt with similar issues. The economic literature on strategic mergers or company restructuring and the corresponding institutional changes in work practices might contribute to this debate although a full exploration of this is outside the scope of this book.

The need for a new definition of cooperation does not imply that previous approaches to international cooperation must be wholly discarded, but rather indicates that they are profoundly incomplete. Several such theories suggested points that frame the context, such as that localised interaction can come to have wider political significance and can participate in integration at other scales. This is a worthwhile point and one that often serves to underpin funding programmes such as Interreg. One manager voiced this explicitly, arguing that federal and national governments had got to the point where they could not afford to see local transboundary cooperation fail:

> *Unbelievable. You wouldn't believe how much political this is [sic]. Believe me. They cannot afford to destroy this cooperation, they cannot afford it, it is impossible, fortunately. They are really bound to approve. So there was a big noise around it and then the Germans said oh my God we cannot afford to get these problems with the French, so our Ministry said we will give you more money. (. . .) So finally after we had a really big battle, it was really awful*
>
> *(Daniel, Naturpark Pfälzerwald, Germany).*

This intrinsically contested nature of the process must be better understood in order for protected area managers to grasp the complexity of situations which are revealed to be 'how much political'. In the next chapter, I discuss some of the myths and assumptions that surround cooperation in transboundary protected areas, further arguing that political and cultural aspects linked to the construction of transboundary spatial entities need to be taken into account.

Notes

1 Griffin, Paul (1999) 'Going Away', Lyon & Lamb : Southwold, p.5.

2 The term *functionalism* as used in this context should not be confused with functionalism in other fields such as sociology or biology (Taylor 1990 : 125).

3 Personal translation from: 'Pour moi, l'amitié franco-allemande, c'est pas la coopération transfrontalière. Quand je parlais hier du mythe de l'amitié franco-

allemande, c'est peut-être de ce qui relève des générations plus âgées que nous, (...) c'est les Stammtisch, et les bouffes, les voyages d'étude, les cérémonies, machin, on se tape sur l'épaule et c'est très bien, et puis on fait des choses qui fondamentalement ne posent pas de problème. On se pose des questions qui ne fâchent pas. Et surtout, on ne pose pas les questions qui fâchent. Pour pas gâcher l'amitié franco-allemande. Une sorte de mythe, comme ça, attention, l'amitié franco-allemande, c'est précieux. (...) Mais le problème, c'est que ça obnubile un certain nombres d'autres questions – et on ne déconne pas, l'amitié franco-allemande fait qu'on reste copains, la guerre tout ça, c'est passé, c'est génial, mais le paysage, l'aménagement, la route que voulez nous imposer, tout ça c'est des conneries. (...) Quand vous nous piquer des idées, c'est des conneries. Et ça, les français on du mal à le dire. Et ça, c'est de la coopération transfrontalière'.

4 Personal translation from: 'Il faut être cohérent. Soit on croit qu'il y là un espace commun naturel où il y a une communauté d'intérêt. C'est comme l'Union Européenne: soit on croit qu'il y a une communauté d'intérêt, soit on n'y croit pas. Si on y croit, à cette communauté d'intérêt, il faut faire en sorte que les politiques, que les actions progressivement soient coordonnées et soient menées en commun. Donc moi j'y crois à cette communauté d'intérêt. Il y en a peut-être d'autres qui pour diverses raisons, y croient moins, quoi'.

5 Personal translation from: 'Nous ici, enfin la structure ici a souhaité ce label, elle n'a pas chigné, ce n'est pas quelqu'un qui est venu le lui imposer. Elle l'a demandé. C'est nous, à l'interne, qui avons monté la proposition mais ils auraient pu aussi refuser en disant 'qu'est-ce que c'est que ça ?'. Donc, c'est différent si tu espères quelque chose, tu vas le faire vivre, que si tu n'en attends pas grand chose et que c'est quelque chose qui est venu à un moment donné. C'est important. Maintenant, peut-être que, dans nos instances, tout n'a pas encore été parfaitement bien compris comme ça mais, en tous cas, ici, au niveau de l'équipe technique, les gens voient bien que ce ne sont pas des lauriers qu'on donne pour ensuite dormir sur ses deux oreilles, c'est vraiment un encouragement et du travail à faire. Rien n'est acquis. La difficulté de mettre en application une réserve de biosphère est la même que pour toutes les idées pour lesquelles tu n'as pas d'autre pouvoir que la persuasion'.

6 Personal translation from: 'Per esempio il discorso della riserva di biosfera, noi, che ci teniamo un po' di più, abbiamo fatto una lettera al Presidente del parco del Mercantour e non ci hanno ancora risposto, cioè gli abbiamo fatto una lettera in cui gli dicevamo: noi abbiamo già preparato il dossier per la parte italiana, voremmo presentarlo quest'anno ma abbiamo visto che non ce la facciamo, quindi dobbiamo andare a finire un altr'anno, però gli abbiamo chiesto qual è il loro parere, perché ci piacerebbe che la riserva fosse transfrontaliera, visto che c'erano state comunque delle adesioni quando avevamo fatto il convegno e volevamo sapere dal punto di vista pratico quali erano le sue intenzioni, cosa intendeva fare, se potevamo scrivere già nel dossier che questa era la prima metà di una seconda metà che sarebbe arrivata in un secondo tempo. Era il minimo che potessimo fare, e lui non ci ha ancora risposto'.

7 In this section, because the comments made relate directly to individuals, the decision to protect individual's identity more stringently has meant that speakers are identified without reference to specific places. However, all these quotes – as elsewhere in the book – are directly taken from the field.

8 Personal translation from: 'Mentre io sono stata direttore del Parco Naturale dell'Argentera, oggi Alpi Marittime, il Mercantour ha avuto cinque differenti direttori ed i responsabili dei diversi dipartimenti (scientifico, communicazioni) sono stati anch'essi cambiati diverse volte. E importante sottolineare questo perché

la collaborazione tra le istituzioni è prima di tutto collaborazione tra le persone: i parchi sono fatti dalla gente che ci lavora'.

9 Personal translation from: 'Quand j'égrène tous les dossiers, au moins dans mon domaine, la protection de la nature, d'abord j'ai un gros problème c'est que j'ai pas d'équivalent dans l'équipe allemande. Le seul avec qui je coopère c'est 'Lukas', un forestier mis à disposition du Naturpark pour le dossier du lynx, et c'est le seul. Il est très compétent, ça se passe très bien. On a fait un voyage d'étude dans le Harz ensemble, on fait notre boulot. Mais c'est limité. Moi je n'ai pas que le dossier du lynx. Alors il n'y a pas de répondant. Alors chaque fois il faut que j'aille chercher les différents niveaux: Mayence, Oppenheim, Neustadt, le Landkreisverwaltung-machin. Je commence à me construire mon réseau'.

10 Personal translation from: 'Lorsque vous vous adresser à un président, à un Geschäftsführer, d'ailleurs un Geschäftsführer c'est différent d'un directeur de parc ... (...) C'est vrai qu'avec les allemands on a découvert tout ça. Qu'il y a un rôle qui est différent entre techniciens et élus. Il y a de la part de l'élu une délégation beaucoup plus grande. Alors qu'en France, les élus, parfois ils revêtent les habits du technicien, quoi. Il y a tout ce jeu, que bon, ils font. Et alors entre la notion de Geschäftsführer (...) et le sous-directeur ... C'est vrai que moi, quand je me présente en allemand, je dis 'Leiter'. Je ne dis pas Geschäftsführer, parce que cette notion, c'est vraiment ... C'est presque le comptable, quoi. C'est le chef du personnel, ou voilà, qui est directeur mais qui est juste là pour que ça marche bien. Alors que moi, je n'ai pas ce rôle. Mais j'ai un secrétaire général qui s'occupe de la logistique, et moi je dirige et je joue aussi un rôle technique, sur des dossiers et sur des projets'.

11 Personal translation from: 'Nous, avec le temps, on a réussi à être réellement un animateur de territoire et un coordinateur, eux ne le sont pas. Résultat: que ce soit une question de conservation de la nature, une question d'éducation, de tourisme, de forêt même, le Naturpark, chaque fois, n'est pas au courant de l'information, soit il n'est pas du tout dans le coup parce que les gens ne voient pas pourquoi ils le mettraient dans le coup, résultat, si nous on demande au Naturpark de se voir pour discuter de ça, ils se demanderont pourquoi on veut les voir puisque, eux, ne sont pas au courant. (...) Moi j'ai déjà fait des réunions en allemand, en Allemagne, avec des Allemands, en en informant mes collègues du ..., mais ils ne sont pas venus parce qu'ils n'ont pas l'info, parce qu'ils sont débordés, parce qu'ils sont une petite équipe, c'est vrai que nous sommes nous une quinzaine mais bon, tu vois, on a été cherché un autre interlocuteur'.

12 Personal translation from: 'Car malgré tout, le fait que le Naturpark Pfälzerwald reste, de fait, porteur de la réserve de biosphère, il est affiché, il est ... il n'a pas la légitimité que pourrait avoir une autre structure. Ils restent dans la fameuse Kistenkultur des allemands, c'est à dire que ça reste ... ils ne peuvent pas sortir de leur boîte Naturpark. (...) Donc là il y a un vrai problème, ce qui fait aussi que dans l'équipe actuelle du Verein, les quatre personnes, il y a des gens qui très clairement sont restés 'Naturpark', et puis des gens comme Daniel qui sont 'réserve de biosphère', donc il y a en plus des problèmes internes'.

13 Personal translation from : 'Je suis intimement convaincu que le Naturpark n'est pas la bonne structure porteuse pour la réserve de biosphère (...). Alors il faudrait une autre structure, reconnu comme tel, avec un statut fort, car les allemands aiment avoir un statut fort derrière eux qui leur donne une légitimité. (...) Deux trois personnes: y'a pas besoin que ça soit une équipe pléthorique'.

PART III
HYBRID BOUNDARIES

Chapter 8

Mapping a Bounded Other

He had bought a large map representing the sea,
Without the least vestige of land:
And the crew were much pleased when they found it to be
A map they could all understand.
'What's the good of Mercator's North Poles and Equators,
Tropics, Zones, and Meridian Lines?'
So the Bellman would cry: and the crew would reply
'They are merely conventional signs!
'Other maps are such shapes, with their islands and capes!
But we've got our brave Captain to thank'
(So the crew would protest) 'that he's bought us *the best –*
A perfect and absolute blank!'

(Lewis Carroll)[1]

(Re)territorialisation and the institutionalisation of space

Maps produced by protected area administrations are useful illustrations of the process of constructing the Other, providing a graphical illustration of one aspect of (re)territorialisation though the institutionalisation of space. Using a series of maps from the case study areas, the changes in the perception of the Other are identified, laying emphasis on the way boundaries are represented graphically. This process further illustrates the rhetoric of integration and distinction, illustrating the politically charged nature of seemingly-objective maps. I argue that the resulting maps have, in certain cases, attained mythical status in their own right, appearing repeatedly as icons representing the success of cooperation and promoted as such. While such maps represent crucial steps in the process of (re)territorialisation, some of the crucial and more problematic transboundary issues are swept aside when they are elevated to the status of a sacred icon. Thus what started out as a planning tool becomes a result in itself. I further argue that in cases where the creation of a transboundary map is identified as impossible by those involved, or when different maps are used by different administrations, this further represents a form of resistance to the creation of a transboundary entity.

Institutionalising space with maps

The poem quoted above slyly points a finger at those who draw maps. Tracing boundaries on a sheet or drawing maps are not innocent activities, and Velasco-Graciet notes that geographers and mapmakers, so involved in the process itself, forgot to study the boundaries they drew as a topic in itself: 'the geographer, too involved in inventing and tracing boundaries, avoided studying the boundaries in themselves'[2] (Renard 1997 : 28 quoted in Velasco-Graciet 1998 : 17). In his analysis of protected area policies in Trinidad, Sletto uses Paasi's idea of the institutionalisation of a spatial entity to link the theorization of boundary making with the social production of bounded protected areas, a process informed by the concurrent processes of differentiation and integration. He notes that 'regions are socially constructed as different and unique, at the same time as regions are integrated from within through narratives of sameness' (Sletto 2002 : 190). This follows the idea of the production of entities as geographical locations and discursive entities (Ò Tuathail 1996 : 5).

This chapter focuses on the maps of transboundary protected areas designed and produced by a variety of actors including protected area administrators, non-governmental organisations, state bodies and international organisations or donors. I argue that cartography and maps take part in this construction of these areas as discursive entities. The maps considered come from the areas covered during the field visits. More maps have been produced by the Franco-German Vosges du Nord/Pfälzerwald and the Italo-French Alpi Marittimi/Mercantour than the Polish-Slovak Tatry/Tatras, the Polish-Slovak-Ukrainian East Carpathians and the Romano-Ukrainian Danube, for reasons linked to budget and policy choices. The number of maps available from each site is thus very variable, and includes both individual protected areas within one country and truly 'transboundary' maps. The number of maps per site ranges from two in the Tatras to the selection of over 10 transboundary maps spanning thirty years, from the Maritime Alps.

The function of conventional cartography is to transform space into a legible, ordered territory, thereby institutionalising it. Maps are therefore part of the technical infrastructure necessary for the governance of space (Ò Tuathail 1996 : 4). However, maps are more than simple tools for the control of space, and Crampton argues that there is a clear moment when cartography ceased to be seen as a straight-forward communication system and became seen as imbued with power relations (Crampton 2001 : 235). Building on Harley's influential work, a map is a representation that belongs to the terrain of the social world in which it is produced (Harley 1989 : 1), creating an 'élite discourse' (Perkins 2003 : 344). Maps are fiat constructions, graphical, reduced pictures of the world responding to a particular need (for a recent review of analytical approaches to analysing maps see Perkins 2003).

A map is necessarily a construction, a 'model representation, that is to say a pertinent and coherent deformation of the body of the earth'[3] (Raffestin 1995 : 244). Firstly, a map is a graphical representation that can only aspire to be an instrumentally reasonably accurate model of reality, which draws attention to some coherent, pre-determined aspect of it. Secondly, it is a representation to scale, since 'it always carries a scale: a relationship of reduction between the real measured distance on the ground and the represented distance. It implies a generalisation, that is to say a loss of information resulting from a compromise'[4] (Ferras & Hussy 1991 : 209). Furthermore, 'rather than being mirrors, maps are cultural texts, which construct the world rather than reproduce it' (Woods 1992 in Paasi 1996 : 20). Thus maps are foremost social and political instruments of power in the division of space that participate in the (re)territorialisation of the area mapped. As Harley further notes, 'much of the power of the map, as a representation of social geography, is that it operates behind a mask of a seemingly neutral science. It hides and denies its social dimensions at the same time as it legitimates' (Harley 1989 : 3). This echoes Latour's arguments on the construction of scientific knowledge: 'Knowledge does not reflect a real external world that it resembles via mimesis, but rather a real interior world, the coherence and continuity of which it helps to ensure' (Latour 1999 : 58). A map ('knowledge') is not neutral, but rather reflects a wider web of relations called on to construct it.

Historically, this legitimisation has taken several forms. Maps have participated in the mise-en-scène of the geopolitical gaze, what Ò Tuathail calls the art of setting a stage scene or arranging a pictorial representation, using a mechanism of theatrical illusion comprising two different stage domains or stage levels: 'on one hand a physical/climatological/geographical/material/spatial/natural (back)ground to the stage, while, distinct from this, we have a human/historical/temporal/political/cultural foreground or surface of appearances' (Ò Tuathail 1996 : 30). This geopolitical gaze was all the rage during the first half of last century, finding favour with authors such as Halford Mackinder, Nicolas Spykman and Karl Haushofer, in a form of Cartesian perspectivalism with a detached viewing subject surveying a worldwide stage. Harley argues that this idea of an all-seeing eye pervades cartography, since 'cartographers manufacture power: they create a spatial panopticon' (Harley 1989 : 8).

This geopolitical heritage in cartography is important in the historical process of reifying maps as explanatory representations and icons, justified as objective by reference to their 'natural' rootedness, in true positivist fashion. Raffestin notes that 'the representation ensures the mise-en-scène, the organisation of the original grasp of power as a show (. . .). The image or model, that is to say any construction of reality, is an instrument of power and has been since the earliest days of mankind (. . .). We have even turned images into 'objects' as such and, with time, have got used to transforming these images, mere simulations of the original objects, rather than transforming or acting upon the objects themselves. Therefore, is it surprising that we manipulate them, have manipulated them and will continue to manipulate them ever more?'[5] (Raffestin 1980 : 130–131). This idea of performance is interesting, and has been amply developed – often confusingly – by a variety of authors following Derrida's ideas of deconstruction (Dewsbury *et al.* 2002), sometimes to the point of oblivion. What I gather from these ideas is that a performance always represents a choice. As Campbell argues on the subject of maps of partitioned Bosnia, 'importantly, these performative practices of representation do not simply "imagine" one assemblage of identity; they also un-imagine another' (Campbell 1999 : 401). In other words, representational practices serve a double function: they bring into being one conception by simultaneously removing another.

The process applied here can broadly be assimilated to deconstruction, in which maps are considered as 'texts'. But while inspired by Derrida's ideas, the more literary turns of deconstruction are avoided, 'engaging not only geopolitical texts but also the historical, geographical, technological, and sociological contexts within which these texts arise and gain social meaning and persuasive force' (Ò Tuathail 1996 : 73). Certainly, as Harley notes, '"text" is certainly a better metaphor for maps than the mirror of nature' (Harley 1989 : 4). If maps are taken to be texts, then Harley argues that it is possible to analyse maps precisely as rhetorical texts. This use of rhetoric feeds into what I described earlier as *spatial* socialization, that is to say the process through which individual actors and collectivities are socialized as members of specific territorially bounded spatial entities and through which they more or less actively internalise collective territorial identities and shared traditions (Paasi 1996 : 8).

While all this sounds fair and well, the next stage is not necessarily so obvious. On a very practical level, how on earth does one deconstruct a graphical object? Crampton fleetingly hints at the possible use of semiotics (Crampton 2001 : 40), an idea developed more extensively by Ferras and Hussy (Ferras and Hussy 1991) in the field they call graphical semiology, analysing the use of points, lines and surfaces by building on the work of the semiologist Luis Prieto. For Harley, the deconstruction of maps 'was a heterogeneous amalgam of approaches' (Crampton 2001 : 241), the objective of which was to 'read between the lines' of a map. Yet Harley's death did not allow him to provide a viable research agenda, or method, for carrying this out. Here, I adopt a largely intuitive method, attempting to tell the stories the maps embody.

Before leading on to a practical application of these ideas, a note of caution is useful. Taking maps to be wider discourses about space is to make a strong assumption about the role of the person actually designing them. Curiously, none of the authors cited above has anything to say about the actual cartographer's role as an artist making choices of colour, layout or graphics. Some of these choices are bound to be arbitrary or simply to reflect a series of artistic choices. It is assumed that these individual choices reflect an elusive social reality, as though the artist were simply a channel through which this is revealed. When a map is produced, the fancy of the cartographer – who may or may not have professional training in the design of maps – is bound to play a role, as are the technical means at his or her disposal. Yet, because maps are not only the result of social reality but also the means of constructing and institutionalising it, these individual choices are integrated into the object itself. While the cartographer may have chosen a thick orangey-red line to represent a border simply because the software package recommended it or because it contrasted nicely with blue lakes, when the end result becomes part of official publications produced by an administration the individual choice is forgotten. The map becomes an icon, recognized by others, constructing and institutionalising a facet of social space.

The mapping of protected areas

Mapping nature, constructing space

Following Harley, I mentioned above that one of the strengths of maps is that they operate behind a mask of a seemingly neutral science. Images, caricatures and maps share the curious peculiarity of mobilising rational means in order to provoke irrational attitudes and behaviour (Raffestin 1995 : 247). Raffestin develops the idea of science and politics using myths as tools when he writes about the use of maps in Germany before and during the Second World War. On one hand, writes Raffestin, 'science has destroyed myths, but has built other new ones; on the other hand, the political realm has constantly created others, mercilessly fighting against those blocking the way. Science has always been based on myths, those of nation or of race, for example, which when melted and moulded together with "naturalistic" elements and Social Darwinism, made up Fascism's original putty'[6] (Raffestin 1995 : 246). According to Raffestin, myths are important since what is initially mediated by myths is subtly transformed into something more immediately tangible.

Mapping protected areas is naturally far different from mapping geopolitical and military campaigns under Fascist rule for propaganda purposes. However, since all maps are intrinsically political, it is not pertinent to distinguish between propaganda and non-propaganda maps. In addition, the naturalisation of political reality under seemingly 'objective' and 'naturalistic' mapping is pernicious and this is particularly true in the mapping of protected areas. Protected areas are intrinsically political objects. The myths alluded to here are sometimes easy to identify, and include themes like borderless nature and natural regions discussed earlier.

Maps of protected areas depict an area of space set aside for a purpose linked to nature conservation. The aim of these maps is initially to define an area, creating an inside and an outside, defining and delimiting one or several boundaries. Biophysical and societal elements are combined on an equal footing in order to show a coherent ensemble, combining the two levels described by Ò Tuathail: the biophysical backdrop and the societal foreground. By subtly equating the two and justifying the coherence of the societal on the basis of the biophysical, maps of protected areas institutionalise particular conceptions of space, and thereby political projects. Territorial identities can be constructed around narratives of sameness based on biophysical arguments: the Other becomes the Same because the two share a landscape or watershed, for example. Although this can be a very positive force for good, and a way of overcoming initial fear of the Other by focussing on shared features, the naturalisation of a societal entity on biophysical grounds leads to a naïve understanding of integration. When things go wrong, and cooperation falters, the disillusion is harder to overcome, as it seems 'unnatural', paradoxically leading to a stronger rejection of the Other, taken to be almost a traitor.

Maps and othering

If maps are social and political instruments of power in the division of space then these reflect the (re)territorialisation of the area mapped. Thus maps are subtle reflections of how an area is defined and how the Other is viewed and can be used as indicators of this changing relationship.

A review of the maps in a given area, both national and transboundary, indicates some of the attempts to portray the Other across the boundary, as well as the first steps in representing a common transboundary identity. Maps are involved in the dual process of integration and distinction and are much more than simple pictures of an area. They reflect the power relations and conceptions of identity and otherness that surrounded the creation of new spatial entities. In all the areas visited, international boundaries had been redefined and demarcated within living memory. This means that the Other was not taken for granted in these areas, at least by older generations. The stretch of Franco-German boundary considered, for instance, was redefined following the Second World War, when France regained the regions of Alsace and Lorraine. Similarly, the Franco-Italian boundary in the Maritime Alps was redefined in 1947, although negotiations and arbitrations linked to land use and access rights were only finally settled in 1954 (House 1957 : 117). These relatively recent changes contribute to making boundaries and boundary definitions highly charged issues in these areas. Identity and sense of belonging are not taken for granted and lines on maps can stir powerful emotions.

Islands and peninsulars : isolate and unify

In 1978, three years after its official creation, the Parc Naturel Régional des Vosges du Nord, in France, published its first map. The map was included in the Charter of 1975 (Figure 8.1). Printed in black and white, it covered the area of the park and some of the surrounding towns. The boundary of the park, represented by a broken line, was no sharper than the boundaries of the two Départements. What the map did not reflect was that not all the local communities within the park had in fact signed up to the Charter and thus were not formally part of it. The strongest boundary and the defining feature to the north of the park area was undoubtedly the international boundary marked in thick black, beyond which there was nothing, a *terra nullius*, containing neither name nor feature. The world, as far as this map was concerned, ended at the boundary and the park might well have been a promontory of land in a hostile sea. Roads leading up to the boundary ended abruptly, going nowhere. There was mention of ruined castles and protected features – it was not indicated whether these were 'natural' or 'cultural' – but no mention of other monuments linked specifically to the boundary.

In 1994, twenty years later, the official map changed in several ways (Figure 8.2). It was in colour and only featured biophysical elements, together with some stars indicating major towns. No roads, no castles, no paths. The graphics used seemed to portray a more homogenous landscape, largely forested and different from the 1975 mosaic. Small strictly protected areas

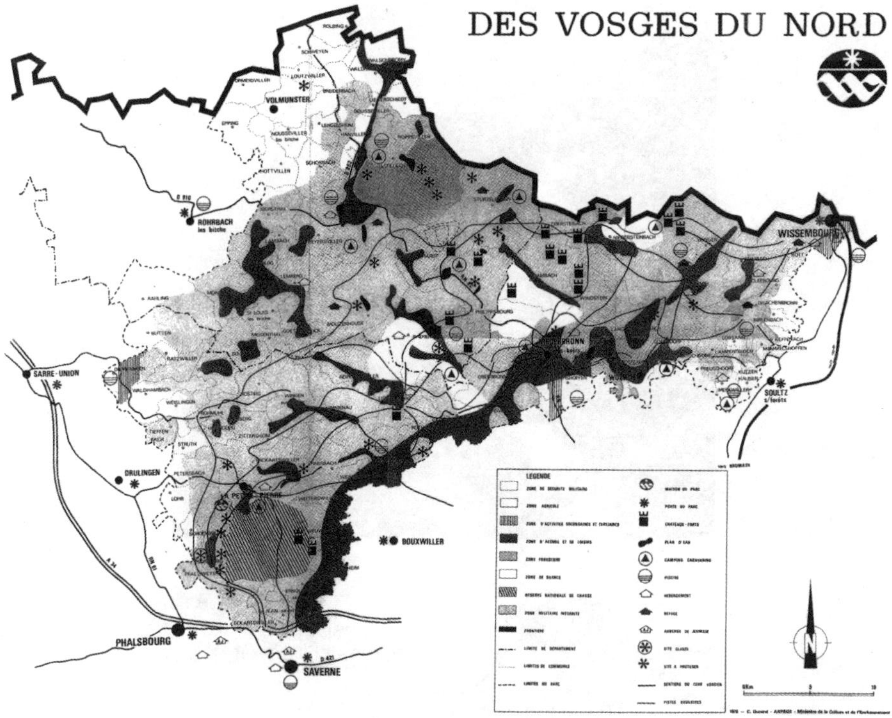

Figure 8.1 Parc Naturel Régional des Vosges du Nord, 1978
Reproduced with kind permission of the Parc Naturel Régional des Vosges du Nord and the Institut Géographique National, autorisation number 70 40038, IGN Paris.

appeared, together with non-protected 'sites remarquables', as did rivers and orographic features. The outer boundary of the park appeared more marked and the homogeneity of the entity reinforced by the lack of internal, communal or departmental boundaries. The boundary of the biosphere reserve buffer zone also appeared, reflecting the new double identity of the entity. The international boundary, while substantially less bold than in 1975, was nevertheless the strongest line to appear on the map, in a separate legend together with the boundary of the park. Rather than a complete blank on the other side, the name of the country, 'Allemagne', was written in French.

This island motif, depicting an isolated, coherent unit, was something of a leitmotiv in protected area mapping, often existing in parallel to official statements and policies of integration within the wider landscape. In 2001, Geographic Information Systems arrived in the Vosges du Nord in a big, colourful way. Map making became a daily habit and something of an industry with the park's administration, with a team of young employees producing GIS-based maps all day long, in a large room under the eaves of the medieval castle housing the park's administration. The 2001 Charter no longer had only

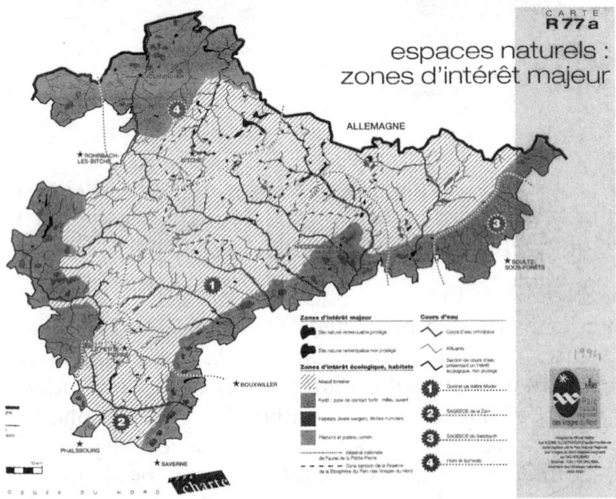

Figure 8.2 Parc Naturel Régional des Vosges du Nord, 1994. The protected area mapped as an 'island'. The promontory above, photographed in the Cinque Terre National Park, Italy, amusingly echoes the pattern suggested by the map. (Photo J.J. Fall)

Reproduced with kind permission of the Parc Naturel Régional des Vosges du Nord and the Institut Géographique National, autorisation number 70 40038, IGN Paris.

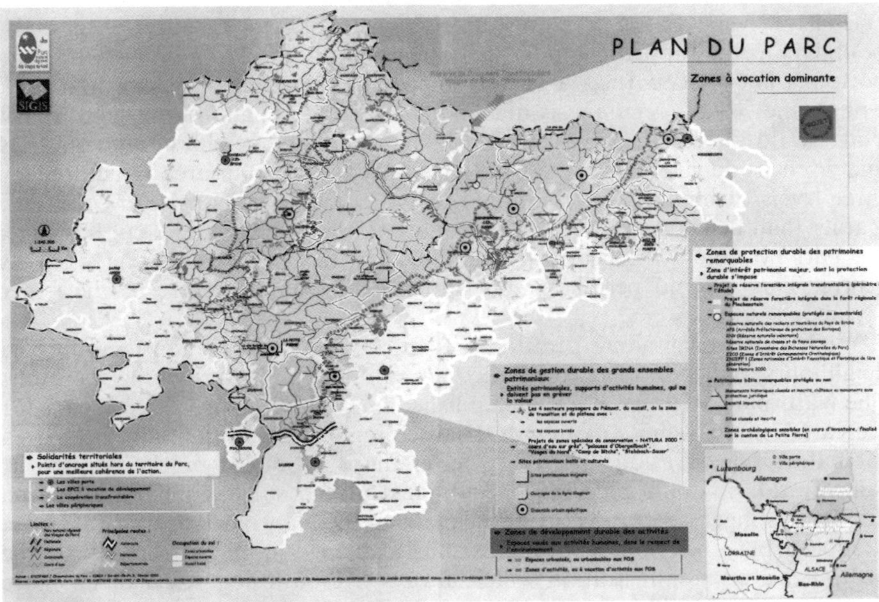

Figure 8.3 Parc Naturel Régional des Vosges du Nord, 2001
Reproduced with kind permission of the Parc Naturel Régional des Vosges du Nord and the Institut Géographique National, autorisation number 70 40038, IGN Paris.

one map, but six sorted according to themes, with one serving as an overview (Figure 8.3).

In a rather sickly rainbow of colours and graphics, this main map summed up the five thematic ones, producing a map of 'zones à vocation dominantes', a zonation system based on major land use. The legends of the map took up practically as much room on the page as the map itself, including explanations for zonation, biophysical (rivers, vegetation cover) and societal elements (roads, towns, heritage sites), as well as a selection of specifically protected areas, classified sites and archeologically sensitive areas. It would have been difficult to put more information into one page. As it was, it was already practically illegible.

A new and interesting category that appeared on this map was territorial solidarity (solidarité territoriale), a term covering 'anchorage points situated outside the territory of the park for more coherent action'.[7] This explicitly involved an arrow symbolically representing a link to the German side of the transboundary biosphere reserve. The name of the country no longer figured explicitly, but rather the 'transboundary' name of the biosphere reserve – the two national names one after another – appeared in the area across the international boundary. The graphical bonanza extended to the different boundaries represented in the appropriate legend: national, regional and communal boundaries, as well as the boundary of the park and, rather curiously, rivers. These were presumably considered to be a form of 'natural' boundary as they appeared under the category of 'limites' or boundaries. In the bottom right-hand corner, a smaller map was represented, placing the park within the wider regional context. In this, the neighbouring Naturpark Pfälzerwald appeared, together with the names of the neighbouring countries and French 'départments', the whole being reduced to an area structured by three levels of boundaries, delimiting 'départements', 'régions' and countries. Rather than being, as in 1975, a promontory of land in the middle of nowhere, the park was now the centre of the world, with circles expanding from its centre. This was no periphery, but rather the dynamic centre of the wider world.

While the French administration was working out its position regarding its neighbours and the wider world, the German managers were doing the same. The Naturpark Pfälzerwald was established in 1958, set up as a green space for recreation in the vicinity of an industrialised region, and not initially as an area for nature conservation. There are fewer maps produced by the Naturpark Pfälzerwald, no doubt reflecting the administration's substantially smaller size, despite its longer history. Many of the maps of the Naturpark Pfälzerwald were produced by other bodies – such as tourist offices – and not directly by the protected area administration which always has been composed of no more than five people. This meant that analysing the existing maps was more problematic, as they reflected other interests than simply those of the park's administration which was itself a 'Verein', or association of local communities, locally elected people and non-governmental organisations.

Around 1996 (Figure 8.4), a map was produced by the Naturpark. This map appeared in various publications, including a small brochure published around 1998 presenting the double identity of the area as a 'Naturpark' and a 'Biosphärenreservat'. Although the map appeared in this brochure, no mention was made on the map of the biosphere reserve and none of the three internal zones appeared. The layout, though in colour, was not unlike the 1994 French map, with the territory appearing as an island surrounded by white space. Disconnected from the surrounding landscape, the area did seem to gain in internal coherence, the defined area essentialised. Although arrows indicated the direction of major German cities outside the park, no mention was made of the neighbouring country. There was no legend indicating the different elements although convention implied that this was a wooded area with many rivers, roads, train lines and small towns. Biophysical and societal elements both featured, as in many tourist maps inviting exploration. A thick pinkish line indicated the difference between the boundaries of the park and the international boundary to the South, but this was less marked than in the earlier French examples. Two international border crossings were indicated along the pink boundary, with roads stopping abruptly. However, despite this snub to the neighbouring country, a dotted path entitled 'Deutsch-Französische Touristikroute' was marked, although this stopped rather curiously shortly before the boundary. It was the only mention of the existence of a neighbouring country.

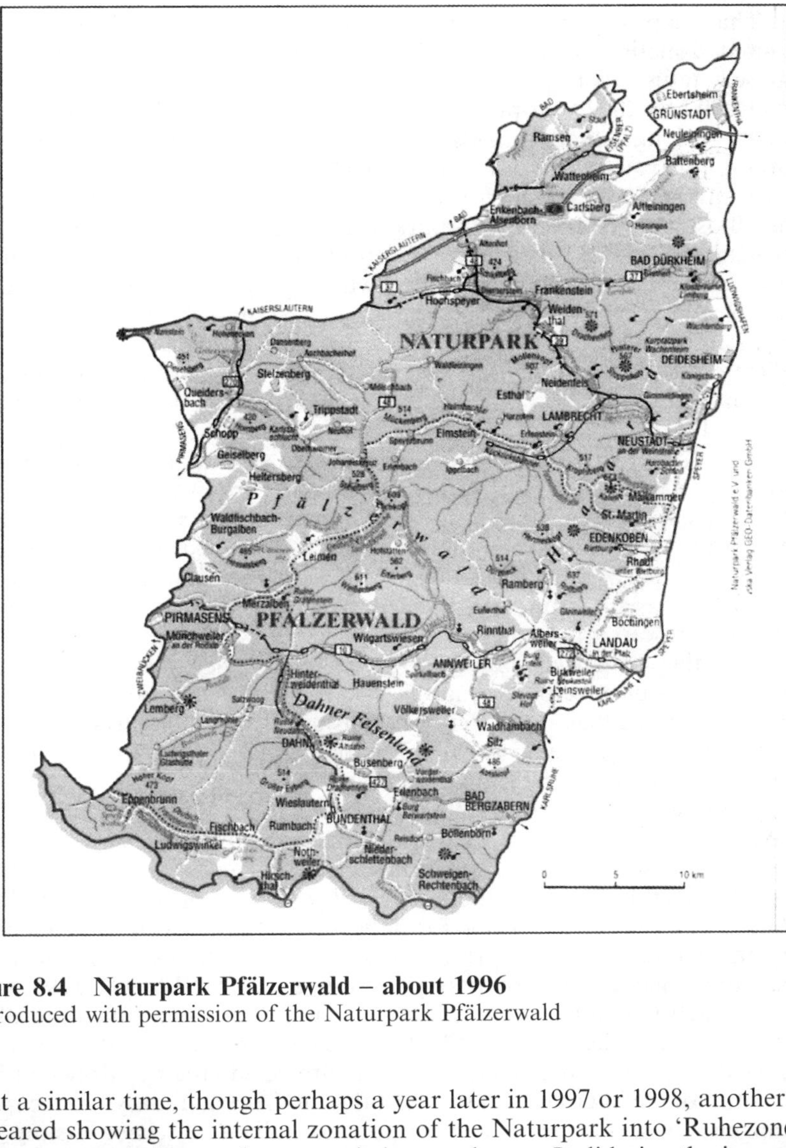

Figure 8.4 Naturpark Pfälzerwald – about 1996
Reproduced with permission of the Naturpark Pfälzerwald

At a similar time, though perhaps a year later in 1997 or 1998, another map appeared showing the internal zonation of the Naturpark into 'Ruhezone', or quiet areas, superimposed on an existing road map. It did give the impression that the Naturpark was part of a larger ensemble. Again, no mention was made of the biosphere reserve on this document and the zonation presented was in fact conflicting with the biosphere reserve zonation presented to UNESCO. At this time, the whole issue of zonation was in fact problematic for the Naturpark. The park administration claimed that it was unclear who was responsible for it and stated that the Ministry of the Environment in Mainz was about to decided on it separately. It was, by all standards, a very sore point. That map was therefore an attempt to enshrine an ill-defined and ill-recognised zonation within an administration whose accountability and recognised status was low.

From 2000 to 2002, the production of maps in the Pfälzerwald protected area was no longer in the hands of the park's administration. Following a reshuffling of responsibilities, the Federal State of Rheinlandpalatinat's Ministry of the Environment took over all responsibilities for the zonation of the 'Biosphärenreservat Naturpark Pfälzerwald', during something of a 'coup d'état' and power struggle which left much bitterness among individuals. During this period, it remained unclear which body was formally responsible for the biosphere reserve, as opposed to the Naturpark, even though both designations covered the same area. In the aftermath of Germany's reunification and following an alignment with the former East German legislation, biosphere reserves had acquired formal legal status within Germany. It was thus no longer acceptable to the Ministry for a non-governmental 'Verein' such as the Naturpark to start making zonations that would hold legal standing. The fact that a working group had been established by the Verein that included local communities, scientists as well as representatives from the neighbouring French protected area became a big issue. This group was immediately dismantled and the Ministry took over. The definition of the zonation was handed over to scientists, 'experts' within various state research bodies.

The map they produced (Figure 8.5) reflected this new rational, science-led approach. The objective of the map was to indicate a zonation that could be given legal standing, based on what was considered sound biophysical evidence of biodiversity and the requirements of nature conservation. Like many of its predecessors, the map represented the Pfälzerwald as an island, linked to the surrounding landscape only by what appeared to be grey roads of different size. While the map referred to the 'Biosphärenrezervat Naturpark' it was clear that only the biosphere reserve zonation appeared here, clearly divided into red core zones, green buffer zone and a pale yellow transition zone. This followed the German legal requirements for biosphere reserves which stipulated a minimum of three per cent core zones, and 20% combined core and buffer zones, requirements that were more prescriptive than UNESCO's 1995 Seville Strategy. Each individual section was named. There was no mention of the 'Ruhezonen', the quiet areas for silent nature contemplation at the core of the Naturpark idea. The outer boundary of the protected area was drawn in blue, with no hint of the international status of the southern boundary. The map was accompanied by a document referring to each individual section, listing their respective biophysical characteristics, as well as the legal protected status of the areas. Nowhere on the map, nor anywhere in the document, did it appear that the Pfälzerwald was part of a transboundary entity.

While this last German map – and the Rheinland Palatinat Environment Ministry – appeared to have firmly turned their backs on their French neighbours, other maps existed explicitly covering both entities. The maps themselves, as well as the forces contributing to their production, subtly reveal further issues in the institutionalisation and contestation of the transboundary entity.

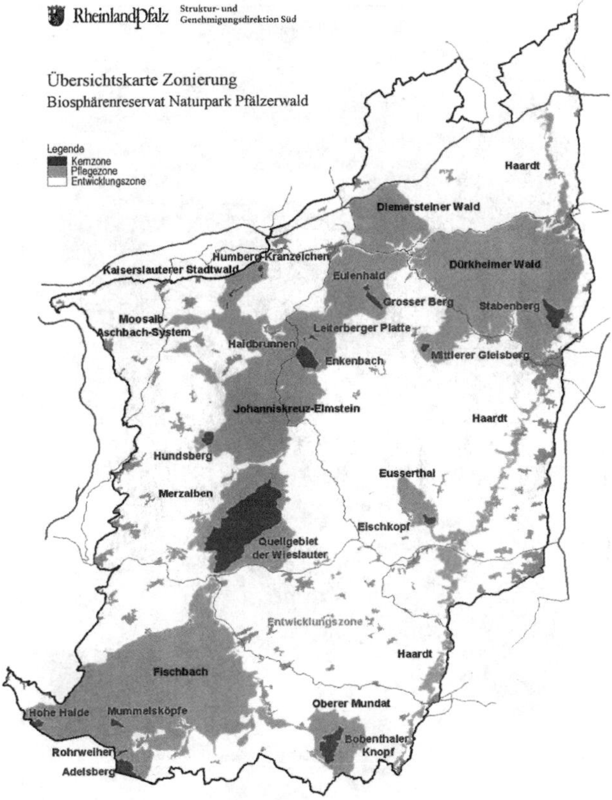

Figure 8.5 The new official map, 2002
Reproduced with permission of the Naturpark Pfälzerwald

Franco-German transboundary maps: institutionalising integration?

Following the designation of the transboundary biosphere reserve in 1992 and as part of the first European Interreg Programme, a large map of the two entities combined was produced, mainly to encourage transboundary tourism. It was seen by both protected area administrations to be a very good, concrete thing to produce that would symbolically reflect the transboundary nature of the area, enshrining it on paper. The map was produced and published in 1996 by Top-Stern Karten, a private German map-making company that also designed the map that appeared on the cover of the folded map (Figure 8.6).

This rather garish-looking map was surrounded by the logos of the French and German protected areas at opposite corners, as well as the European flag and the MAB logo representing biosphere reserves. At the centre of the page, a sharp green line delineated the biosphere reserve as one transboundary entity, with no hint of the international boundary running through it. Karlsruhe and

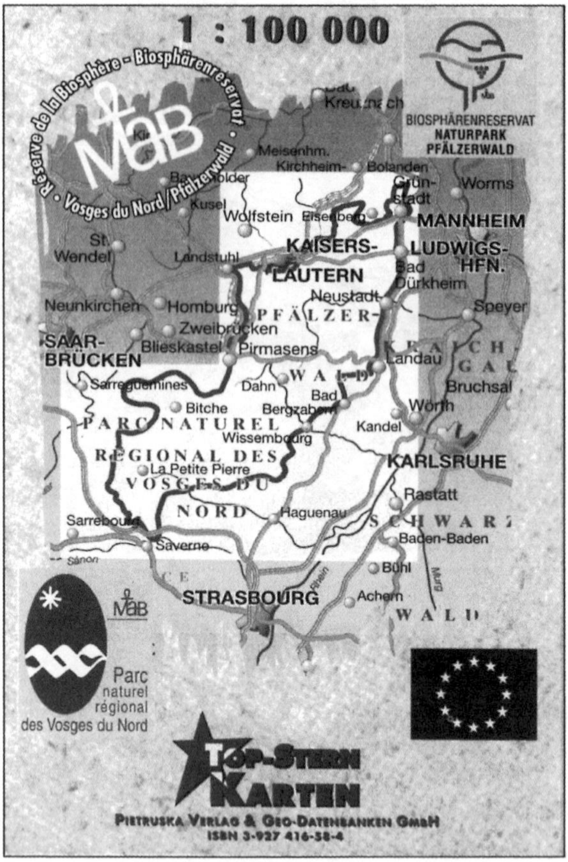

Figure 8.6 First 'transboundary' map, 1996
Reproduced with kind permission of Pietruska Verlag & GEO-Datenbanken GmbH.

Strasbourg could have been in the same country. The international boundary did however appear on the inside folding-out tourism map. This common map was considered a great success and sold well. The protected area administrations were proud of it and distributed it widely. During the course of the fieldwork, I got handed this map several times, in both countries and it was often hinted that these were some of the last copies as stocks had run out.

Building on the success of this common map, a more ambitious programme was set up for creating a transboundary resource-management Geographic Information System, with databases covering both countries. This was funded in part by a LIFE project, another European Union programme. This US$ 640,000 project was overseen by the German protected area. It came to an end in 1999. A press conference in a site close to the boundary was organised in order to present the results, which included a map based on a satellite image –

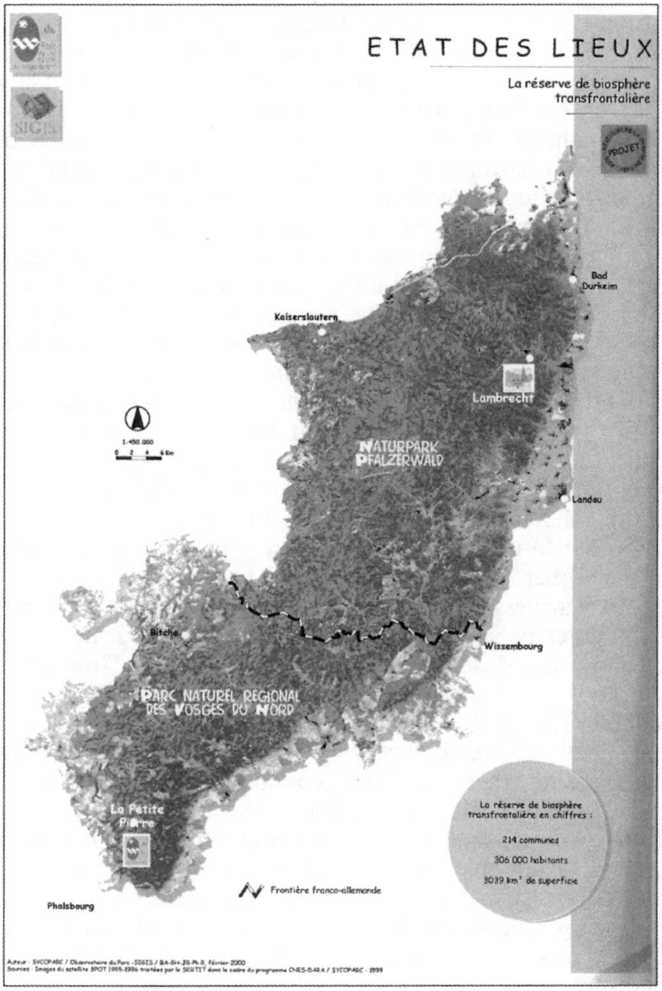

Figure 8.7 Overlaying transboundary projects on a satellite image, 2000
Reproduced with kind permission of the Parc Naturel Régional des vosges du Nord and the Institut Géographique National autorisation number 70 40038, IGN Paris.

the only concretely 'transboundary' result of the whole project (Figure 8.7). This map was an interesting example of the process of integration and distinction, indicating to what extent the two were inextricably linked.

The area portrayed was exclusively the territory of the two protected areas, again figuring as an island in a white sea. This map appeared both in black and white in several publications, as well as in colour on a large poster. The entity appeared coherent and were it not for the clear black and white line representing the international boundary, it would be impossible to distinguish one country from the other. That, indeed, was the whole point of the map which naturalised the transboundary entity, distinguishing it from the surrounding landscape yet integrating it as one unit. This representation was internalised by several protected area managers, such as Théo who implicitly referred to this when he stated that 'you only need to look at a satellite picture to realise that there are a certain number of shared issues'[8] (Théo, Parc Naturel Régional des Vosges du Nord, France). Yet, especially in black and white, the most striking feature of this map was the boundary chopping the unit through the middle: a black and white squiggly line, superimposed on the background satellite picture.

This ambitious programme of a transboundary GIS system marked a turning point in the cooperation between the French and the German protected areas. Although results were glossily presented to the press, it was clear that all actual GIS databases only covered the French area, and the satellite map was in fact the only 'transboundary' result. Further concerns over financial matters and allusions of the mismanagement of funds further embittered the situation. The map was thus the result and reflection of a contested process and as such incarnated the inherent contradictions of the project: the integration of a professed biophysical unity coming up against the daily realities of confrontations with the Other, leading to a bitter, paradoxical result of increased distinction and wariness. Yet the image of unity that the map reflected and performed did enter people's minds, shaping a different scale of planning.

In 2002, following measured success the previous year, four transboundary farmer's markets were set up within the two protected areas, repeatedly bringing together local producers from both countries in one location. Unlike the announcements the previous year which only featured names of places, the brochure advertising these in 2002 contained a map (Figure 8.8) that represented the two protected areas joined together.

This did not explicitly mention the transboundary biosphere reserve, despite the emphasis on sustainable development, preferring to focus on the common organisation by the two individual protected areas. The graphics in the brochure – the page split into yellow on one side and red on the other, with the logos of the two protected areas facing each other – furthered this idea of two entities organising one thing together. There was thus no mention that they might have one umbrella identity. The map used the 'island' graphical metaphor again, linking up the transboundary entity to the surrounding landscape by mentioning key large towns, as well as the Rhine river (the French spelling was used) and a red and white line representing the international boundary. The villages where the markets took place appeared in red, with their respective national spelling.

When certain practical and legal details were overcome – the Germans, for instance, were horrified to discover the French were planning to sell cheese and meat on non-refrigerated stalls, something not allowed under German law – the markets proved a great success and were well attended. This was a recognised success, improving morale and trust amongst the two protected area administrations. Yet it was a relatively non-controversial project that did not stir up too many deeper issues while remaining well within the realm of the

Figure 8.8 Transboundary markets, 2002
Reproduced with kind permission of the Parc Naturel Régional des Vosges du Nord and the Institut Géographique National, autorisation number 70 40038, IGN Paris.

continuing 'amitié franco-allemande' by focussing on the idea of two separate, distinct entities carrying out something together. The map thus reflected the distinction between the two, allowing for measured integration during and around one common project.

The map that appeared shortly afterwards in the brochures and materials for the official inauguration of the transboundary biosphere reserve in 2002 was more problematic. Despite being recognised as a transboundary biosphere reserve by UNESCO since December 1998, there had never been an official ceremony with a handing-over of formal certificates. Therefore, partly as a public relations exercise and partly as a *de facto* relaunching ceremony after two years of turmoil within the Franco-German cooperation projects, an official 'creation' or 'foundation' ceremony was organised (respectively 'céremonie de création' or 'Gründung'). Thus on the 23rd February 2002, on both sides of the boundary in succession, a selection of key figures gathered. The German Land's Minister, the French Minister of Land Planning and the Environment, the President of the French MAB Committee, as well as Dr Peter Bridgewater, the Director of UNESCO's Division of Ecological Sciences – complete with his characteristically cheerful flowery buttonhole – turned up among others for official speeches, photographs and cocktails. The official invitation heralded 'the end of boundaries within nature' between the two countries in catchy bilingual slogans: 'Entre la France et l'Allemagne, la nature n'a plus de frontière' or 'Zwischen Frankreich und Deutschland hat die Natur keine Grenze mehr'. Yet, revealingly, the map attached to the brochure showed a different picture.

The purpose of the ceremony was to launch the transboundary biosphere reserve. The map, however, did not indicate where this might lie and no further map appeared in the accompanying literature. Instead, the map (Figure 8.9) indicated a section of a schematic road-map, largely centred on the German side in a small insert. Key towns and cities were indicated, as well as the two locations of the ceremonies on either side of the orange dotted boundary. The map did not refer to the protected areas or to the biosphere reserve and simply indicated a rough circle defining the area within which the meetings would proceed. It was almost as though the public-relations' coordinators had shot themselves in the foot, highlighting the absence of a shared transboundary

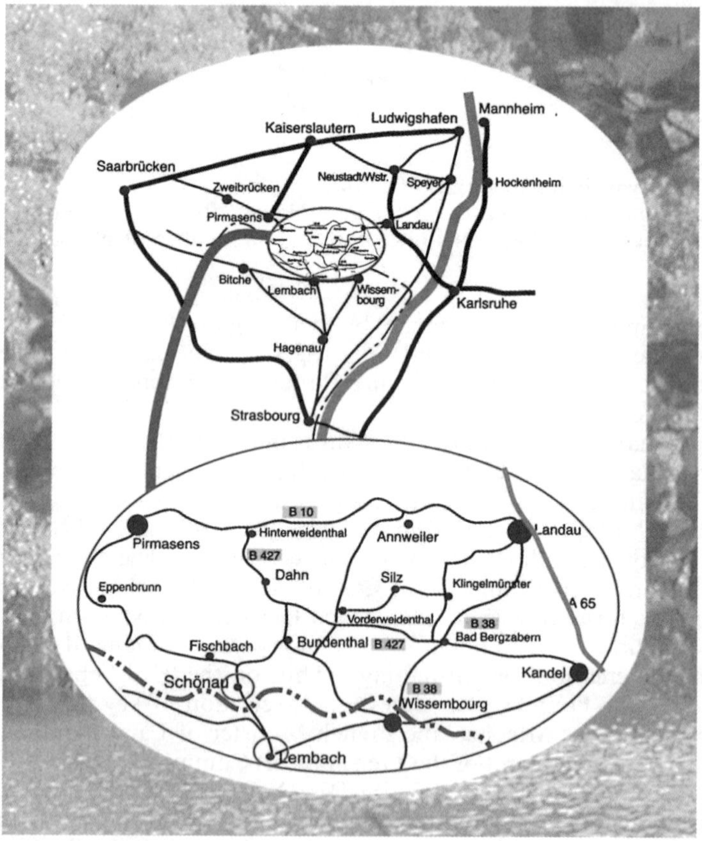

Figure 8.9 An alternative definition of the area, 2002
Reproduced with kind permission of the Parc Natural Régional des Vosges du Nord and the Institut Géographique National, autorisation number 70 40038, IGN Paris.

entity rather than its creation. This would not have been be so meaningful if other maps had been produced, and this could have been discarded as no more than a pragmatic 'how to find the meeting' road map. But the issues of shared zonation and shared management plans were such loaded themes to both partners and the disagreements on these issues sorely felt by both sides, that the non-existence of other maps was more than circumstantial.

Despite the ambiguous message that this event sent out, other maps in the area went further in promoting the integration of the transboundary entity. The most powerful map was paradoxically only present in one place, printed on a wooden table in the Vosges du Nord Visitor's Centre, located in the castle housing the park's administration in La Petite Pierre (Figure 8.10). This simple yet effective map painted in bold green on the grainy surface of the table, brought together the transboundary entity in a simply and effective way. Little wooden posts that could be picked up off the table contained information about each separate protected area, yet the construction of the whole emphasised unity. The international boundary appeared as a black line but was not particularly striking. The graphics conjured up the idea of a web of different biophysical and societal features linking up the various dimensions. The names of Alsace and Lorraine – historically charged for local people – were shown on equal par with the term for Germany. Rivers flowed across the international boundary and roads joined up the two sides. The transboundary

Figure 8.10 A wooden museum map in the dungeon of a castle
(Photo J.J. Fall)

entity was naturalised as one unit, yet connected to the surrounding landscape. It was a visually striking creation, all the more so for being in the old dungeon of a medieval castle built in days when the boundary needed the protection of such fortresses.

Yet this map, so clearly depicting many of the ideas actively promoted by the protected areas, may have gone too far in projecting integration. The entity certainly looked integrated but many issues remained for the managers from the two countries. As long as the transboundary unit was taken to be one homogenous green lump, then the more controversial issues could be swept aside. But as soon as the reality was more subtly recognised to be a combination of internal zonation issues, with differing levels of authority and accountability, as well as culturally constructed work patterns and cultures, then the myth of integration collapsed. No map representing that complexity had yet been produced.

Trials and tribulation of transboundary zoning

Because maps participate in the process of institutionalising spatial entities, they are both instigators and indicators of the process. The Franco-German case discussed above concerned one transboundary biosphere reserve, hinting at some of the issues surrounding the establishment of a common zonation. Because biosphere reserves are based around the idea of dividing up space into a series of zones, they are by definition involved in the mapping of it. Yet while zoning – and mapping – a biosphere reserve is not simple in itself, the zonation becomes even more complex when applied on a transboundary level.

Biosphere reserves are based around the idea of defining a series of zones divided by boundaries within a more-or-less defined whole. Thus the zonation exercise is often crucial to the success of the project, defining from the outset which partners are to be involved – be they protected area administrators, scientists, local people, local authorities, experts and employees from different Ministries and a variety of possible levels of political decision-making. In transboundary situations, this naturally becomes a more complicated process, involving at least two separate administrations in different countries, but very often more. The production of a common map is often considered a first crucial step, a process likely to be the first substantial activity carried out jointly. It is politically charged, since differences in administrative structure, ways of work and levels of authority and accountability come to the surface. Because mapping is an act of power and control it has historically been the responsibility of states. This implies that mapping at a transnational level is politically and symbolically transgressive.

The reliance on zoning to determine different functions and uses for various areas relies on the idea that these fit in with existing national categories of protection, as well as with non legislated land use. It is not therefore necessarily obvious what transboundary zonation exactly entails when legal and administrative systems differ. The definitions of core, buffer and transition areas can be found in the Seville Strategy, but the two latter especially remain relatively broad, subject to interpretation within national legislation. This means that while the actual legal status of the three zones can be left to individual countries and adapted to local conditions and legislation, it does create a challenge in a transboundary context.

According to the Seville Strategy, 'core areas are securely protected sites for conserving biological diversity, monitoring minimally disturbed ecosystems, and undertaking non-destructive research and other low-impact uses. The buffer zone usually surround or adjoins the core areas, and is used for cooperative activities compatible with sound ecological practices, including environmental education, recreation, ecotourism and applied and basic research. A flexible transition area, or area of cooperation, may contain a variety of agricultural activities, settlements and other uses and in which local communities, management agencies, scientists, non-governmental organisations, cultural groups, economic interests and other stakeholders work together to manage and sustainably develop the area's resources' (UNESCO 1995 : 4). Within these definitions, no specific allowance is made for transboundary situations.

For some actors, the ultimate objective of a transboundary biosphere reserve is to have one level of coordinated work, often taken to mean one 'management plan', in which case it is important for the various zones to hold the same legal status in the different countries. Yet for others, such a level of integration is threatening and to be avoided at all cost. The extent to which such understandings differ within a country and within different levels of authority – park directors, scientists or Ministries for example – is considerable. In all cases, however, it is necessary to decide to what extent zonations should match each other. Does it matter if the 'buffer' zone in one country only has the equivalent legal status of the 'transition area' in its adjoining neighbour? This question is in fact rarely explicitly addressed in the existing sites, although core areas practically always benefit from the highest level of protection available in each country – except in the Vosges du Nord, in France, where medieval castles with millions of visitors each year are included in the core zones, much to the despair of the Germans. The question is more acute in the transition areas and large variations exist as to their status and to the role – or existence – of local communities within them.

Interestingly, although the Seville Strategy prescribes a 'flexible' boundary to the transition area, in practice all the areas visited prefer a sharply defined line on a map, even in cases where this does not correspond to any particular legal demarcation. Fuzzy lines are difficult to depict on maps and to explain to people. In all sites visited protected area managers wanted a clearly defined transition area, defining an inside and an outside, and a limit to their authority. Differing status in land management authorities in the various countries was also a regular problem. Similar terminology did not necessarily imply similar authority or accountability for managers, such as directors and employees of 'national parks' or 'natural regional parks'. In these cases, cooperation within each country – for example between protected area officials and foresters who concretely manage the land – was as much a challenge as cooperating across the border. In the situations considered here, it was only within the Central and Eastern European older national parks (Polish Tatras, Polish Bieszczady,

Ukrainian Danube Delta) that park directors exercised total decision-making power over the land, as well as in the core areas of French national parks (French Mercantour). In all other situations, the protected area administrations had to negotiate, coax and coordinate to get their zonation respected.

Jigsaws and patchworks in the Carpathians

The idea of establishing a transboundary protected area in the East Carpathians had been around for over twenty years, although it really started to take form in 1990 during the UNESCO-MAB meeting in Kiev when the idea was formally presented by members of the Polish MAB Committee. Despite a series of setbacks in getting the project formally drafted by the governments concerned, the project was inscribed in the World Network of Biosphere Reserves in two stages, with the Slovak and Polish applications being registered in 1993, joined by the Ukrainian one in 1999. This led to the formal designation of the whole unit as one biosphere reserve, called the 'East Carpathians Biosphere Reserve (Poland/Slovakia/Ukraine)' on the official certificate. This entity was more complicated than three identical pieces of jigsaw. The Polish entity was made up of two protected landscape areas and a national park; the Ukrainian entity combines one national park, one protected landscape area and one regional landscape park; while the Slovak entity was made up of one national park. Each of these entities was governed by a single administration, or was directly managed by another body such as the forestry service. At the same time, all of these somehow came together under one transboundary biosphere reserve that had neither independent funding nor any additional staff. Under such circumstances, it was hardly difficult to imagine the challenge of producing a common map.

Although on paper the transboundary tri-lateral entity appeared coherent, creatively combining existing protected areas into one transboundary biosphere reserve, the political and administrative complexity was tremendous. In order to understand this, it is useful to initially only consider one entity at a time. The Slovak side was constituted by one national park. This seemed the most straightforward scenario, with one apparent administration governing one area. Yet a short look at a map of the area showed up a much more complex reality. The Poloniny protected area (Figure 8.11) in Eastern Slovakia was a useful starting point to illustrate the problematic nature of mapping the transboundary biosphere reserve in the East Carpathians. The map was produced in 1998 by the protected area administration, funded by a joint project with the American Peace Corps and the United States Information Service. This meshing together of institutions illustrated the complex web of actors involved both in the process of zoning and in the subsequent production of maps.

The map merged biophysical elements such as lakes and rivers, with societal factors including human settlements and roads, as well as protected area boundaries – both external and internal. Lakes and rivers were concrete objects with a firm link to reality. Nobody would even think of doubting their

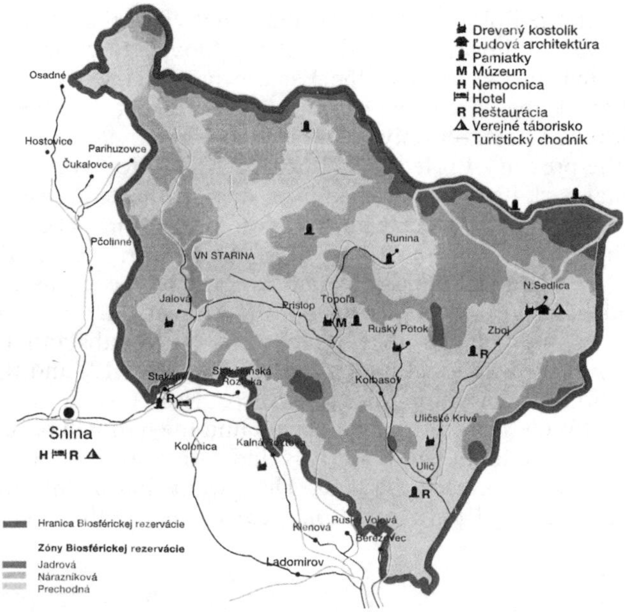

Figure 8.11 Poloniny National Park, 1998

existence on the actual site. At the same time, the three zones of the biosphere reserve also featured, in different colours. The outside boundary of the protected area was very heavily marked in red (Hranica Biosférickej rezervácie) and formed the dominant feature of the map, structuring the ensemble. A selection of pictograms representing key points of interests for potential visitors also featured, both on the inside and outside of the protected area. There was also a walking trail marked in yellow, which seemed to follow the boundary of the protected area. The various features appeared to indicate that although the protected area was very heavily defined as an entity, it was nevertheless connected to the surrounding landscape. On this map, the biosphere reserve zonation appeared to structure the ensemble in a coherent fashion, almost echoing the higher ground that one could conjure up by looking where the rivers flowed.

Yet not all these elements were of a similar standing. The zonation did not reflect 'natural', tangible zones and while these had instead been defined as part of the political process of seeking recognition from UNESCO as a biosphere reserve, they did not have any practical application on the ground. Furthermore, despite appearing the most solid feature on the map, the outside boundary to the protected area was substantially redefined in the year preceding the publication of this map. At the time of the fieldwork in 1998 and 2000, this was not demarcated in any way on the ground. In addition, the title of the map did not tell the whole story. The biosphere reserve was in fact coextensive with the Poloniny National Park that was officially created on the 1st October 1997, following a decision of the Slovak Environment Ministry. This was created on the basis of the Eastern half of the Vychodné Karpaty Protected Landscape Area, which stretched further to the West. The administration was however only set up on the 1 of January 1998, by splitting the staff of the previous Protected Landscape Area into two.

The national park itself was divided into 'small protected areas', 'core areas' and 'buffer zones', with decreasing levels of protection. The national park zonation plan still required the approval of the Ministry of the Environment (Act 287/1994 : 11), something that had not yet happened as the park was awaiting guidelines on methodology from the Ministry. During the interviews and discussions, it emerged that there were quite substantial differences between the projected national park zonation (Figure 8.12) and the biosphere reserve zonation. In addition, because there was a somewhat conflictual relationship between the forestry department and the national park authorities, the zonation of the national park was indicative at best as the forestry department actually managed the land. This meant that neither the biosphere reserve nor the national park zonation were legally established in any way.

Figure 8.12 Poloniny National Park, about 2001

However, it was clear that in order to obtain biosphere reserve status, such a zonation had to be carried out, even if this only appeared on paper. This was an interesting example of what Albert termed 'new medievalism', a situation in which there is an overlapping of various authorities on the same territory (Albert 1998 : 53), with confusion about who is accountable to whom for what. This appeared particularly clearly on different incompatible maps (Chapter 5).

So what initially appeared to be a relatively straightforward map of an area masked a much more heavily charged political reality. At the time this map of the biosphere reserve was published, it was not at all clear to UNESCO what areas were actually part of it (Robertson, 1998, *pers. comm.*), as they appeared different to those designated as such in 1993. What was left out of the map was equally revealing. While the area appeared joined up to the road network on the West and South, the Northern and Eastern areas seemed empty. The fact that this protected area bordered Poland and Ukraine was thus omitted, even though at the time this area was one of the partners in the much-heralded trilateral biosphere reserve. This map thus played on several levels, operating behind its mask of seemingly neutral science: it put biophysical and societal elements on the same level, it made a strong statement about its neighbours by firmly turning its back on them, and it enshrined a largely immaterial zonation in a very convincing way.

Similar comments could be made about each entity within the tri-lateral biosphere reserve, although specific maps for the Ukrainian protected landscape area and regional landscape park did not exist at the time the fieldwork was carried out in the region (1998 and 2000). This fact, in itself, was revealing and is returned to later. However, in the case of the Polish entity, maps for the whole area including the landscape protected areas were available although it seemed that these had been produced by the national park – an administrative body that may well not have had the formal authority to do so. Nevertheless, despite the difficulties faced by the individual protected areas, transboundary maps showing the whole area did exist.

The one referred to primarily in 2000 was rudimentary, composed of little more than black lines on a white background (Figure 8.13). Despite being graphically rather basic, this undated map produced around 1999 was considered by the local managers to be the most pertinent representation of the transboundary entity. In 2004, it still is (Niewiadomski, 2004, *pers. comm.*). One of the issues when setting up the tri-lateral entity had been to get the Ukrainian partners to put down on paper a defined area that could be integrated into the whole, as well as defining precisely the Slovak and Polish internal zonation. Various backlogs in this process led to the Polish and Slovak areas being initially declared a transboundary biosphere reserve, joined by Ukraine six years later in 1999. Thus actually having a map at all was a result in itself, however imprecise it might have been. When I was shown the map, the main concern of the Polish manager was that some of the Polish sections were out-of-date, with local changes to the boundaries of the national park likely to come about in the near future due to plans for extending the park. The exact location of the boundaries, even on a map of this scale, was taken to be crucial.

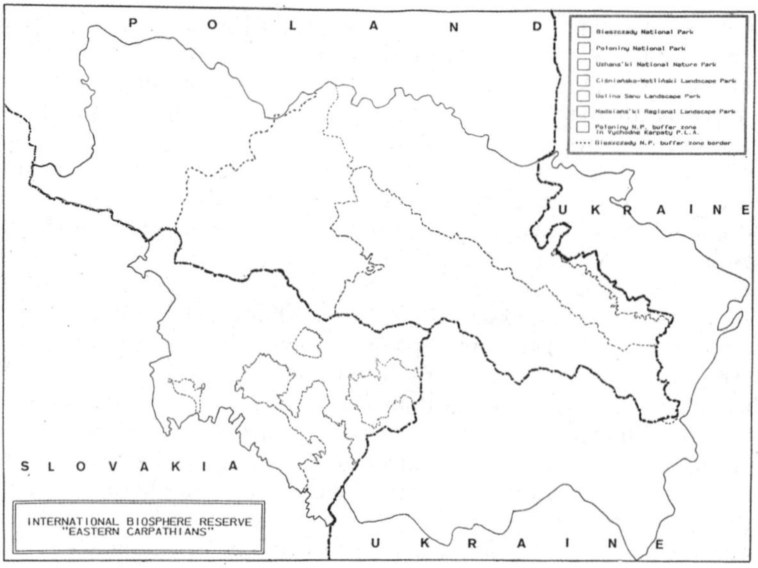

Figure 8.13 East Carpathians, 1999
Reproduced with kind permission of Mr Zbigniew Niewiadomski

Maps as necessary or sufficient proof of a spatial entity

Because of the iconic nature of the 'first' trilateral biosphere reserve, these maps have been reproduced many times in publications discussing trans-boundary cooperation (for example UNESCO 2002 : 137). The map was thus considered a *necessary* justification for the existence of a tri-lateral entity. This could be taken solely as the positive recognition of the role of maps within protected area planning. I believe it was more complex than that. Problems arose because these maps were implicitly taken as *sufficient* proof of the existence of a transboundary spatial entity. The risk in this case was that the maps enshrined a representation which did not have any concrete administrative or political existence. The map thus became a *fin-en-soi*, a result in itself of cooperation, rather than a contributing factor in (re)territorialisation and regional transformation. Having achieved a common map, no easy task in any case, the temptation was to consider that the process of cooperation largely occulted the complexity of building common projects.

Attractive maps of transboundary protected areas are used as tools in public relations, subtly providing this concrete proof of cooperation. UNESCO has liberally reproduced the East Carpathian map within its publications (Figure 8.14), redrawing it for graphical clarity and aesthetic appeal. It has become an icon in itself. Likewise, a glossy publication on the national parks and biosphere reserves in the Carpathians, partly funded by IUCN and WWF,

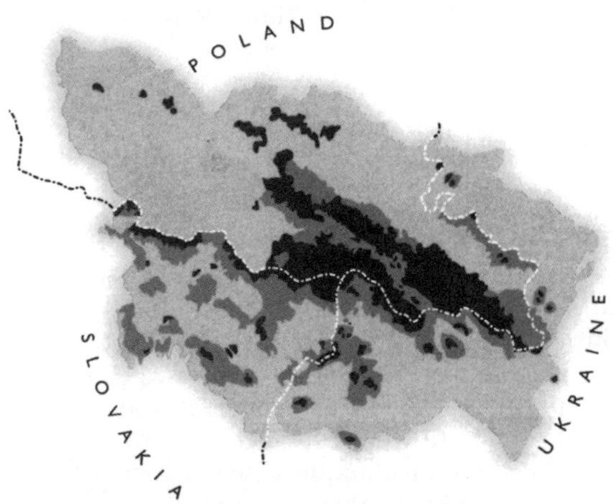

Figure 8.14 UNESCO 2002
Reproduced with kind permission of the UNESCO Secretariat in Paris.

produced transboundary maps by creatively binding together existing zonations (Figure 8.15). The result is attractive, giving the impression that all protected areas throughout the region enjoy coherent transboundary zonation. Yet these maps very much precede such coherence on the ground, subtly suggested by the substantial differences between them. They are forms

Figure 8.15 IUCN and WWF 2000
Reproduced with kind permission of Dr Ivan Voloscuk.

of spatial propaganda, promoting a certain politicised representation of an area, subtly suggesting a desired territoriality.

Maps and resistance

Passive resistance: the absence of maps

Such an act of power does not go unchallenged. It is not only those promoting such visions who recognise the persuasive force of maps. Those opposed or resistant to the idea of transboundary cooperation also have ways of getting themselves heard. The simplest technique is passive resistance. This involves not cooperating in producing a map that promotes an unwelcome vision (see Chapter 7). Below, I discuss to what extent this approach has been adopted in the Danube Delta. The second technique, discussed subsequently in the case of the Tatras, is more active and involves drawing the map for the entire entity without consulting the partner protected areas, thereby actively promoting a desired spatial scenario.

In the Danube Delta, despite the professed desire of several senior managers to produce a common map and despite a large transboundary project dedicated to doing so funded by a LIFE grant, no transboundary map had been produced. The official explanation was that, at the last stage, money ran out to have the map printed. The only existing transboundary map was rudimentary at best (Figure 8.16), and was drawn not by the protected areas but rather by the Ukrainian and Romanian National MAB Committees, based not in the Danube Delta but far away in their respective national capitals.

Although managers on the ground within the protected areas professed the desire to cooperate with their neighbours, they had not produced a useable common map. Rather, like earlier examples within France, Germany and Italy, their own maps firmly turned their backs on their neighbours (Figure 8.17 and 8.18).

Both these maps named the neighbouring country yet stopped short of indicating any link between their entity and the other official half of the transboundary biosphere reserve. Zonation criteria did not correspond and there was no agreement on whether this was likely to change in the future, nor how or whether it should be addressed.

This was hardly surprising in an area where the international boundary was still contested, leading to political tensions between Ukraine and Romania, notably linked to the presence of hydrocarbon deposits beneath the Black Sea. Crossing the boundary within the biosphere reserve was nearly impossible and involved at the time a private boat trip with 'fees' of fifty dollars – a princely sum in those parts – requested by the Ukrainian border guards on entrance and departure. Medical examinations on arrival in both countries were mandatory 'due to the sanitary conditions of the neighbouring country'. A clear sense of humour when carrying out field work in such conditions does come in handy. Yet people dealing with this context day to day, as part of their daily work, may find that a sense of humour only goes so far. Clearly, in this case, it would

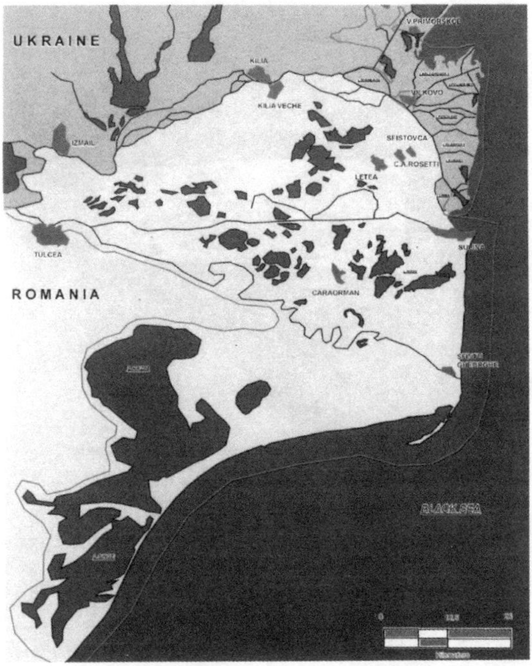

Figure 8.16 Produced by Ukrainian and Romanian MAB Committees, 2000
Reproduced with kind permission of the Ukranian UNESCO MaB Committee.

have been politically unwise as well as unlikely for protected area managers to produce transboundary maps. Passive resistance can therefore be explained as a coping strategy within a politically charged situation. It can either denote resistance to regional transformation or else more subtly reflect a wider political hostility to the idea of transboundary contact and cooperation, stemming from different levels of authority.

Active resistance: mapping the Other without consent

Another strategy that can be applied in transboundary situations is a more active form of resistance. The resistance, in this case, happens in situations where one partner is more eager or more able to envisage cooperation. Active resistance can be likened to a form of aggression, although remaining on a highly controlled level. Thus in cases where adjacent protected areas cannot agree on the creation of a transboundary map, for whatever reason, one partner can decide to produce a transboundary map alone. In the case of the Polish Tatras, in the face of administrative restructuring and confusion on the Slovak side, the Polish administration initially drafted a map alone. This covered the entire territory of the transboundary biosphere reserve (Figure

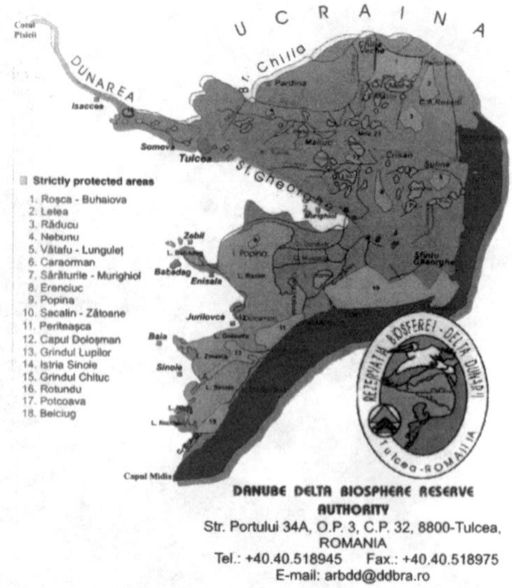

Figure 8.17 Danube Delta Biosphere Reserve, Romania
Reproduced with kind permission of Dr Angheluta Vadineanu.

Figure 8.18 Danube Biosphere Reserve, Ukraine
Reproduced with kind permission of UNESCO-MAB Ukraine.

8.19), yet it was clear that the Slovaks initially had scant input in its production.

At the time, following a major reshuffle of authority between the forestry service and the national park, it was unclear to the Poles who was in charge in Slovakia. Although this map was in circulation within the Polish administration, the manager who handed it to me wrote 'concept of changes to make 17.05.2000' on the document, in blue ink. It was therefore recognised that this was not an official document accepted by all. Yet at the same time, this map was distributed, promoting a desired scenario for the transboundary entity. As in the case of the East Carpathians, the map of the transboundary entity predated its effective existence.

This subtle form of cartographic imperialism is far from innocent and breaks many taboos. In situations where neighbouring countries remain on cool terms at a governmental level, officials would be very wary of neighbouring countries mapping the territory outside their jurisdiction. Transboundary maps rest on the assumption that this taboo is overcome through a process of negotiation, with no irredentist claims on either part. When this is carried out unilaterally, such an understanding is broken. Paradoxically, therefore, rather than contributing to further the ideals of transboundary cooperation, such an

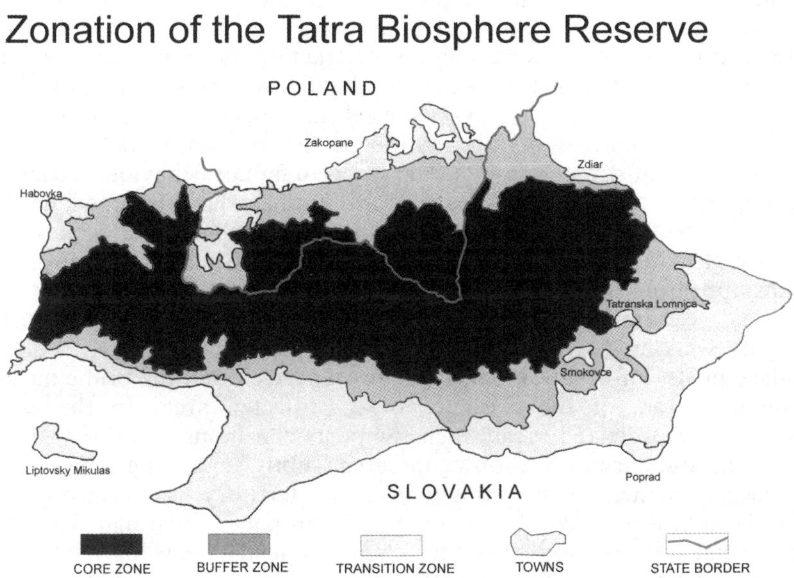

Figure 8.19 Produced by Tatra National Park, Poland, 2000
The map was first published in Kot, M.; Krzan, Z.; Siarzewski W.; Skawinski, P.; Voloscuk, I.; (2000) 'The Tatra Mountains Biosphere Reserve', in *Biosphere Reserves on Borders*, Breymeyer, A and Dabrowski P. (eds); Warsaw, The National UNESCO-MAB Committee of Poland, Polish Academy of Sciences. Reproduced with kind permission of Dr. Zbigniew Krzan.

action can backfire. In the case of the Polish and Slovak Tatras, the situation was at a standstill. The map produced, however, indicated an integrated situation, with a high level of coherence between the two entities. The boundary between the two, marked in red, appeared much less striking than the stark black core zone. Interestingly in this situation, the Polish administration undertaking the mapping had responsibility for a much smaller area than that ascribed on the map to the Slovak side. Thus in addition to promoting their ideal scenario, the map allowed the Polish national park to gain symbolic control of a much larger area. This was at a time when the Polish park was considered by the management to be reaching saturation point due to extremely high visitor numbers: a situation in which expansion, even on a purely symbolic level, had clear advantages. The promotion of a transboundary entity, in this context, was worth the risk of breaking some taboos. Subsequently, however, this map did become the accepted zonation of the biosphere reserve, reproduced later in the year in a book published by the Poles (Breymeyer *et al.* 2000 : 108), in collaboration with one Slovak author.

Maps as icons of regional transformation

There are several ways of bringing about regional transformation. One attempt to induce (re)territorialisation is expressed in the idea of 'raising awareness' – a term much favoured in conservation agencies and applied to a variety of themes including species loss, habitat destruction or pollution. Awareness raising is usually carried out by means of publications and literature, more directly by environmental education specialists, rangers and guides or else by directly locating sources of information in the landscape, for example on signposts placed at strategic points. The aim is to induce behavioural change by increasing the amount of appropriate information available to individuals. The particular aspect of 'awareness raising' that is examined here relates to the direct use of maps on signposts or information panels aimed at visitors.

Maps appearing on information panels usually indicate the location of the visitor: as 'you are here' icons clarify the position of the onlooker, usually suggesting walking trails or paths for exploration. In addition to signs and boundary posts, information panels of this type are often the prime means of informing visitors of the existence of a protected area. In the case of transboundary protected areas, such signposts can be used 'raise awareness' of the transboundary dimension of the area, subtly suggesting a new scale of experience, whether or not this additional territory is effectively easily accessible to visitors. Like their counterparts on paper, such maps contribute to the social construction of space, acting as instruments or mediators in (re)territorialisation, this time directly aimed at users of protected areas or at local inhabitants.

In certain officially designated transboundary protected areas, there are varying indications of the transboundary dimension (Figure 8.20 and 8.21) on the panels themselves. These two panels were located on opposite ends of a car park that crosses the Franco-German boundary in the Vosges du Nord/Pfälzerwald. In the French example, not only was no mention made of the

Figure 8.20 Parc Naturel Régional des Vosges du Nord
(Photo J.J. Fall)

German protected area, but additionally the map seemed to indicate that the onlooker was quite a way from the boundary itself. The German panel, however, showed both names although no mention was made of the French protected area on the map itself. On neither panel was there any indication that the areas considered were part of a biosphere reserve.

Ignoring the location of the panels on the ground would be a mistake. Since they were to all intents and purposes facing each other across a stretch of gravel, they were implicitly part of a boundary landscape, with the boundary passing between them. Nearby, ruined artefacts of the formerly controlled boundary lay as rusting posts and old stripy barriers. The message was therefore ambiguous, reflecting the concurrent trends of distinction and integration. The Other's existence was symbolically acknowledged to the point of coordinating the location of signposts, yet the maps did not reflect this recognition. Silently ignoring each other head-on, the maps proclaimed alternative scenarios for either side of the boundary.

In other cases, maps have been used to enshrine both the national protected area and the additional transboundary level. In the Polish Bieszczady, the maps of each coexist on the panels proposed to visitors, hinting at the multi-levelled administrative structures. In the Ukrainian Uzhansky National Park, on the other side of the boundary, the panels tell a similar story (Figure 8.22).

Figure 8.21 Naturpark Pfälzerwald
(Photo J.J. Fall)

The recognized importance of this shared vision is illustrated by the photograph taken during a tri-lateral meeting: three of the men appearing on the photo are senior representatives from the three countries constituting the East Carpathians biosphere reserve. During the meeting, the delegates were specifically taken by the Ukrainian national park director to view this panel, thereby demonstrating his commitment to the cooperation. The panels were no longer only used to inform visitors but attained the status of icons physically confirming the adherence to a common agenda. Doubtless there would be other examples of the use of maps within the landscape but these two short cases illustrate some of the ambiguities and complex issues raised by maps. By physically locating maps outdoors, visible to all, a certain spatial scenario is performed. This subtly manipulates onlookers, further promoting a certain conception of space. This is rarely unambiguous, oscillating continually between integration and distinction.

Figure 8.22 Uzhansky National Park
(Photo J.J. Fall)

Conclusions

This chapter suggests that the maps of transboundary protected areas have ambiguous and politically charged stories to tell. Using a series of examples, I have argued that the study of maps assists in unveiling some of the problematic aspects of (re)territorialisation in which desired constructions of space are promoted and contested. In certain cases, these maps have attained mythical status in their own right, appearing repeatedly as icons representing the success of cooperation and promoted as such. Examples indicating forms of resistance, both passive and active, to such imposition of power in the landscape have been developed, indicating further dimensions of the struggle for regional transformation.

Creating a 'transboundary' map is always a loaded political project. In the Vosges du Nord and the Pfälzerwald, this politicised process came up against latent problems and tensions involving the different levels of accountability of each administration, as well as the resistances and practical barriers that appeared among a multitude of actors involved in cooperation in various capacities. The institutionalisation of the 'transboundary' level of planning was a multi-faceted process, implying more than simply drawing a map across a boundary. These surveys of sites constructing transboundary maps have indicated some of the challenges of creating shared representations of space. Simply declaring that 'nature has no more boundaries' did little to inform the process and rather appeared unhelpfully naïve. Furthermore, by negating the symbolic aspects of the process, such a slogan actually ended up being counter-productive. These examples also showed to what extent maps – and the absence of maps – could be used to inform the political analysis of an area and the processes taking place within.

While such maps represent crucial steps in the process of (re)territorialisation, their transformation into sacred icons is counterproductive in the long run. Considering maps to be the end result of cooperation, rather than a tool for achieving this, sweeps more problematic issues aside. Instead of being celebrated as sacred icons, maps should be critically studied as indicators of otherness by those involved in cooperation. By more fully grasping how both the Self and the Other are represented on maps, a more complete picture can emerge, acting as an initial and ongoing diagnosis. The absence of a transboundary map should thus be identified for what it is: a failure to construct space cooperatively on a symbolic level.

Notes

1 Lewis Carroll, 'The Hunting of the Snark', The Works of Lewis Carroll, edited by R.L. Green, Hamlyn, Prague, 1965.

2 Personal translation from: 'le géographe trop occupé à inventer, à tracer des frontières se priva de l'étude même de ses frontières'.

3 Personal translation from: 'représentation modélisée c'est-à-dire une déformation pertinente et cohérente du corps de la terre'.

4 Personal translation from: 'elle porte toujours une échelle: rapport de réduction entre la distance réelle mesurée sur le terrain et la distance représentée. Elle implique une généralisation, donc une perte d'information qui résulte d'un compromis'.

5 Personal translation from: 'la représentation assure la mise en scène, l'organisation en spectacle de l'emprise originelle du pouvoir (. . .). L'image ou modèle, c'est-à-dire toute construction de la réalité, est un instrument de pouvoir et ce depuis les origines de l'homme (. . .). Nous avons même fait de l'image un "objet" en soi et nous avons pris l'habitude, avec le temps, d'agir plus sur les images, simulacres des objets, que sur les objets eux-mêmes. Dès lors, faut-il s'étonner que nous les manipulions, que nous les ayons manipulées et que nous les manipulerons toujours davantage?'

6 Personal translation from: 'D'un côté, la science a détruit des mythes, mais en a construit des nouveaux; de l'autre la politique en a constamment engendré, pour s'acharner sur ceux qui lui barrent la route. Toujours, elle a fonctionné à coups de mythes, ceux de la nation et de la race, par exemple, qui, amalgamés et fondus ensemble avec des éléments "naturalistes" et darwinistes sociaux, ont préparé la "pâte fasciste"'.

7 Personal translation from: 'Points d'ancrage situés hors du territoire du Parc, pour une meilleure cohérence de l'action'.

8 Personal translation from: 'Il suffit de regarder une photo satellite, pour se rendre compte qu'il y a un certains nombres de problématiques qui sont extrêmement communes'.

Chapter 9

The Myth of Boundless Nature

Naturalizing spatial metaphors in a borderless world

Protected area managers place themselves in a paradoxical position: on one hand many adhere to a belief in the immutable 'boundlessness' of nature, yet on the other they face stark difficulties when applying such ideas to practical management when people, not nature, need engaging with.

> *It is difficult to make people accept there are no more borders*
> *(Gheorghe, Danube Delta Biosphere Reserve, Romania).*

> *It's true that this European park, it's good, because it's true that the boundary, it's stupid, really, a boundary between people, well, the vegetation is the same, but I don't know, in the current state of things, (...) I'm not too sure ... Well, what bothers me is that everyone, but it's the same as the single currency, is that everyone persists in their own identity and, finally, 'we are all equal' when we aren't really all equal*[1]
> *(Thomas, Parc National du Mercantour, France).*

Building on previous chapters discussing the definition of boundaries, I take a critical look at the process of transboundary zonation. This concrete example of defining boundaries across two or more national systems is described and analysed through the discourses of protected area managers. Societal or cultural differences, including differences in accountability, management structure, legal status and budget as well as the often-identified differences in language, 'culture' and environmental traditions are repeatedly identified by managers as obstacles conspicuously conflicting with the myth of a boundless world.

It is more than a coincidence that discourses on 'borderless nature' have gained favour at a time when economic arguments about a 'borderless world' have become ubiquitous. Terms such as 'globalisation' are pervasive and therefore largely unquestioned. This latest terminology replaced metaphors of the Global Village which have become rather passé, yet remain within the same type of imagery. Similarly, the image of the globe from outer space has lost its novelty yet remains omnipresent, having originally caused a fundamental shift in thinking about the relationship between humans and their environment in the 1960s, following the Apollo space programme (Hajer 1995 : 8, see also Cosgrove 2001). At the same time as pictures of the Blue Planet were being used to sell anything from washing powder to protected areas, the emergence of a condition of postmodernity over the last four decades brought other changes. This coincided with 'a dramatic materialist and ideological *deterritorialization* of the geopolitical order established under American hegemony after World War II. This process of deterritorialization refers not to the creation of a borderless world, but to the loosening of the spatial order' (Ò Tuathail 1996 : 228–229). The myth of a borderless world is therefore not simply an adjunct of globalisation but rather becomes paradoxically both necessary and contested within this context.

Despite the pervasive image of a borderless world, it is banal to say that boundaries continue to be drawn and to have concrete effects. Contrasting myths coexist and conflict. Here, I link transboundary zonation to institutional issues, discussing the irruption of culture and difference in the zonation process. I conclude by suggesting that approaches to 'social nature' offer the promise of transcending the practical problems posed by the stark confrontation of biophysical and societal conceptions of boundaries. An explicit deconstruction of discourses on nature points a way forward from the impasse that protected area managers end up in when seductive myths collapse.

Defining common boundaries

The call to create transboundary zonation raised interesting questions on the practical applications of the borderless world and boundless nature myths. It appeared surprisingly paradoxical to state that a borderless world could be strengthened by drawing more boundaries, yet this was apparently never questioned. Instead, the idea was that common boundaries defined by accepted and shared criteria would replace other less natural ones. This was implicitly suggesting erasing *fiat*, political (international) boundaries with others modelled on *bona fide*, 'natural' ones, reorganising territorial networks in line with 'nature'. Yet however much managers identified this as desirable, it remained difficult to do in practice. All managers suggested that the creation of a transboundary zonation was a central aim when creating a transboundary entity. It was explicitly required by UNESCO in the case of transboundary biosphere reserves. None of the case study areas had achieved this to the satisfaction of all:

> *There is no common zonation of the transboundary biosphere reserve at the moment*[2]
> *(Théo, Parc Naturel Régional des Vosges du Nord, France).*

> *Even if the zonation is a bit unclear at the moment, the principle is the same in both countries*
> *(Nikolaï, Danube Biosphere Reserve, Ukraine).*

> *There is no common zonation map in the Danube Delta. But we have been working on a common vegetation map*
> *(Nicolae, Danube Delta Biosphere Reserve, Romania).*

As mentioned earlier in the chapter on transboundary maps, it was not altogether clear what transboundary zonation actually implied (Chapter 8). At the very least, it involved managers recognising that the boundaries of specific zones defined by adjacent protected area administrations were compatible or complementary with those within their own area. It involved defining not only boundaries, but also the status and management objectives of the zones. One manager explained what had happened in this particular field in the last few years, stressing the practical difficulties of making two different systems compatible, when both worked within separate timeframes:

> *So, many things and at the same time few things [have been achieved]. We would in fact like it to go as quickly as possible more in depth but we realised that (...), transboundary issues, and on one hand transboundary zonation, at the moment, is almost impossible, because you can see that we, I, have already handed my proposal to UNESCO, we had also to renew the MAB label. So every time we had to renew the MAB label, in 99, my proposal was sent off, it's not for lack of trying with [the coordinators], we tried to see how to construct this transboundary zonation, it was spoken about on both sides. But in fact, let's not forget that the term transboundary doesn't mask one thing and that is that we have two radically different administrative organisations*[3]
>
> *(Hugo, Parc Naturel Régional des Vosges du Nord, France).*

Interpretations of the three biosphere reserve zones varied. Correspondingly, the definition of what constituted transboundary zonation did too. Managers were quick to point out when zonations did not match, often despite the existence of shared maps. UNESCO, understandably, was keen to enshrine the concrete existence of transboundary zonations, even if this meant publishing maps they knew to be out of date, as in the case of the Vosges du Nord/ Pfälzerwald map that appeared in several publications despite having been discarded by local managers as inappropriate (UNESCO 2002 : 139). The strength of the myth 'made flesh' within a map and its corresponding competitive value among international conservation organisations was too strong to ignore. Although managers were often proud of the official sanction UNESCO gave to their work, they were sometimes perplexed at the lack of critical appraisal of what was recognised by some to be no more than improvised suggestions for a common zonation:

> *[Théo] This explains that in December '98, the International Committee of UNESCO created the 'Vosges du Nord/Pfälzerwald Biosphere Reserve', without their really being either a common team or a unique support structure, and without a common zonation. And from the point of view of UNESCO, I think that in this case, well ... I was surprised.*
>
> *[Juliet] The common zonation, it appears on the maps, there, next to the one you have just drawn.*
>
> *[Théo] It's a juxtaposition*[4]
>
> *(Théo, Parc Naturel Régional des Vosges du Nord, France).*

This manager appeared surprised that the lines he had drawn on the map were taken at face value. He seemed disappointed that there had not been more of a critical appraisal of what transboundary zonation actually implied. Lines on maps took on different meanings for different people or institutions. Discourses of boundaries were rarely shared in adjacent protected areas.

Boundaries, I suggested earlier, are not only always political but are also always contested. This contested nature was masked on an official level by homogenizing discourses, with boundaries taken to be unproblematic objects of equal status despite appearing in a variety of contexts. This homogenizing discourse negated the political nature of boundaries that was necessarily context-dependent. There was not one universal type of boundary dividing one universal nature. Rather, different meanings and statuses were assigned to lines on maps in different countries:

> *Let's consider the zonation: what is going on? When we draw a line on this map, when we decree that the forest is the buffer zone, forested, ok, exactly, they know that this line belongs to an informal domain, that the Park's Charter only indicates it within a series of guiding rules, but they don't get upset if we explain clearly what lies behind that, the clauses and there is no need to build up a specific legislation. In Germany: it's impossible! In Germany, when a line is drawn on a map, this line needs to be supported by ... a distant ... As soon as we started discussing zonation, we came up against this problem. When we said that all our core zones are sites that wouldn't have a different protected status, nature reserves, biological reserves, ... 'Arrêtés de Protection de Biotope', well, I'll not mention all the tools used but some are linked to special rules and some are not. For the Germans, that is incompatible. A core zone is only a zone that has a ... for example, that corresponds to a specific criteria within the law. So, already at this stage, it really wasn't on the same basis*[5]
>
> *(Hugo, Parc Naturel Régional des Vosges du Nord, France)*

Hugo had no problem in considering lines-as-boundaries as elements defining spatial entities of different status and legitimacy, belonging in certain cases to nothing more than the 'informal domain'. However, he suggested that the German attitude was substantially different. These differences in attitude to legislation and protocol were recognised to be a problem by the German administration, especially as they were struggling internally to determine who was formally responsible for carrying out zonation. There, lines on maps were never informal. The German protected area was caught between the German MAB committee and the Land's relevant Ministry. This complicated the drafting of a transboundary zonation to the point when it was simply dropped as an immediate objective. There were therefore a whole host of different understandings of what boundaries meant on legal, ecological or symbolic levels:

> *First we made a common proposition in the early Nineties. Look at your map. It was the same zonation as the Vosges du Nord, like in the Palatinate forest. And then our national committee says that this German-French zonation it doesn't obey the rules. 'You have to make a new zonation that obeys the national rules'. And so the cancellation [sic] was torn apart and now we have two different zonations. The French zonation and the German zonation. And that is a problem now. And here the association wants to get a common zonation but the State says 'no, we want to have these national zonation rules which says 3% must be out of human management, must be wilderness'. So they made this little plan you see here and this plan is now in discussion and I am sure this plan will come. And the Land will make a rule for the German part of the*

common biosphere which will say that the common zonation must be done . . . and so we have a little problem with our French partners

(Lukas, Naturpark Pfälzerwald, Germany).

Part of the issue here was a different understanding of what constituted nature and therefore what required protection (see Chapter 9). Mirror images on both sides of the boundary were not seen to be absolutely necessary, but managers recognised that compatible or complementary zonation needed to be achieved, although the actual practicalities of this remained unclear. In several sites, managers mentioned that the transboundary zonation was simply based on separate zonations, with no specific shared interpretation of the status of each zone (Danube Delta; East Carpathians; Tatras). In the Tatras, it was obvious that these interpretations were substantially different, particularly regarding the (in)existence of permanent human settlements. In the cases where achieving increasingly compatible zonations was identified as an objective (East Carpathians, Vosges du Nord/Pfälzerwald), managers recognised that some form of negotiation would be necessary. They were not however always equipped to carry this out, either through lack of pertinent skills, legitimacy or institutional capacity:

The Ministry for environment and transport of Rheinland Palatinate, they are in charge of this. While on the French side they already have made a decision on their zonation and now we have to go into the process of getting compatible . . . complementary zonation systems on both sides, which is a big task. (. . .) The only place where we may have one joint core area is a nature forest reserve 200 hectares on the German side, and one hundred on the French. This is going to be designated now, and this will be the first German-French core area

(Daniel, Naturpark Pfälzerwald, Germany).

The creation of one shared core area was not straightforward, yet had iconic status. It was viewed in this case as symbolically important and indicative of progress along the road to greater compatibility. It also allowed a variety of individuals to get to know each other while the formal process was being undertaken, fostering further contacts. Any such attempt involved negotiation and discussion. Not surprisingly, the nitty-gritty of attempted but failed or abandoned negotiation peppered the managers' depiction of the process in all the sites. This was often seen to be a struggle for legitimacy between a selection of different administrative bodies and individuals. Certain people or administrations were more legitimate tracers of boundaries, while others were deemed legally unfit or unauthorised to draw lines on maps. The most acute example was in the German Pfälzerwald:

Well, the mistake was on the German side, because the French were always ready, (. . .) they showed us all that maps and all the planning, and came to Germany, and presented it to all the politicians etcetera. One representative of the German Ministry came to La Petite Pierre [the French administration], they showed him everything, and then finally, some of the guys at the Ministry level, they decided to do a German zonation, and not a transboundary zonation. (. . .) We don't have the right . . . We had created a working group on the zonation, the transboundary zonation, with French and German representatives, and [the French director] came and he presented the French ideas and everything and I translated. And then the Ministry came and said this is not your task. You don't have the right to do that. (. . .) So we say okay, we accept it, but we want you to include the French. As we did. But they were completely excluded. They promised several times, and they excluded them. And I don't know why

(Daniel, Naturpark Pfälzerwald, Germany).

When two interpretations conflicted in the absence of any formal negotiation, managers often suggested what would happen if their own interpretation was applied to the 'other side'. Lukas hinted at this in the quote presented earlier ('*the Land will make a rule for the German part of the common biosphere which will say that the common zonation must be done . . .* '), suggesting that one (German) interpretation could be applied or imposed, on the (French) Other. Similarly, over the boundary, Hugo suggested what would happen if the French criteria were applied in Germany, nevertheless implying that this would not happen:

But well, a common zonation, evidently, will not be set up. If they had applied our zonation, all the forest would be in the buffer zone, inside that there would be the Naturschutzgebiet (protected areas) that they already have and that would be all their network of protected areas which is in the Biotopkartierung (biotope mapping), and then the village would be in the transition zone, and the rivers would also be in the transition zone. But they didn't do that, they wanted to apply criteria which are quite strict on the German Land's side, 3% of the territory, red zones . . .[6]

(Hugo, Vosges du Nord, Pfälzerwald).

This setting down of positions was interesting in that it illustrated how particular interpretations and spatial discourses were confronted and negotiated discursively, on both a concrete and symbolic level. Managers confronted Self and Other, as well as their respective interpretations of zonation in a ritualised confrontation of discourses. Here, this was acted out within the interview, yet such verbalised confrontations also took place during informal discussions and mealtime conversations. These performances of differences and similarities verbally examined the possibilities for integration or distinction.

Different interpretations of zonation reflected different strategies for dividing up protected areas into a series of coherent zones. In the Franco-German case, the French landscape within the Parc Régional was seen to be very varied, including open landscape, forests and extensive agricultural zones. The German Naturpark was largely forest. The same could be said of all the transboundary sites, reflecting differences in history, settlement and previous land management. Although nature itself may be 'boundless', the resulting landscapes reflected human management and political choices. This was particularly true in the East Carpathians where populations had been forcefully removed from large tracts of land on the Polish side following World War II, while the Ukrainian and Slovak areas were still relatively densely populated in several parts of the biosphere reserve.

The criteria chosen to define each zone were thus site specific, stemming from the local conditions within each protected area. Applying the interpretation emerging from one side onto the Other's different landscape was seen by managers to lead to an inappropriate balance between the three zones. The definition of what was appropriate for each zone was therefore heavily context-dependent. Within the core zones, for instance, what constituted the most valuable feature of an area (the 'wildest' bit) could not be defined satisfactorily for the other national contexts. In other words 'wilderness' was not a universal, value-free (or apolitical) thing. Partly because of this, none of the biosphere reserves handed over to the neighbouring entity exclusive responsibility for one type of zone. Indeed, in the absence of three identified zones, a biosphere reserve could not be presented by a national MAB Committee for formal designation by UNESCO, as designation was still organised on a national level even for transboundary sites.

This difficulty in translating zonation interpretations was compounded in the German case by the additional strict spatial quotas for each zone laid out by the German MAB Committee. These would not have been fulfilled if the French zonation criteria had been applied to the German Naturpark. In the Mercantour/Alpi Marittime, where two adjacent protected areas were examining the possibility of seeking designation as a biosphere reserve, the existing zonations were identified as being substantially different. Again, this was seen to be a delicate issue that required addressing, particularly as the French park had no formal jurisdiction over its own buffer zone:

> *So we started doing this proposal [of a transboundary biosphere reserve]. Also because when we speak of a core zone, well this is the parks, for us it is the whole park for them only the 'central zone', but on the contrary the buffer zone and the transition zone are harder to define, especially if it has to be a precise delimitation*[7]
>
> *(Alessia, Parco Natural Alpi Marittime, Italy).*

> *They don't have the same system. First of all because it is a natural park so they don't have the same system, they only have one zone. So there is no identified peripheral zone. It's the communes [communities] of the park but it's not identified as we are in the peripheral zone with a line, with legally defined boundaries. (. . .) The peripheral zone: we don't have the same authority as on the central zone, no, not at all. That is to say we have very few jurisdictional means to intervene in the peripheral zone. (. . .) Let's say that we have . . . I think the law was very intelligent at the start to make a zone of strong protection and a zone that they called the 'pre-park', but there have not been concrete means to reflect on what this means for specific planning within this zone*[8]
>
> *(Chloé, Parc National du Mercantour, France).*

Chloé identified the differences between the Italian and French parks principally in terms of an absence of a peripheral or buffer zone on the Italian side. However, as she progressed in her reflections, it appeared that while she considered the French side to be more coherent, in practice the peripheral zone was largely inoperative. Thus the differences she initially stated as an impediment to transboundary zonation collapsed. Nevertheless, as she continued to discuss the issue, this difference was taken to be fundamental.

Boundaries, even if largely inoperative on the ground, held symbolic strength and were seen to entrench differences.

In this section, I discussed the problematic issues surrounding the establishment of a transboundary zonation. While managers suggested that the creation of a transboundary zonation was a central aim when creating a transboundary entity, this proved poignantly difficult to do. The cases examined here all involved creating a transboundary biosphere reserve in which three types of zones were defined. However, the individual (national) interpretations of these three zones differed considerably. Criteria appeared to be strongly context-dependent and therefore not unproblematically applicable in other contexts, even when areas were adjacent. There was furthermore no 'universal' agreement on the significance of a protected area boundary, compounded by differences in legislation as well as by different attitudes to legislation. Thus negotiations around transboundary zonations were confrontations around differing spatial discourses in which independent administrations sought to promote their own particular interpretations, reflecting their individual strategies. These discourses concerned space, but also inevitably had temporal dimensions.

Divergent spaces, divergent times

As developed in Chapter 2, boundaries are not only spatial and territorial but also temporal phenomena. Human activities are regulated, organised, controlled both within space and time – yet just as spaces differ, so too do people's relations to time. If space and time are constructed simultaneously, then they can also be constructed differently. In the protected area administrations, the speed with which decisions were taken differed, deadlines were different and were not treated in the same way in separate contexts. When common zonation was attempted, independent time-frames created specific problems:

> *So for us the park's plan had to be produced and finalised, you see. And so we decided by ourselves. We could have waited a bit for the Germans, but in this case they hesitated for two years over their zonation, so we quite quickly gave up the idea of making our zonation fit in with a common zonation*[9]
>
> *(Théo, Parc Naturel Régional des Vosges du Nord, France).*

In the Mercantour/Alpi Marittime, managers identified the need for a common zonation but were unsure of how to proceed as priorities differed. At the time of the fieldwork, the French park had decided to lay emphasis on seeking World Heritage status, while the Italian park was more interested in establishing a biosphere reserve. The decision made here was to distribute different roles, although it remained unclear quite how these would be combined in the future:

> *The [biosphere reserve] project is currently being followed up more by the Italian park, we rather split up roles so they are making progress on the biosphere reserve and we are working on the World Heritage project. We hope to present a proposal before the end of*

> *the year. (. . .) So we haven't worked on the zonation, we worked with the idea that the two zones of the park would be the biosphere reserve but now, within that, we need to establish a zonation. (. . .) I think that Italy will (should) make a proposal. (. . .) She [the director] has already made a suggestion for her side and depending on what they come up with, the degrees, we will adapt ours. Let's say that it works sufficiently well for them to tell us something and for us to adapt to it and vice versa. That doesn't create any problems*[10]
>
> *(Chloé, Parc National du Mercantour, France).*

This very positive comment painted a very unproblematic picture of equal contact between the two administrations, with one making constructive comments to the other. However, when I discussed this with Italian managers they expressed surprise. It was far from clear to them that the French expected suggestions. Rather, they had gathered that their neighbours were not too keen to discuss the idea at all, as there had been little contact in previous months and letters remained unanswered. Thus the process was left to drag on. While the French park professed an interest in the possibility of common zonation, they did not pursue it actively. In any case, for them, designation as a World Heritage site seemed to offer more potential, more prestige and less problematic zonation issues.

Negotiation, frustration and divergent deadlines

When an attempt at transboundary zonation failed or stalled irredeemably, frustration inevitably appeared among those involved, colouring wider interactions. Enthusiasm among managers inevitably lagged as timeframes were redefined. Divergent deadlines in national contexts meant that a drawn out negotiation could not take place.

> *What do you want us to say? For us, our zonation is approved on the 2nd October and is set. We stopped and fixed our zonation in January 2000. Full stop. And then they come on the 22nd August, with a project. We don't know who defined it, and who was consulted. I didn't go to the meeting on the 22nd August, on one hand because I couldn't, because I had other things to do* [slang term], *but also on the other hand to indicate that us French, we didn't have anything left to do with their zonation. (. . .) I tell myself that if in 10 years time we have a common zonation, at the time of the next revision of the biosphere reserve, well then that will be very very good. That is the objective we have to set for ourselves*[11]
>
> *(Théo, Parc Naturel Régional des Vosges du Nord, France).*

Part of the point of seeking official designation as a biosphere reserve was the official sanction this gave transboundary work. There was a fine line between managers wanting this, and using it as a bargaining chip or a carrot to convince unwilling partners to cooperate. As mentioned in Chapter 7, cooperation was not always an equal exchange between partners of the same status or dedication. Several times, managers noted that official transboundary designation came too soon, before serious engagement and negotiation had taken place between adjacent protected areas. Once official designation was made, pressure to cooperate paradoxically lessened:

> *It would have been better if UNESCO had not recognised the transboundary biosphere reserve so quickly. Because it is not really transboundary*
> *(Nikolaï, Danube Biosphere Reserve, Ukraine).*

> *[JF] So in fact it would have helped you more if UNESCO had said no?*
> *[Théo] Yes, absolutely. UNESCO should have said to us 'we encourage you, we support you in your project of establishing transboundary cooperation, we are behind you, we think this is really interesting, it's great, but for there to be a biosphere reserve, you need a project, a sort of Charter, a guide for assisting management, call it what you will, the Germans call it Raumplankonzept, and you have to show us that there is common management' (...) Common zonation does not exist. And even less a Raumplankonzept. So here, I was a little bit ...*[12]
> *(Théo, Parc Naturel Régional des Vosges du Nord, France).*

This was a fine line for UNESCO and managers to draw. UNESCO's policy called for promoting as many satisfactory sites as possible, increasing the number of transboundary biosphere reserves around the world as a badge of success. Similarly, managers also gained from official recognition and so were unlikely to adopt strategies to delay it. They nevertheless realised they therefore lost an important bargaining chip once this had been obtained. When I mentioned this to a senior manager in the UNESCO Secretariat, she expressed surprise that managers would be resistant to receiving quick recognition, and firmly wished that this should not appear in any report, equating it to UNESCO 'shooting themselves in the foot'. Even managers who said that recognition had come too quickly expressed ambiguous feelings:

> *After all it's also good to say that there is a transboundary biosphere reserve, the first in Europe, it's rather like receiving a laurel wreath, but concretely on the ground, this biosphere reserve has no framework. Nothing is transboundary about it: this has to be said*[13]
> *(Théo, Parc Naturel Régional des Vosges du Nord, France).*

This initial survey of transboundary zonation indicated that however 'boundless' nature may be seen to be by managers, tracing boundaries through it in different countries was problematic. The creation of homogenous transboundary zones remained an identified objective, yet its significance was unclear, as were its practical execution and consequences, both spatially and temporally. The meaning assigned to boundaries was heavily context-dependent, linked to national legislation, planning practices and traditions. There was not one universal form of boundary, just as there was not one universal understanding of nature. Furthermore, legitimacy in tracing boundaries was contested in many sites, as different national administrations debated their respective roles, further complicating understanding on a transboundary level. Timeframes were different. Furthermore, managers suggested that official recognition should be strategically timed, acting alternatively as a stick or carrot. Boundaries existed more or less formally within legislations or sometimes not at all. The tracing of them, however, was always linked to specific institutional structures. This is discussed more in detail in the following section.

Bounded institutional spaces

When managers discussed boundaries and zonation they inevitably mentioned corresponding institutional structures. The central question concerned determining who or what administration was legitimate in drawing boundaries and establishing a zonation, both nationally and across the international boundary. Legitimacy was not always clearly established, particularly as biosphere reserves did not always formally appear within national legislations. Thus identifying the Other with whom to engage was not straightforward, as discussed in the chapter on cooperation (Chapter 7):

> *All this means that, in fact, we can sort of come and discuss with our German partners saying that we are really the coordinators and that, when we discuss things, we know that we have already spoken to the people and that it is us who manage our zonation, of our biosphere reserve, in inverted commas. On the other side, it's not the Naturpark but the Ministry of the Environment and Forests in Mayence, so the Land took it in hand and said 'we will do this', and they will indeed, as you saw, and they are doing these red, green and other perimeters. The Naturpark is not saying anything, it is watching and will see what is thrust upon it. In any case, it's a power relationship between the environmentalists of the Ministry and the foresters, because almost everything is in the forest. (...) We were here, the park, working directly on our zonation while over there, it's the Ministry of the Environment and Forests*[14]
>
> *(Hugo, Parc Naturel Régional des Vosges du Nord, France).*

> *The French organisation, the Sycoparc, says 'who is responsible in the German part for the biosphere reserve? Is it the association, or is it the State?' And up until now it seems that it is the State that is responsible. It does not improve our work with the French*
>
> *(Niklas, Naturpark Pfälzerwald, Germany).*

In addition to institutional confusion, the lack of legislative recognition for biosphere reserves was repeatedly identified as a problem in Poland, Slovakia and France. This was seen to have a direct impact on the ability of managers to carry out coherent zonations that were compatible with existing designations. In the absence of clear guidelines, different institutions and administrations projected responsibility on others, further perpetuating the confusion:

> *The biosphere reserve still has no legal status, and in Poland we have a conflict with the Deputy Minister. In 1997, I asked about possible ministerial support for the work of the biosphere reserve but he told me that it was not within the interest of the Ministry of the Environment. The relationship between the MAB Committee and the Ministry is difficult, and the MAB Committee is seen as a source of possible funding. They think it should finance its own activities*
>
> *(Andrzej, Bieszczady National Park, Poland).*

This confusion lead to surreal situations in which different administrations found themselves drawing alternative boundaries in space, constructing coextensive but conflicting territorial entities. While this created problems on national levels, it further created confusion in transboundary situations:

> *[Maximilian] I think we have a chance in the next ten years to become together [sic] with the French organisation and I hope it but there are a lot of problems based by the Federal State. And we don't know if the biosphere reserve will be taken by the Federal State and they say 'we are doing biosphere reserve and you are doing Naturpark'.*
> *[Juliet] In the same area?*
> *[Maximilian] In the same area. That is a problem*
> *(Maximilian, Naturpark Pfälzerwald, Germany).*

> *The biosphere reserve is mainly an international agreement. Each national park must have a management plan, but this does not exist in the Poloniny national park yet because we are waiting for the methodology from the government regarding issues like zonation and so on. It is not yet clear how the national park and the biosphere zonation fit together*
> *(Dominik, Poloniny National Park, Slovakia).*

The institutional confusion, with zonations as puzzles that diversely 'fit together', was compounded by cultural differences in work practices. The recognition of the role institutions or administrations played and their corresponding status also differed considerably in different contexts. Identifying differences in work practices was widespread among managers: identity was defined relationally and constructed discursively. In the French Mercantour, for example, Alexandre said that working with the Italians was difficult because they were too laid back, while the French needed frameworks to function. In the Alpi Marittime, comments by Alessia stressed the overt emphasis on protocol that existed within French bureaucracy. In fact, Italian managers describing the French perfectly echoed what the French were saying about the Germans in a different location ...

'Legitimacy' came to have very different meanings in different countries (see Chapter 7). Differences in recognition awarded to advisory or coordination committees created surprisingly problematic situations, when paradoxically the initial objective of these was to bridge institutional differences. In order to work around the delicate distribution of levels of legitimacy, managers suggested a variety of transboundary structures that could assist in defining broad zonation principles. Joint scientific committees were often envisioned to be the most 'objective' and therefore appropriate advisory groups, removed from political processes. After all, the thinking went, these functioned well within national contexts and could therefore be extended to cover the whole transboundary zone. However, creating them as transboundary bodies was problematic for institutional, financial and practical reasons. These reasons easily explained their collapse, as in the case of the East Carpathians or the Danube Delta where fuel costs and visa fees made practical communication difficult and meetings unlikely. This was not to say that other less tangible issues were not equally at fault: they were simply not mentioned by managers facing overwhelming practical issues. In addition, it was not always clear who was responsible for setting up transboundary committees. One manager mentioned that it should be the responsibility of the relevant Ministry, not the protected area's administration:

> *There is a need for a cooperative committee for the one biosphere reserve, which should include a number of people from each side, to organise the practical cooperation. At the moment, there are only personal contacts between individuals. (...) It should be created on the level of the Ministry but since the first signed agreement, there have been at least three different Ministers of the Environment. The current one is not so excited about the idea of the biosphere reserve. The biosphere reserve is not written into Slovak law, but it is coordinated by the Academy of Sciences*
> *(Dominik, Poloniny National Park, Slovakia).*

In addition to the practical arguments relating to the establishment of such bodies, comments on divergent institutional practices occasionally cropped up, indicating that creating a shared entity implied more than simply defining a group of people:

> *For the working group on biodiversity, they wanted to have a clear status. You know, the Germans really like to have a clear status and to be officially legitimated, very officially, for participation in a simple working group otherwise they don't function. We work in a very informal way within a working group that exists, exists no longer, we don't panic, but for them this is impossible and unthinkable*[15]
> *(Hugo, Parc Naturel Régional des Vosges du Nord, France).*

This was corroborated on the German side by managers who mentioned struggling for a clear definition of their role in relation to other existing structures. It was interesting to note that while both partners displayed a clear understanding of the needs of the Other, differences were still seen to create practical problems. A negotiated, in-between mode of functioning had not yet emerged.

This section discussed the links between boundaries, institutions and the search for alternative shared management bodies. The fact that each of these notions covered very different things in separate countries emerged very strongly. Obtaining legitimacy for drawing boundaries, and determining what institution should be responsible for this, was far more problematic than managers initially imagined. It raised a series of context-dependent understandings and practices related to boundaries that were frequently poorly identified, weakening the likely emergence of a negotiated, 'third way' (or 'fourth way', in the case of the tri-lateral East Carpathians). This sudden irruption of 'culture' in what was thought of the domain of 'nature' is further considered in the next section.

Identifying difference, constructing the Other

Dealing with difference: the irruption of 'culture'

There was often a turning point during interviews when following a relatively straightforward description of local features, landscapes or wildlife, a comment about 'culture' would irrupt and change the flow of the discussion. Suddenly, in what was a conversation about 'facts', an emotion marking the boundary between Us and Them emerged. The degree to which this was then welcomed as a further discussion point depended. Some managers did not wish to elaborate, perhaps feeling that such ideas were inappropriate, while others were more inclined to explore issues. Comments differed, but included the following exclamations:

> *Maybe we are the ones who are wrong, but that's how we are*[16]
> *(Manon, Parc Naturel Régional des Vosges du Nord, France).*

> *I ask myself many questions: I had lived believing in the myth of German efficiency. So I don't know if it is a lack of willpower ...*[17]
> *(Hugo, Parc Naturel Régional des Vosges du Nord, France).*

> *The French are all shits!*[18]
> *(Marco, Parco Naturale Alpi Marittime, Italy).*

The Italian manager who made the last far-from-politically correct comment about the French did not wish to elaborate and refused to discuss the topic further, perhaps feeling that he had gone too far. In this case, the identification of the symbolic boundary between Us and Them lead to a closure of communication. This avoidance of difficult topics was not the only strategy managers adopted to deal with the irruption of difference. In many of the interviews, 'cultural' difference and related issues were swept aside. Questions related to such subjects were received with surprise and repeated attempts to explore the subject were met with scorn or answered with short replies stating that transboundary communication posed no problem. This lack of engagement or refusal to address the issue creates methodological problems: how do you analyse something that is not there? Were 'cultural' issues really a non-existent issue for some managers? Or did the refusal to address them point to the fact that transboundary interaction had not reached the point when they appeared and started to challenge myths of boundlessness? Or was this simply a methodological side-effect indicating that I was asking non-pertinent or inappropriate questions?

In a business manual on management across cultures, Schneider develops the argument that all management traditions are culturally based and are therefore often substantially different, even between two institutions in the same country. She notes that there are two potential traps: 'to assume similarities and to assume differences' (Schneider 1997 : viii). She further states that 'we need to recognize that these underlying, and often hidden, cultural assumptions give rise to different beliefs and values about the practice of management. These assumptions are also manifest in the behavior of managers and employees, as well as in our everyday working environment, from the designs of the buildings we enter, the interior office, to the very design of job descriptions, policies and procedures, structures and strategies' (Schneider 1997 : viii). Théo clearly identified the importance of being aware of such differences between administrations, both in institutional structure and in work patterns, although he noted that raising such topics was often considered taboo by both partners. Much of the literature on this topic was developed by and for private companies wishing to do business in other countries, yet this was largely ignored by protected area managers, more likely to be versed in the natural sciences than in theories of management. Théo noted that cultural differences were taken more seriously in the private sector:

> *Private companies, where there is a question of money, say, well they train themselves much more on intercultural issues, to know how to deal with a partner or a Japanese client, or a German one, in order not to get things wrong, for instance apparently you never shake hands with a Japanese. But that, if you don't know it, well you're stuck. And then, that's it really, so for private firms, that immediately means markets closing and market values going 'pop!'*[19]
>
> *(Théo, Parc Naturel Régional des Vosges du Nord, France).*

In the case of protected areas and despite what the protected area literature may claim, there was often no direct economic gain from cooperation, although there was funding secured specifically for cooperation projects. When differences were acknowledged and taken into account in the planning process, the real transformative potential of increased interaction appeared in all its multiplicity and destabilising complexity, mediated in some cases by coordinators. Increased knowledge did not however lead unproblematically to increased interaction and effective contacts as might have been expected. Initially the opposite appeared to be the case. However, by identifying differences, myths of homogeneity were replaced by more pragmatic objectives:

> *I think we must accept not to be too ambitious, we won't have a common team, a common place overnight ('c'est pas demain la veille'). So that, pfiiiouu, those are beautiful ideas I had had at the start. It's unthinkable now. But we have to move on with concrete ideas, practical ones which involve partners. On the ground. So then we will build something, but in the very very long term ...*[20]
>
> *(Hugo, Parc Naturel Régional des Vosges du Nord, France).*

Myths and ideals of rapidly creating a shared space with shared institutions collapsed, only to be replaced by other pragmatic ideas and practical projects. Increased engagement with the Other did not therefore lead linearly step by step to unity, but rather led to the collapse of initial myths. In some cases, identifying differences and discursively constructing identities in explicit contrast to the Other was actually detrimental to common work. Myths may be counterproductive in the long run yet while they last they can be surprisingly potent. When managers described the failure of certain transboundary projects and described their stark confrontation with practical difficulties, abandoned myths continued to haunt discussions, rather like half-forgotten, distant friends who were fondly referred to in passing. Identifying differences was one thing, but moving on and constructing new ways of work – new myths? – was another. Several managers had clearly undertaken the first step, being willing to acknowledge differences and identify similarities. However, while this served to 'define' the boundary between the two, it was a high risk process with an uncertain outcome.

> *(How do I define) cultural differences? Needs and desires and rhythms of work are not at all the same between France and Italy. And, also, available funds are not at all the same either, and how these funds are used is even less so. When there are funds, people don't necessarily have the same wishes to apply them to the same places, at the same time and with the same standards*[21]
>
> *(Thomas, Parc National du Mercantour, France).*

In such a context, identifying what priorities might be shared was not straightforward since identifying a problem was itself a culturally-informed process. Hajer has stated that an environmental problem 'should not be conceptualised as a conflict over a predefined unequivocal problem with competing actors pro and con, but it is seen as a complex and continuous struggle over the definition and the meaning of the environmental problem itself. Environmental politics is only partially a matter of whether or not to act, it has increasingly become a conflict of interpretation in which an increasingly complex set of actors can be seen to participate in a debate in which the terms of environmental discourse are set' (Hajer 1995 : 15). While Hajer applied his analysis to the global 'new environmental conflict', such a statement was equally pertinent on a local scale where sets of actors extended not only to the protected area administrations involved, but also to other international and national actors.

Faced with a collapse of the myth of boundless nature, the French and German managers decided to address what seemed like profound 'cultural' differences head-on in order to grasp why ways of work and priorities were not the same. UNESCO funded this as a pilot-project, allowing the two parks to invite a cross-cultural facilitator to assist them in identifying differences and similarities as well as setting a programme for future work. This process, though highly effective in achieving its objectives, partly served to entrench differences. One manager (Théo) mentioned that some of the staff were shocked to discover the extent to which the two protected areas were different, both on institutional, 'cultural' and personal levels. Contacts turned into quasi-conflicts and the boundary between Self and Other became even more of a trench. The question, naturally, was what happened after such a process.

> *Some came to understand why the German colleague reacts and acts like this, or the French colleague like this. In some cases they just use it as an excuse, they say well this is all cross-cultural, you can't do anything about it. (. . .) So identifying the problem did not help in those cases because it just supports the prejudice. Some are prepared to*

> *say, well, I had a wrong understanding, or there was prejudice, I am ready to change it, I got new insight, and the others will say, well, I always knew that. You can't help that. Germans are just like this, French are just chaotic ...*
>
> *(Daniel, Naturpark Pfälzerwald, Germany).*

Identifying differences was one thing, instituting change another. Managers were often loath to address how this might be carried out, other than suggesting vague ideals of finding new ways. There was however little evidence that this had been applied anywhere comprehensively. Instead, it appeared to be a long, slow process. Managers would speak about the training they received, going over some of the things that they had learnt, gaining ownership of it and reinterpreting it in their own context. One manager explicitly linked this up to what had happened during specific meetings, discursively negotiating Self and Other by confronting behaviours:

> *So we try to take this quite seriously, you see. In a structured way, in order to ask real questions. And so, it's funny, because then we had these training sessions, we had a few workshops on intercultural communication and the trainer said to us 'well, there are the Germans and the French', and he explained to us for example that the French communicate in an implicit way, between the lines, some things aren't said, and body language indicates to the person listening what he is supposed to understand without it being said explicitly. This is the opposite of the Germans. Explicit. And so they tell you. That is to say if something isn't going well, 'vlaaaa'. And often, at the beginning, we would have meetings with the Germans, and they would speak amongst themselves, and they were insulting each other. The things they said ! We were devastated. We thought they would never be able to work together again. But they were saying things as they were. 'I think that here you have fucked things up', 'there, we had committed ourselves, your colleagues are hopeless', 'it's written here, what have you done about it? You've done nothing'. But really, it was what is never said in France. In France, in any case, once the minutes of a meeting are drafted, they are forgotten*[22]
>
> *(Théo, Parc Naturel Régional des Vosges du Nord, France).*

Some managers noted that the identification of differences had forced them to change their behaviour directly, modifying the way they interacted in order to communicate more effectively with the Other. This was either expressed as a loss of innocence, as a need to face reality, or else integrated and appropriated, as individuals became almost evangelical in their desire to convert others to their level of self-awareness:

> *I said Mr [Maximilian], you know, it is very important to maintain contacts and to ... I don't know how to express this in English ... 'Kontakte pflegen' ['nurturing contacts'] ... we cannot just do it on the phone or by email, you have to meet people, and many French people they are very, can I say, person-oriented? Germans are more 'Sach-orientiert' ['object-oriented']. I said it is very important if we have good French colleagues we have to meet them regularly, we have to talk, we have to exchange, we cannot always go there with an agenda and say today we want this and this, no, we have to have an open discussion, we have to go to lunch together and all these kinds of things, it is very important. And I think that meanwhile he understands that*
>
> *(Daniel, Naturpark Pfälzerwald, Germany).*

In this section I have explored what happened when myths of homogeneity and boundlessness were confounded by the perceived irruption of 'culture' and difference. The stark appearance of heavily symbolic boundaries dividing work patterns, modes of communication and outlook destabilised managers. The responses to this were far from linear and predictable. Instead, the speed with which differences were identified varied, as did the responses developed to deal with them. In one case, an explicit attempt to identify differences and bridge them pro-actively in training sessions was carried out. Again, the response to this was not linear, and increased knowledge of the Other did not lead directly to increased interaction. Instead, reactions of individuals varied. Existing prejudices were as likely to be reinforced as they were to be transcended. Furthermore, identifying what 'problems' managers faced was shown to be culturally dependent, a theme that is further explored in the next chapter.

Counter-myths of heterogeneity

It would be wrong to suggest that all managers moved from a belief in boundless, homogenous nature to a realisation that boundaries existed. This would be wrong partly because it would suggest a 'point zero', a time when ideas and practices were set in stone before cooperation and interaction began. Instead, it was clear that a whole host of opposing and conflicting myths or dominant stories coexisted within administrations. Instead of such an unrealistic, uniform situation, a variety of myths coexisted and were created discursively along the way by managers, often directly in contrast to each other, yet all part of the dynamic process of negotiating identity.

Not surprisingly, 'national' identity was one of the most potent counter-myths. It was indicative that most of the managers referred to their colleagues using the shorthand of nationality. The state as a fundamental territorial reference-point was sometimes felt by managers to be directly under threat from transboundary cooperation. Power necessarily fostered resistance. Inevitably some strong resistance emerged when territorial changes were felt to be imposed 'from above' or from the 'other side'. Just as some managers were resistant to abandoning their belief in boundless nature, others refused to question their identity founded on national qualities. Not surprisingly, considering the multi-scaled, interlocking face of identity, these categories of managers were not mutually exclusive. There was an opposite dynamic to 'homogenous nature collapsing in the face of a heterogeneous reality': the rejection of homogenising transboundary nature in favour of national identity. This dynamic relation between integration and distinction permeated every discussion concerning the creation of transboundary spaces (Figure 9.1).

Figure 9.1 Relations between dynamics of spatial integration and distinction

Crucially, when differences were seen to be negated in favour of the imposition of a greater myth, that of a 'European', transboundary entity, managers responded defensively:

> *The guards [rangers] say 'we don't want to, we are French and European, but French before all else', it's logical, we haven't yet got this ... Well, of course it is necessary to start now to have the European parks in 10 years time, but it shouldn't ... it needs to be done little by little, that's for sure, but I don't think it should be ...*[23]
>
> *(Thomas, Parc National du Mercantour, France).*

One of the heavily symbolic difficulties linked to national identity that appeared when confronting the Other was the question of language. Not surprisingly it emerged repeatedly as the most obvious symbolic boundary, yet its actual relevance was debated. Managers often used the example of language as a metaphor (or rather a synecdoche) for cooperation as a whole. If languages were mutually comprehensible, so they suggested, then cooperation was easy.

In the East Carpathians, for example, several managers said that the three languages were mutually comprehensible if all spoke their own language clearly, adding that local dialects were particularly close. In any case, the Polish manager was fluent in Ukrainian and could serve as a go-between when necessary. What was interesting in the East Carpathians and the Tatras, was that all older managers would have learnt Russian at school, but this, again due to the highly symbolic nature of language and national identity, was never identified as a practical option. Individuals confessed to speaking Russian, but suggested that this belonged to a distant era and one that was better forgotten. Younger managers, however, were more likely to have learnt English as a foreign language and this was also occasionally used, though more rarely or else only within the context of international meetings with 'outsiders'.

In the Danube Delta, several of the senior Romanian managers spoke good Russian and communicated by phone largely in Russian with their Ukrainian counterparts. For written and email communication, English was preferred. The Ukrainian director was making a very concerted effort to learn English, continuing to take weekly lessons after the World Bank funding had run out. This did not stop several Romanian managers from mentioning however that while the Romanian side had members of staff who spoke Russian, the reverse was not the case. One manager noted that '*they should have some Romanian speakers*' (Grigore, Danube Delta Biosphere Reserve, Romania). Language lessons had in fact been offered to a selection of staff members when the World Bank project was carried out, but when the grant ended many of these individuals lost their jobs, taking their newly-learnt skills with them. In the wider Danube Delta area, where several meetings hosted by the Council of Europe included Moldovans in an attempt to widen transboundary cooperation along the river, English was officially designated as the working language. Many of the emotionally potent aspects of language were alluded to in the other two cases, as one French manager in the Vosges du Nord clearly expressed:

And so there are differences that are totally cultural. And this also means saying that I am sick of having to muddle along in German. Me, I speak French. And if you want to cooperate with me, you learn to understand me. To establish a real relationship of transboundary cooperation which would be the same as international cooperation. We have to go through a certain number of stages of getting to know each other, and that is difficult[24]

(Théo, Parc Naturel Régional des Vosges du Nord, France).

Likewise, a German manager noted that a lack of engagement with the neighbour's language was taken to mean a lack of personal motivation for the process as a whole. Additionally, misplaced pride related to lacking linguistic talents compounded cultural misunderstandings.

Also they [the German managers] sometimes find it difficult to communicate with the French because they don't speak French. They never made an effort to learn French. And then you go to a meeting and I say I can sit next to you and translate everything and they say it is not necessary I understand everything and I tell you they understand twenty per cent maximum. And then you imagine what happens ... So then Théo speaks German very well but sometimes he speaks German with the Germans but sometimes when we are in France he wants to speak French and he is right. Germans cannot expect to come to France and the French speak German all the time

(Daniel, Naturpark Pfälzerwald, Germany).

Now I only work with people who either speak French or else English. So obviously this selects. So if someone only speaks German, I will not go any further. But you always find people ... When you want to communicate, you find ways[25]

(Hugo, Parc Naturel Régional des Vosges du Nord, France).

In the Alpi Marittime/Mercantour, managers adopted the method favoured in Switzerland for multi-lingual meetings: everyone spoke in their native tongue. This naturally assumed that everyone understood the other language correctly, something which was not necessarily to be taken for granted, as was hinted at by Daniel in the Franco-German case quoted above. In the Franco-Italian protected areas, however, this seemed to work out quite well:

Meetings take place in both languages, they speak Italian, we answer in French. That's it. (...) But it's different for [the Italian director] who can speak both languages without any problem[26]

(Chloé, Parc National du Mercantour, France).

This last section has highlighted that spatial myths were never unique but rather coexisted with other conflicting ones, alternatively constructing or performing homogenous or heterogeneous spaces. Tensions appeared between different myths, as for example when 'boundless nature' confronted strongly held ideas of 'national identity'. Language was used as an example of such highly-emotive issues that created both symbolic and practical boundaries between managers in different countries. The different responses to the existence of linguistic difficulties again illustrated the multi-faceted, non-linear responses to the identification of 'cultural' boundaries between protected areas.

Conclusions

I started this chapter by referring to the widespread naturalising metaphor of 'boundless nature' that was used in transboundary protected area projects and literature. I suggested it was largely taken to be unproblematic by protected area managers. However, a finer analysis of the discourse of managers suggested on the contrary that political issues frequently appeared in a domain initially considered to concern environmental matters. Thus the 'political' inevitably erupted within the 'natural'. Because of this, transboundary protected areas were useful examples for reintroducing the political into discourses of nature, unmasking taken-for-granted distinctions separating nature and society.

The first section discussed attempts to create effective compatible transboundary zonations. This proved difficult as the criteria for defining boundaries (and therefore zones) were strongly context-dependent and not unproblematically applicable in adjacent areas. There was no 'universal' agreement on the significance of boundaries, compounded by differences in legislation as well as by different attitudes to legislation. Negotiations around transboundary zonations were confrontations of spatial discourses in which administrations sought to promote their own particular interpretations and strategies. This problematic drawing of boundaries played a part in governing the shifting understanding of what was Inside and Outside as it reflected on 'society' and 'nature' being territorialized as distinct ontological domains (Whatmore 2002 : 61).

Legitimacy in tracing boundaries was contested in many sites, as different national administrations debated their respective roles, further complicating understanding on a transboundary level. Boundaries existed more or less formally within legislations or sometimes not at all. The tracing of them, however, was always linked to specific institutional structures. Obtaining legitimacy for drawing boundaries and determining what institution should be responsible for this, was far more problematic than managers initially imagined. It raised a series of context-dependent understandings and practices related to boundaries that were frequently poorly identified.

It was clear that myths of homogeneity and boundlessness were confounded by the irruption of 'culture' and difference. The response to this was far from linear and predictable. Increased knowledge of the Other did not lead directly to increased interaction. Instead, reactions of individuals varied and existing prejudice was as likely to be reinforced as it was to be transcended. Identifying what 'problems' managers faced was culturally dependent. It would be wrong to suggest, however, that all managers moved from a belief in boundless, homogenous nature to a realisation that boundaries existed. Instead, it was clear that a whole host of opposing and conflicting 'myths' or dominant stories coexisted within administrations, such as that of 'boundless nature' and 'national identity'. Language was used as an example of a symbolic and practical boundary between managers. The different responses to the linguistic difficulties illustrated the multi-faceted, non-linear responses to the identification of 'cultural' boundaries.

Notes

1 Personal translation from: 'C'est vrai que ce parc européen, c'est bien, parce que c'est vrai que la frontière, elle est stupide, réellement, une frontière entre les peuples, bon, la végétation est la même mais je ne sais pas trop, dans l'état actuel des choses, (...) je ne sais pas trop ... Bon, moi ce qui m'ennuie c'est que chacun, mais c'est la même chose que la monnaie unique, c'est que chacun persiste dans son identité et, finalement, on est tous pareils alors qu'on n'est pas tous pareils'.

2 Personal translation from: 'Il n'y a pas de zonage de la réserve de biosphère transfrontalière en ce moment'.

3 Personal translation from: 'Donc, à la fois beaucoup de choses et en même temps, peu. On voudrait que ça aille effectivement le plus rapidement en profondeur mais on s'est rendu compte (...), le transfrontalier, d'une part que le zonage transfrontalier, à l'heure actuelle, est quasiment impossible, parce que tu vois que nous, moi, j'ai déjà rendu mon dossier à l'UNESCO, on avait nous aussi un renouvellement du label MAB. A chaque fois on avait un renouvellement du label MAB, en 99 donc, mon dossier est parti, c'est pas faute d'avoir essayé avec [les coordinateurs], d'essayer de voir comment œuvrer ce zonage transfrontalier, ça a été évoqué des deux côtés. Mais en fait, n'oublions pas quand même que le transfrontalier ne masque pas une chose, c'est qu'on a deux organisations administratives radicalement différentes'.

4 Personal translation from: 'Ce qui explique qu'en décembre 98, le Comité International, là, de l'UNESCO, à crée la réserve de biosphère des Vosges du Nord/Pfälzerwald. Sans qu'il y ait vraiment d'équipe commune, sans qu'il y ait de structure de support unique, sans qu'il y ait de zonage commun. Et du point de vue de l'UNESCO, moi je trouve que là, moi j'ai été surpris, quoi. [Juliet] Le zonage commun, il apparaît sur les cartes, là, avec celle que vous avez dessiné. [Théo] C'est une juxtaposition'.

5 Personal translation from: 'Prenons le zonage : qu'est-ce qui se passe ? Quand nous on a un trait sur cette carte, quand on décrète que la forêt est zone tampon, forestier, ok, exact, ils savent que ce trait est du domaine de l'informel, que la Charte du Parc ne l'indique que dans une vision réglementaire, mais ils ne s'offusquent pas si on explique bien ce qu'il y a derrière, les clauses, et il n'y a pas besoin pour ça de construire une législation particulière. En Allemagne, c'est impossible ! En Allemagne, quand on met un trait sur une carte, il faut que ce trait derrière, il soit ... loin. Dès qu'on a commencé à discuter zonage, on s'est confronté à ce problème. Quand on disait toutes nos aires centrales sont des sites qui n'auraient pas de statut différent de protection, réserves naturelles, réserves biologiques de ..., des Arrêtés de Protection de Biotope, enfin, je passe les outils utilisés mais tu en as des réglementaires et des non-réglementaires. Pour les Allemands, c'est incompatible. N'est aire centrale qu'une zone qui ait une ... par exemple, qui répond à un critère de la loi, etc. Donc, déjà là, c'était vraiment pas sur les mêmes bases'.

6 Personal translation from: 'Mais bon, le zonage commun, à l'évidence, il n'y en aura pas. S'ils avaient appliqué notre zonage, toute la forêt serait en zone tampon, à l'intérieur il y aurait les Naturschutzgebiet qu'ils ont déjà et qui seraient tout leur réseau d'espaces protégés qui est dans le Biotopkartierung, et puis le village serait en zone de transition, et les ruisseaux aussi seraient en zone tampon. Mais ils n'ont pas fait comme ça, ils ont voulu appliquer des critères qui sont assez strictes côté Land allemand, 3% du territoire, les zones rouges ...'.

7 Personal translation from: 'Allora abbiamo cominciato a fare sto dossier, però. Anche perché quando si parla di zona centrale, vabbè sono i parchi, per noi è il parco per loro è la "zone centrale"; invece la buffer zone e la transition zone sono iù complicate da definire, soprattutto se ci deve essere proprio una perimentazione precisa'.

8 Personal translation from: 'Ils n'ont pas le même système. D'abord parce que c'est un parc naturel donc ils n'ont pas le même système, ils n'ont qu'une zone. Donc la zone périphérique n'est pas identifiée. C'est les communes du parc mais c'est pas identifié comme nous en zone périphérique avec un tracé, des limites définies par la loi. (...) La zone périphérique, on n'a pas la même autorité que sur la zone centrale, non, pas du tout. C'est-à-dire qu'on a très peu de moyens juridiques d'intervention sur la zone périphérique. (...) Disons qu'on a, je trouve que la loi était très intelligente au départ de faire un zone de protection forte et une zone qu'ils appelaient "le pré-parc", mais il n'y a pas eu les moyens de réflexion au niveau d'un aménagement spécifique de cette zone'.

9 Personal translation from: 'Donc nous, le plan du parc, il fallait qu'il sorte, quoi. Et donc qu'on décide tout seuls. On aurait pu attendre les allemands un peu, mais là comme ils ont tergiversé deux ans pour leur zonage, on a laissé tomber assez rapidement l'ambition de caler notre zonage sur un zonage commun'.

10 Personal translation from: 'Le projet (de réserve de biosphère) est actuellement plus suivi par le parc italien, on s'est un petit peu partagé les rôles donc eux avançaient plus sur la réserve de biosphère et nous sur le patrimoine mondial. On espère déposer un projet aussi d'ici la fin de l'année. (...) Alors on n'a pas travaillé le zonage, on a travaillé sur le fait que ça soit les deux zones du parc qui seraient réserve de biosphère mais maintenant, à l'intérieur, il faut qu'on établisse le zonage. (...) Je pense que l'Italie devrait faire une proposition. (...) Elle a déjà proposé pour sa partie et, en fonction de ce qu'ils ont obtenu, les degrés, on adaptera sur la notre. Disons que ça fonctionne suffisamment bien pour que eux nous disent une chose et que nous on l'adapte, et vice versa. Cela ne pose pas de problèmes'.

11 Personal translation from: 'Qu'est-ce que vous voulez qu'on dise? Nous, notre zonage il est approuvé, le 2 octobre il est calé. Nous, en janvier 2000 on a arrêté notre zonage. Point. Et eux ils viennent le 22 août, avec un projet. On ne sait pas qui l'a défini, avec quelle concertation. Moi je ne suis pas allé à la réunion du 22 août, d'une part parce que je ne pouvais pas, car j'avais d'autres choses à foutre, mais d'autre part aussi pour marquer que nous français, on n'avais plus rien à faire avec leur zonage. (...) Moi je me dis que si dans 10 ans on a un zonage commun, lors de la prochaine révision de la réserve de biosphère, et bien ça sera très très bien. C'est l'objectif qu'il faut se fixer'.

12 Personal translation from: '[Juliet] Donc en fait, ça vous aurait plus servi que l'UNESCO dise non? [Théo] Oui, tout à fait. L'UNESCO aurait du nous dire "on vous encourage, on vous appuie dans votre projet de coopération transfrontalière, on est derrière vous, on trouve ça hyper intéressant, c'est super, mais pour qu'il y ait une réserve de biosphère, il faut un projet, une sorte de Charte, un guide d'aide à la gestion, on appelle ça comme on veut, les Allemands disent Raumplankonzept, et ils faut que vous nous montriez qu'il y a un management commun". (...) Zonage commun: il n'y a pas. Et Raumplankonzept, encore moins. Donc moi, là j'étais un peu ...'.

13 Personal translation from: 'Après tout c'est aussi bien de dire qu'il existe une réserve de biosphère transfrontalière, première en Europe, il y a un côté laurier qui est pas mal, mais concrètement sur le terrain, cette réserve de biosphère n'a aucun cadre. N'a rien de transfrontalier: il faut le dire'.

14 Personal translation from: 'Tout ça fait qu'en fait, nous, on peut à la limite venir et discuter avec nos partenaires allemands en disant qu'on est réellement coordinateurs et que, quand on discute, on sait que derrière, on a déjà, au préalable, discuté avec les gens, et que c'est nous qui menons notre zonage, entre guillemets, de réserve de biosphère. De l'autre côté, ce n'est pas le Naturpark mais le Ministère de l'environnement et des forêts de Mayence, donc le Land qui l'a pris en main et qui a dit "c"est nous qui nous en occupons', et qui va, tu l'as vu, et qui fait ces périmètres rouges, verts et autres. Le Naturpark ne dit rien, il regarde et il verra ce qu'on lui impose. De toute manière, c'est un rapport de force entre les environnementalistes du Ministère et les forestiers, puisque tout est en forêt pratiquement. (. . .) Nous on était là, le parc, à travailler directement sur notre zonage tandis que là-bas, c'est le Ministère de l'environnement et des forêt'.

15 Personal translation from: 'Pour le groupe de travail sur la biodiversité, ils voulaient avoir un statut clair. Tu sais, les Allemands aiment beaucoup avoir un statut clair, et être légitimé officiellement, très officiellement dans une participation à un simple groupe de travail sinon ils ne fonctionnent pas. Nous on bosse de manière très informelle dans un groupe de travail, qui existe, qui n'existe plus, on se panique pas, mais eux c'est impossible et impensable'.

16 Personal translation from: 'Peut-être que c'est nous qui sommes faux, mais bon on est comme ça'.

17 Personal translation from: 'Moi je me pose des questions: j'ai vécu dans le mythe de l'efficacité allemande. Alors je ne sais pas si c'est un manque de volonté . . . '

18 Literal translation from: 'I Francesi sono tutti stronzi!'.

19 Personal translation from: 'Les entreprises privées, là où il y a une question vitale de fric, quoi, et bien elles se forment beaucoup plus à l'interculturel, pour savoir comment on aborde un partenaire ou un client japonais, ou allemand, pour pas faire d'impers, car il paraît par exemple qu'on ne sert jamais la main à un Japonais. Mais ça, si vous ne le savez pas, vous êtes planté, quoi. Et là, c'est radical, donc pour les entreprises, c'est tout de suite des marchés qui se ferment, des chiffres d'affaires "pouf"!'

20 Personal translation from: 'Je crois qu'il faut accepter de ne pas être trop ambitieux, que c'est pas demain la veille qu'on aura une équipe commune, un lieu commun. Alors ça, pfiiiouu, ce sont des belles idées que j'avais eu au début. C'est impensable, maintenant. Mais il faut partir sur des idées concrètes, pratique, qui impliquent des partenaires. Du terrain. Alors on construira, mais alors à très très long terme . . . '.

21 Personal translation from: '(Comment est-ce que je définis) différences du culture ? Les besoins et les envies et les vitesses de travail ne sont pas du tout les mêmes entre la France et l'Italie. Et, en plus, les crédits ne sont pas du tout les mêmes non plus et l'utilisation des crédits, encore moins. Quand il y en a, les gens n'ont pas forcément les mêmes envies de les appliquer aux mêmes endroits et aux mêmes heures et aux mêmes finitions'.

22 Personal translation from: 'Donc on essaie de mener ça de manière un peu sérieuse, quoi. Un peu structuré, pour poser les vraies questions, quoi. Et du coup, c'est marrant, parce que du coup on a eu ces formations, on a eu ces quelques formations à l'interculturel, et le formateur nous a dit "bon, il y a les allemands et les français", et il nous a notamment expliqué que les français avaient une forme de communication implicite, sous-entendue, des choses qu'on ne dit pas, et puis gestuelle qui fait que l'interlocuteur il est censé comprendre sans qu'on l'exprime explicitement. Au contraire des allemands, où c'est le contraire. Explicite. Et puis ils vous le disent. C'est à dire que si quelque chose ne va pas, "vlaaaa". Et souvent,

au début, on avait des réunions avec les allemands, et ils discutaient entre eux, et ils s'engueulaient. Des trucs. Nous on étaient catastrophés. On s'est dit qu'on ne pourraient plus jamais travailler ensemble. Mais ils se disaient les choses crûment. "Je pense que là vous avez merdés", "là, on s'était engagés, vos collaborateurs sont nuls", "C'est écrit, qu'est-ce que vous avez faits? Vous n'avez rien fait". Mais vraiment, c'était ce qui ne se dit jamais en France. En France, d'ailleurs, le compte rendu, dès qu'il est rédigé, il est oublié'.

23 Personal translation from: 'Les gardes disent "on veut pas, on est français et européen, mais français avant tout", c'est logique, on n'a pas encore cette ... Bon, c'est sur qu'il faut commencer maintenant pour avoir les parcs européens dans 10 ans, peut-être, mais il ne faut pas, il faut y aller petit à petit, c'est sûr, mais je pense qu'il ne faut pas aller ...'.

24 Personal translation from: 'Et donc il y a des différences qui sont tout à fait culturelles. Et c'est aussi de dire que moi j'en ai marre de baragouiner en allemand, quoi. Moi je parle français, quoi. Et si vous voulez coopérer avec moi, vous apprenez à me comprendre. Donc établir un vrai rapport de coopération transfrontalière, qui serait le même que la coopération internationale. On est bien obligé de passer par un certains nombres de phases de connaissance mutuelle, et ça c'est difficile'.

25 Personal translation from: 'Je ne travaille plus qu'avec des gens qui soit parlent le français, soit l'anglais. Alors c'est clair que ça sélectionne. Alors quelqu'un qui ne parle que l'allemand, moi je ne vais pas trop regarder. Mais on trouve toujours des gens ... Quand on veux communiquer, on trouve les moyens'.

26 Personal translation from: 'Les réunions se passent dans les deux langues, ils parlent en italien, on répond en français et voilà. (...) Si ce n'est [la directrice] qui, elle, peut parler les deux langues sans aucun problème'.

Chapter 10

Drawing Lines in Hybrid Spaces

The dividing line between nations may well be invisible; but it is no less real. How does one cross that line to travel in the nation of animals? Having travelled in their nation, where lies your allegiance? What do you become?

(Montgomery 1991 in Whatmore 2002 : 15).

Naturalising the debate on Otherness

In the previous chapter, I suggested that transboundary protected areas were useful examples through which to reintroduce the 'political' into discourses of 'nature', unmasking taken-for-granted boundaries separating nature and society. This suggestion allows further exploration of what appears to be a fundamental change in distinguishing two ontologies.

Previous chapters have discussed boundaries, the construction of space, negotiated identities and cooperation. I have argued that the establishment of a transboundary entity on the basis of several existing protected areas implied processes of reterritorialisation, that is to say a common reinvention and redefinition of both social and spatial practices. This was seen to involve forms of negotiated identities between an increasingly complex set of local, national and international actors. I dwelt on the concept of boundaries as socio-spatial phenomena, the redefinition of which had an impact on collective territorial identities. I discussed the role of boundaries in the construction of space (Chapter 2), focussing on the notion that identities were constructed by symbolic and material boundaries that defined Self and Other.

As the discussion progressed, I explicitly (re)introduced 'nature' into the equation by discussing the changing faces of protected areas. In the chapter on protected areas and boundaries (Chapter 3), the way such areas were planned suggested intermingled notions of boundaries, 'nature', 'culture' and Otherness. Boundaries were seen not only as political constructs but also as conceptual barriers defining ontologies. The 'biophysical' and the 'societal' were seen to collide in mutually incomprehensible understandings of boundaries. In analysing certain elements of protected area managers' discourses that sought to naturalise spatial entities, I suggested that 'politics' (re)appeared in a domain initially considered to be unproblematically 'natural'. I described attempts to move beyond this sterile clash, suggesting that the key lay in considering nature as intrinsically social. Thus in order to understand the world, the sharp distinction between nature and culture had to be replaced by the idea that nature 'was nothing if not social'.

The argument progressed by focussing on the intrinsically contested nature of boundaries (Chapter 5) that defined an Inside and an Outside by reifying power in space. From the idea that all political and territorial identities were, in a sense, fictional and connected with imagined communities I suggested with Paasi (Paasi 1996 : 15), that the construction of space was based on a dialectic between two languages: the language of difference and the language of integration. This grounded the discussion on the construction of transboundary spaces (Chapter 6). Building on this, I suggested that the notion that 'cooperation' both stemmed from such constructions and actively participated in constructing them needed to be examined critically and redefined (Chapter 7). The next chapter went further in deconstructing prevalent myths of boundless nature and bounded culture by suggesting that taken-for-granted boundaries between 'nature' and 'society' were no longer pertinent (Chapter 8) and that reactions to the irruption of difference were unpredictable.

Transboundary protected areas were therefore useful examples for reintroducing the political into discourses of nature. This problematic drawing of boundaries, both symbolic and concrete, played a part in governing the shifting understanding of what was Inside and Outside as it reflected on 'society' and 'nature' being territorialized as distinct ontological domains (Whatmore 2002 : 61). Rather than negating the physical materiality of the world, this constructivist position supposed that nature could only be known through culturally-specific systems of meaning and signification. This implied that human representations of nature were not simply 'mirrors of nature', but instead were 'cultural products freighted with numerous biases, assumptions and prejudices' (Castree & MacMillan 2001 : 209). This has important consequences in transboundary situations where 'nature' and 'identity' are intertwined.

Part of the issue is identifying how these 'representations of nature' come about. When writing about a conflict over the protection of a forest in British Columbia, Braun and Wainwright suggest that the 'object' of conflict could not be taken for granted. They note that 'struggles over nature, land, and meaning are simultaneously struggles over identity and rights' (Braun & Wainwright 2001 : 59), adding that 'in the crucible of environmental politics all manner of identities and relations were remade' (Braun & Wainwright 2001 : 59). This particular conflict served to construct the varied positions of the individuals and coalitions involved but also, they argue, the forest itself. The idea is that since all claims about nature are discursively mediated, there is no pre-existing unproblematic 'forest' at the outset. Instead, such a 'forest' is discursively constructed. This is different from saying that there is no material reality, no trees, grass and soil; rather this implies that they are constructed as objects of knowledge (for further discussion of the distinction see Demeritt 2002). 'Knowledge and language are the tools we use to make sense of a natural world that is both different from us and yet which we are part of. There is, therefore, no objective, nondiscursive way of comprehending nature "in the raw". We have to live with the fact that different individuals and groups use different discourses to make sense of the same nature/s' (Castree 2001 : 12).

I am not suggesting here that since discourses vary, all may be equally valid. These constructions of nature are 'non-innocent': they carry in them certain (disavowed) political commitments that need identifying. Instead of relativism, I suggest a more active deconstruction of these discourses: ' "deconstructing" these knowledges therefore entails "denaturalizing" them: that is, showing them to be social products arising in particular contexts and serving specific social or ecological ends that ought to be questioned' (Castree 2001 : 13). The key here is to change the very terms in which interactions with and struggles over nature are understood. 'To argue that environmental disputes merely reflect competing "interests" assumes that what is perceived as natural is self-evident, and exists *external* to the domain of power and politics that geographers and political ecologists set out to study' (Braun & Wainwright 2001 : 42) (emphasis in original). Instead, 'environmental politics' are always entangled with a cultural politics of knowing, because 'the very thing that is taken to be the object of environmental studies and politics – namely "nature" – is an effect of power' (Braun & Wainwright 2001 : 41).

The construction of transboundary protected areas and cooperation within these entities are enabled by a set of discursive practices through which what counts as 'nature' is made intelligible. The 'social nature' approach contains real potential for moving beyond the 'cultural' sticking points that were seen as obstacles to the creation of transboundary protected areas. In the previous chapter (Chapter 9), I discussed the collapse of the pervasive myth of 'boundless nature' that underpinned much idealised theory about transboundary management. There is a countervailing yet coexisting myth: that 'cultural' differences are somehow expressed 'within nature'. Nature and wilderness are intrinsically discursive constructions: 'the notion of wilderness being fleshed out here is a relational achievement spun between people and animals, plants and soils, documents and devices in heterogeneous social networks which are performed in and through multiple places and fluid ecologies' (Whatmore 2002 : 14). This performative conception of nature not only has relevance to examining the boundary between the wild and civilized, human from animal but also has wider implications for the study of spatialised difference.[1]

Nature, identity and the Other

Social natures

Throughout the fieldwork, little comments made by protected area managers mentioned or hinted at how difficult it was to cooperate with managers and institutions in the neighbouring country due to different 'management philosophies'. Underpinning many of these comments was the notion that some management strategies or techniques were 'more natural' than others. Since the neighbouring country used 'less natural' techniques, the argument went, it made cooperation difficult or impossible. The idea of 'managing nature' is fundamentally about human interference in biophysical processes. How, then, did some forms of this come to be viewed as more or less natural?

What effect did this construction of seemingly irreconcilable, 'naturalised' differences between protected area administrations have on transboundary cooperation? How did these discourses construct the boundary between Self and Other in connection with that between 'nature' and 'culture'?

Writings on nature and the Other have followed different trends, and are only starting to really emerge. Whatmore, for example, argues that nature itself has been constructed as the archetypal Other (Whatmore 2002). This leads her to call for a new approach, a new 'hybrid geography' where such distinctions are revisited. Drawing on Said's 'Orientalism' approach, other authors such as Gregory have instead linked representations of other 'natures' to representations of the Other as fundamentally different (Gregory 2001 : 85). This has led, for example, to analyses of travel literature in which largely European writers gaze on foreign landscapes, projecting and constructing an other nature (as barbaric, sensual, lush, primitive etc.), in contrast to a standard (usually temperate) 'reference-point' nature. Here, I follow in Gregory's steps in suggesting that the (human) Other is constructed simultaneously to the (non-human) Other.

Several sections within other chapters have discussed issues of zonation within protected areas. I return to the issue of transboundary zonation in order to examine it as a double negotiation: the definition of boundaries between Self and Other and between 'nature' and 'culture'. Chapter 3 identified changes in the way boundaries were comprehended within the protected area movement as a whole, leading to a discussion of the definitions of who and where nature as the Other was. In contrast, this chapter provides a more practical discussion of what protected area managers have said on the matter, discussing how this ties in with the difficulties of engaging with an Other within transboundary entities.

Concentric zonation was designed to define boundaries staking out areas of varying degrees of 'naturalness'. This assumed a pre-existing definition of what was considered more or less 'wild' or 'natural'. In transboundary situations, substantially different views and understandings appeared. This confrontation revealed often profound differences, even within the supposedly value-free scientific community. Science was therefore revealed to be far from neutral but rather founded on a series of political choices and definitions. As Castree has suggested, an understanding of this supposes a new way of apprehending physical reality, not negating its existence, but rather recognising the contingency of social practices. It requires 'an insistence that the physical opportunities and constraints nature presents societies with can only be defined *relative to* specific sets of economic, cultural, and technical relations and capacities. In other words, the *same* "chunk" of nature – say the Amazon rainforest – will have *different* physical attributes and implications for societies, depending on how those societies use it. In this sense, the physical characteristics of nature are *contingent* upon social practices: they are not fixed' (Castree 2001 : 13) (emphasis in original).

In the Mercantour/Alpi Marittime, where transboundary zonation had not yet been attempted, the future process was viewed as unproblematic by one

French manager. Different zones were simply taken to reflect different levels of protection and levels of 'sustainable' management:

> *I think that [the biosphere reserve] is a way of making a sustainable model of planning, this assumes first of all a profound questioning about the different zones and what can be attributed to them as challenges and as objectives and I think that it is something that can be made operational. And, additionally, if we make a biosphere reserve together, this will allow for a much stronger and more interesting core inside. It is something that is in line with the objectives of national parks, that is to say to have a zone really very strongly protected and on the other hand thinking about the planning of the peripheral zones*[2]
>
> *(Chloé, Parc National du Mercantour, France).*

The wildest part here was described as 'strong' (fort) and 'interesting' (intéressant), reflecting the idea that values are contained in nature itself (strong) or are attributed by those beholding it (interesting). In French national parks, clear guidelines lay out that the protection of wilderness is the main objective. There are, for example, no human settlements in the strictly-protected core zone. This manager therefore assumed that what was 'wild' would be defined in the same way in both countries, as it would follow a 'scientific' definition. In the high alpine ecosystems of the Maritime Alps, this was quite likely to be the case. However, it was not the case in all transboundary situations. In the Vosges du Nord/Pfälzerwald, for instance, managers attempted to establish a strictly protected forest reserve spanning both sides of the international boundary as a form of shared core area. In this case, one manager attached to a forest research station working on the project said that it was problematic to agree on a common definition of what the 'climactic' vegetation was:

> *It was not so easy to agree on what the 'natural vegetation type' is*[3]
>
> *(Kathrin, Pfälzerwald, Germany).*

This was interesting, as it indicated that scientists within different countries had distinct definitions of what constituted the 'wildest' form of nature in one area. This was compounded by further difficulties in deciding what constituted 'nature' in both countries. This therefore concerned defining the boundary between 'nature' and 'culture'. In most parts of Europe, centuries of human management have determined the resulting patterns of biodiversity. As a result, actively managed landscapes are often correspondingly 'richer' in fauna and flora or at least are strongly valued as natural/cultural constructs. When protected area managers therefore defined the most valuable parts of the landscape that deserved to be defined as core zones, such culturally-specific definitions came into play. In transboundary situations, these distinct representations of nature conflicted. In the Vosges du Nord, following the redrafting of the Parc Naturel Régional's Charter, a new zonation was introduced with distinct changes in how core areas were defined:

> *The novelty is that we took the obviously natural zones, and we put all the historical monuments, in particular the castles that are in the middle of the forest. We considered that around them, in any case, the zone where they are and the castle itself were carriers of nature ('porteurs de nature'), we are in fact in Europe which is strongly modified by man [sic], where nature and culture are strongly interlocked. If you don't take into account nature that has been modified or has been planned in some way, there is not much nature left that you can take into account, unlike in zones where there is more space like the Siberian North or some sectors of the United States or even Australia. So, we added the historical monuments*[4]
>
> *(Hugo, Parc Naturel Régional des Vosges du Nord, France).*

These decisions were profoundly puzzling to the German managers. Their own conception of nature was substantially different. In any case, the main principles for zonation were laid down on a national level by the German MAB committee. These included the requirement for a strict proportion of 'wild', untouched and strongly 'protected' nature:

> *There are different understandings, for example in Germany, referring to what our national MAB Committee set up as guidelines, a core area could never be an area where you have for example a medieval rock castle with one hundred thousand visitors each year, whilst in France, they also want to protect also all cultural heritage in core areas, so in their core areas you may find such castles – but this is impossible in Germany*
>
> *(Daniel, Naturpark Pfälzerwald, Germany).*

These differences in conceptions of nature led to practical problems when defining common zonations. It also led to a form of 'naturalized' rejection of the Other. What is the point of working together, the argument went implicitly, if they don't even know what nature is? The issue here further challenged the myth of boundless nature discussed in the previous chapter. In many ways, the problem was that there were two different natures, two different forests being constructed simultaneously. In order to transcend these fundamental different ways of defining the boundary between 'nature' and 'culture', a change of awareness would have to take place. Castree noted: 'we have to live with the fact that different individuals and groups use different discourses to make sense of the same nature/s. These discourses do not reveal or hide the truths of nature but, rather, *create their own truths*. Whose discourse is being accepted as being truthful is a question of social struggle and power politics. Furthermore, many nature discourses become so deeply entrenched in both lay and expert ways of thinking that they themselves appear natural' (Castree 2001 : 12) (emphasis in original). The confrontation of two different representations of nature was indeed a struggle, further participating in the negotiated construction of respective identities. However, the process is heavily impeded if both feel that what needs negotiating is in fact non-negotiable: if expert ways of thinking appear natural and differences are naturalized, then defining common criteria is impossible.

One of the French managers was aware of this, and added a further twist to the tale. Earlier (Chapter 7), I suggested that it would be a mistake to consider protected area administrations to behave as one subject when engaging in cooperation, uniformly adopting a level of enthusiasm. Instead, I suggested that individual responses varied tremendously among managers. Likewise, when discussing representations of nature, differences among managers appeared. While it was perhaps pertinent to speak of 'national' differences or patterns in the representation of nature, it would be wrong to suggest that these were shared by all within one administration:

> *But it is not in the French spirit, or the Latin spirit, to have zones left to nature because man [sic] wants to intervene too much. He believes that if he leaves nature wild, it is a loss of power and influence, you know these problems for sure, while we are more divided. In the team, some people are 'Kultur' and there are some people who believe that we should do things like the Germans, set the standard higher, and put in the core zones that are really ... who would get rid of the castles, even if ... (...) I think they [the Germans] would be wrong to change: I am able to recognise when the Germans do things well. No, it's true. I believe their zonation is well thought out*[5]
>
> *(Hugo, Parc Naturel Régional des Vosges du Nord, France).*

These differences surfaced when both administrations attempted to create a shared forest reserve. This required choosing a common location and defining its boundaries together. One manager indicated that this process of negotiation was far from straightforward as it implied a spatialized confrontation of these divergent spatial discourses:

> *We wanted a different location for the transboundary reserve. In France this place is very touristy. That was a really massive problem for us. On the French side of the reserve there is a ruin which attracts some visitors. The French are more concerned with active management than we are. We wanted a strict reserve with no visitors. This still has to be resolved*[6]
>
> *(Kathrin, Pfälzerwald, Germany).*

The different representations of nature were in stark contrast, implying different choices in setting the boundary between natural and cultural ontologies. If the confrontation was starkest in the core areas, it trickled down into the other zones. The French and German managers also had very different conceptions of what should feature in the buffer zones:

> *We added the traditional orchards which are clearly nature-culture zones also, in sharp decline, and for which we have an action programme. We thought it was logical then to have them in the buffer zones, I would say the zones that are a bit natural and still have strong heritage value but which are in any case already being managed by man [sic], on which alternative practices must be experimented. The transition zone on the other hand has become really small: it's all the communities ('communes'), all the zones of activities, the towns, and finally the urban zones. (...) That's it for the zonation*[7]
>
> *(Hugo, Parc Naturel Régional des Vosges du Nord, France).*

> *They (the French park) also said, 'we take all our forest and put it into the buffer zone', in our case it is different too. Because we also want to have open land. So there are some very different views*
>
> *(Daniel, Naturpark Pfälzerwald, Germany).*

What was interesting in this case was that two of the French managers used the German landscape terms to express these differences, referring to the 'Kulturlandschaft' (cultural landscape), versus the 'Naturlandschaft' (natural landscape) positions. This reflected a clear understanding of the differences between the two approaches, as the French managers used 'the neighbour's' term to describe their own situations. Differences were thus identified and negotiated, to the extent that appropriate terms were chosen within the Other's language:

> *And the 'park' tool [natural regional park] started from there, really. (. . .) It was really 'Kulturlandschaft', as the Germans say, but the concept was there, the idea was there*[8]
>
> *(Théo, Parc Naturel Régional des Vosges du Nord, France).*

However, despite the ability to identify different representations of nature, these were still seen to indicate wider, more fundamental differences in work patterns and habits:

> *And the buffer zones are zones . . . but there is still a logic after all, listen, the Germans adopted a certain logic and logic is the backbone of the Palatinate (Pfälzerwald)*[9]
>
> *(Hugo, Parc Naturel Régional des Vosges du Nord, France).*

This transition is interesting: the manager slipped from discussing differences in conceptions to nature to wider cultural differences. Relations to nature, therefore, were one further element in essentialising cultural differences. This link between nature and the Other was explored by Gregory, in his writing on postcolonial nature. He related how degrees of normalness of nature were created, exploring this link between nature and identity in the corresponding creation of imaginative geographies. In travel writing, this was for example apparent in descriptions: 'in rendering temperate nature as "normal" nature, colonial discourse simultaneously constructed nontemperate nature as radically other and thereby established an essential distance between "normal" nature and its excesses' (Gregory 2001 : 98). This meant that the construction of almost 'pathological' natures created a further boundary defining Self from Other. Degrees of normalness and naturalness were compounded by differences in the role assigned to humans.

If nature was more or less natural, this implied varying roles for human action. This action was seen to need regulating. The need to create a common zonation was therefore directly associated with the need to protect the 'same (boundless) nature' on both sides of the boundary, meaning that human action would have to be regulated in the same way. However, because differences were recognised, some form of negotiation was seen to be necessary:

> *Well, my personal point of view is that is that if you want to establish a system with stepping stones, core zones of protected areas on both sides of the border, then we need a complementary zonation system. We have the situation where on one side of the border ecosystems will be destroyed while on the other side they will be protected,*

> *because of different legislations and so on. That is one aspect. (...) There is still quite a difference in forest management philosophies between France and Germany*
> *(Daniel, Naturpark Pfälzerwald, Germany).*

The apparent destruction of ecosystems was here directly associated with human action. On the French side, however, human action was used specifically to maintain certain (cultural) ecosystems and landscapes. Although active management of the environment was intrinsically anthropic, it was constructed as more or less close to nature:

> *I think the German forest practices are nearer to the nature development, we have a concept that is called Naturnahwaldwirtschaft, and in France they have a – in our point of view – an older concept with clear cuts, I think, and something like that, and that is history in our Federal State, since more than ten years now*
> *(Lukas, Naturpark Pfälzerwald, Germany).*

This manager noted that not only were the Others less close to nature, they were also 'backward'. Thus here being close to nature was constructed as progressive and rational. In a twist of the usual myths, 'reason' rested no longer on the side of culture and civilisation, but rather on that of nature. Being close to nature meant being modern. Ontological boundaries were redrawn, identities constructed and negotiated.

The examples in this section have highlighted further difficulties in creating transboundary zonations. I have argued that definitions or representations of nature were contingent on social practices and culturally-specific definitions. There was a wide variety of conflicting definitions of what constituted 'nature', both within and between protected area administrations. The boundaries between the distinct ontological domains of 'nature' and 'culture' were defined and negotiated differently in different contexts. Thus individual managers constructed different 'natures', creating their own truths. This is akin to constructing physical objects of knowledge. The definition of ontological and physical boundaries constructs objects of knowledge implying that these are not pre-existing. Instead, a spatially-defined protected area is 'not something that existed *independently* from the maps, tables, techniques, and practices that made it available to forms of economic and political calculation' (Braun & Wainwright 2001 : 52) (emphasis in original). Rather, protected areas are constructed discursively by these different elements, constituted by the different relations and links within heterogeneous social networks that include both human and non-human actors.

Specific icons of difference

In order to continue to explore these links between nature and the Other, this section deals with the practical identification by managers of environmental problems. The protected area literature reviewed earlier (Chapter 3) suggested that 'shared problems lead to shared management'. However, if it is accepted that objects of knowledge are constructed, then the definition of what constitutes a problem is also contingent on social practices (Hajer 1995 : 6). During the fieldwork, in addition to the iconic figures of 'boundless nature' such as the Ibex wandering through the mountains, specific icons incarnating 'difference' were also referred to. Bark beetles, small insects that attacked coniferous trees, were referred to repeatedly in the Tatras:

> *There is a problem with the management of the bark beetle. In Slovakia, they cut the trees when they are infected, but in Poland we have decided that in the strict nature reserves we will not do this because we want to observe the process. There is a strong conflict about this. The Slovaks say we are breeding the beetles*
> *(Jurek, Tatra National Park, Poland).*

While this was recognised to be a practical problem, others did mention that it created opportunities for learning from different management techniques. Scientists, in particular, were interested in observing the effects of the different tree-cutting or bark-stripping regimes:

> *But we have cooperation, for example we exchange information about the bark beetle – Ips typographicus – because our management methods are different. The Slovaks cut infected trees, but we do nothing and we monitor what happens*
> *(Tadeusz, Tatra National Park, Poland).*

These different choices were linked to what was defined as 'natural'. In Slovakia, where the forests were managed by the Ministry of Forests, bark beetles were defined as pests. They were considered, if not actually unnatural, then certainly 'out of place'. The sheer number of them meant they were no longer defined as natural. The accusation that the Poles were breeding the beetles was interesting: it assumed that they had symbolically domesticated them, bringing them into the realm of the cultural, removing them from wild nature. These cultural differences were not necessarily nation-specific but rather overlapped or were compounded by divergent professional practices. In many countries, there was a more or less overt conflict between nature conservationists and foresters, as both shared radically different views of what constituted 'nature' and therefore what was appropriate management. However, while the differences between individuals may well have been substantial, these were not referred to as emotionally as those occurring with neighbouring countries.

> *Forestry is much more intensely managed in Slovakia. Also, they shoot deer and feed them in the strictly protected areas. In Poland we do not do this. Also one of the bears disappeared when it went over into Slovakia ...*
> *(Krzysztof, Tatra National Park, Poland).*

These tacit accusations of malpractice reflected obviously different definitions of nature. Feeding animals, hunting deer and bears were seen to be perverse cultural incursions into the domain of nature. Poles were accused of breeding beetles and bringing wild pests into the realm of culture; Slovaks were accused of denaturalising wild deer. In both cases, the different tales concerned different definitions of the boundary between 'nature' and 'culture'. The story of the bear was particularly interesting in that it cropped up several times in different places and in different guises. In the East Carpathians, the same story was told of a bison wandering into Ukraine. The animal was implicitly considered to belong to 'us', to the side of the person telling the tale, specifically placed in relation to the Other. In a delightfully moral twist, the implied transgression of the animal that crossed the boundary to the Other Side ultimately led to its death.

More than anything else, this tale of transgression served to illustrate the naturalized differences between the two sides. In suggesting that 'nature' was 'less natural' on the other side, it served to further define Self and Other. In the Alpi Marittime/Mercantour, a similar story was told of the lammergeyers – a type of vulture – being progressively reintroduced in the mountains. One bird was released each year, in alternate countries. French managers repeatedly noted that the 'French' birds, bearing French names, inevitably went to live in Italy. Although this was always told tongue-in-cheek, the recurrence of the tale indicated its symbolic strength. The different versions hinted either at the 'nature-knows-no-boundaries' myth (the birds ignore political designations) or else, more tellingly, as 'boundaries-reflect-different-natural-conditions' (boundaries are 'natural'). For the French managers, this implied that the neighbours had 'stolen' the French birds; for the Italians this meant that the birds preferred to live in Italy because 'nature was more natural' there.

Birds, bears and beetles were not the only animals used as icons of difference. In the Mercantour/Alpi Marittime, the wolf was also returning, migrating into France from packs on the Italian side. Again, this iconic animal served to highlight differences in the way 'nature' was constructed. As mentioned previously, the animals were symbolically attributed a nationality:

> *It's difficult when there are prickly subjects such as the wolf that we have in the parks, because this isn't an easy subject and so is really quite polemical. (. . .) And in addition the wolves came from Italy, so it's not easy*[10]
>
> *(Chloé, Parc National du Mercantour, France).*

Thus not only were the wolves a problem in themselves, but they were additionally problematic because they were associated with 'the Other side'. This comment referred to the difficulty in convincing French shepherds that the wolves were 'natural' and were not part of a 'foreign' invasion. The idea that they had been reintroduced, like the lammergeyers, was pervasive among shepherds and politicians, with one local mayor referring to the threat the wolf posed to 'indigenous fauna' ('faune indigène').[11] The usual accusation was that 'ecologists' had covertly released the wolves. This tale served to 'denaturalise' them, making them legitimate targets for destruction in an area were hunting was prohibited. There was a clear recognition by this manager that while actual conditions differed, representations of nature were culturally-contingent and based on collective myths:

[Juliet] In this kind of thing, are the local people's attitudes similar on both sides of the boundary? [Chloé] Regarding the wolf, no, they aren't at all. On the Italian side they don't have the same problem at all in as far as they have many less flocks [of sheep] and also they have in their culture a cult of the wolf, Remus and Romulus, so it's the female wolf who raised Remus and Romulus, it's a different culture, we have Little Red Riding Hood, it's not exactly the same! Well, there is also the fact that shepherds have always had wolves so they kept small flocks and they guarded them, while here the economic aspect and the ease in keeping flocks prevailed, they had larger flocks, so the conditions have changed, of course[12]

(Chloé, Parc National du Mercantour, France).

These different definitions of 'nature' were a problem when transboundary zonation was attempted. They additionally invariably led to differences in legislation within each zone. In certain cases, these differences were specifically put forward on common signposts. In other areas, individual signposts coexisted on each side of the boundary, as in the 'transboundary' car parks in the Vosges du Nord/Pfälzerwald. In the Mercantour/Alpi Marittime, attempts to limit signposts at the top of mountain passes led to the creation of common ones, incarnating differences within one representation. For practical reasons, however, at the time of the fieldwork these had not yet been installed:

On the signposts there is a presentation, a common map of the two areas, a presentation of everything we have done together, a presentation of all the information points in the two parks and so signs in French, in Italian, but it's exactly the same text and we explain the legislation on both sides because that also, on a legislative level, we didn't have exactly the same rules on one side or another so there is at least an explanation of these differences in order for the person at the mountain pass to know the differences a bit better[13]

(Chloé, Parc National du Mercantour, France).

This short section has discussed how the identification of problems is linked to definitions of nature. The use of iconic figures such as beetles, birds, bears and wolves served to construct a variety of tales further defining boundaries between nature/culture and Self/Other. Certain animals were repeatedly identified as being boundless, while others were specifically assigned a symbolic nationality and place. This locating of animals was linked to contingent 'natures' defined by the boundary between what was natural and what was cultural. Animals such as bark beetles and wolves were variously constructed as wild (and therefore natural and worthy of protection) or domesticated (and therefore cultural and requiring extermination).[14] Animals such as the bark beetles and wolves were therefore 'denaturalised' by a variety of actors in order to be managed.

Initial conclusions

The examples in this section have highlighted further difficulties in creating transboundary protected areas. Because what constituted 'nature' was contingent on social practices and culturally-specific definitions, the boundaries between the distinct ontological domains of 'nature' and 'culture' were defined and negotiated differently in different contexts. This meant that individual managers constructed substantially different 'natures'. Furthermore, this meant that transboundary protected areas considered as objects did not exist *independently* from the maps, tables, techniques, and practices that constituted them. Instead, they were constructed discursively by these different elements, by the different relations and links within heterogeneous social networks that included both human and non-human actors. These elements surprisingly but explicitly included non-humans, particularly in the form of more-or-less charismatic animals. The locating of animals was linked to contingent 'natures' defined by boundaries between what was considered 'natural' and what was considered 'cultural'. Some were repeatedly identified as being boundless, while others were specifically assigned a symbolic nationality and place. Because of the fluidity of the changing definitions of boundaries, these were in effect hybrid animals, contributing to create hybrid geographies in which ontological boundaries no longer followed strict dichotomies.

At the beginning of this chapter, I mentioned the boundary between Self and Other and the boundary between society (or 'culture') and nature. Whatmore joins these elements together to illustrate the creation of the 'Commons'. In a little twist, I suggest that such a figure can be adapted to illustrate the creation of transboundary spaces (Figure 10.1).

This figure expresses the multi-faced dimensions of the creation of transboundary spaces that implies negotiation between two distinct forms of boundaries: that dividing nature/culture (society) as well as that between Self (inside)/ Other (outside). The purifications refer to choices made by the individuals involved in the process who offer binarist accounts of their actions in line with their interests and capabilities. This process of purification means that the resulting spatial object (the protected area) is dubiously remade as *either* social or natural. Yet, fundamentally, it is a hybrid. Whatmore suggests that this process 'marks the spatial inscription of the mutually informing and contested contours of modern purifications of the inside/outside boundaries of political community through the spatial practices of sovereignty and of the social/natural boundaries of living associations through the spatial practices of property' (Whatmore 2002 : 114). Thus the process is one of hybridity in which the indissoluble links between all the elements (things, people, places, animals) forge the resulting identity of the object.

What is the meaning of this hybridity? Is it something concrete or is it just another return of the chimera that has been haunting me? On one of the days I was writing this chapter, I took a short walk to the Botanical Gardens in Durham, looking for a change of scene in the early Spring sunshine. Having spent my days musing about nature/culture, Self/Other and the chimeric ambition to merge biophysical and societal ontologies, I suddenly stopped. Bemused. Excited. I had found a chimera! It was a small bush, little more than a flowering stick.

The signpost informed me that this was officially called a 'chimæra': the surprising and rare result of a graft in which the tissues of two botanical

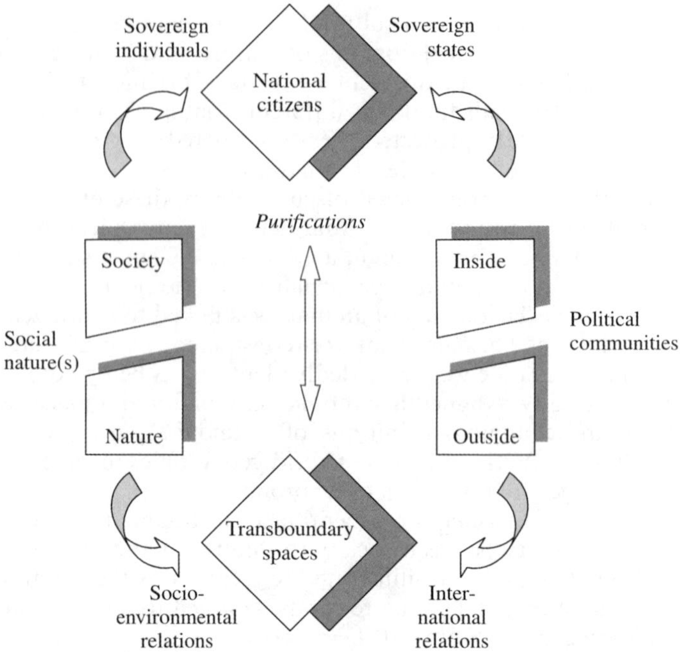

Figure 10.1 De/re-territorializing transboundary spaces
Reproduced and adapted with permission of the author (from Whatmore 2002 : 115)

specimens had merged during grafting; not one species, but two. It was alleged to produce three sorts of flowers, illustrating its multiple identities, different from a simple hybrid. Had this plant transcended the boundaries of nature and culture, additionally merging Self and Other within one entity? It was created by human intervention, yet was undoubtedly part of nature. It was both Self yet intrinsically Other at the same time. I was entranced – and devastated! If chimeras existed, then how could I pretend that they should not be chased?

Hybrid natures or chimeras?

She was the mother of Chimaera who breathed raging fire,
A creature fearful, great, swift footed and strong,
Who had three heads, one of a grim-eyed lion,
Another of a goat, and another of a snake, a fierce dragon;
In her forepart she was a lion; in her hinderpart, a dragon;
And in her middle, a goat, breathing forth a fearful blast of blazing fire.
(Hesiod)[15]

Figure 10.2 'Laburnocytisus Adami'. It originated in the garden of Monsieur Adam,[16] near Paris, in 1825
(Photo J.J. Fall)

Although the chimera in the Botanical Gardens did not actually breath fire (or at least not in my presence), she certainly existed. Like Haraway's cyborg (Haraway 1985), my lonely chimera appeared to do the impossible: to merge elements previously thought of as distinct. Mixing lions, dragons and goats was one thing; merging unbreachable ontologies was quite another. If the promises of the 'social nature' approach were to be taken seriously, then what road should be followed next? If protected areas were remade as *either* social or natural because of purifications made by the individuals who offered binarist accounts of their actions, then what did this actually imply? If nature as 'fleshed out here is a relational achievement spun between people and animals, plants and soils, documents and devices in heterogeneous social networks which are performed in an through multiple places and fluid ecologies' (Whatmore 2002 : 14), then how on earth should it be divided into distinct spatial entities? Was this, in fact, a pertinent question?

I stated earlier that attempting to merge ontologically distinct types of boundaries was barking up the wrong tree, suggesting that the solution was of a profoundly different type. Part of the appeal of the social nature approach is its ability to unmask taken-for-granted distinctions such as that separating nature and society. This means that nature, rather than being separate from the societal, is always social and political. Nature must no longer be considered 'in itself', but rather as an intrinsically discursive construction. Two different consequences stem from this. On one hand there is a need to develop appropriate spatial concepts to describe such hybridity; on the other there is a need to continue to engage with discourses that posit this distinction as fundamental. For if such distinctions are swept away in one ambitious academic swoop, then how are researchers supposed to make sense of discourses that assume a chasm between the societal and the biophysical? The solution lies in understanding the nature and purpose of the purifications that happen during the process of constructing space.

Hybrid spatial entities

Earlier, I reviewed the historically constructed nature of a series of created and recreated spatialized categories (Chapter 2). The concepts of region and bioregion, like that of territory, as well as the boundaries defining them need to be explicitly revisited in the light of this hybridisation of the biophysical and the societal. These coexisting discourses indicate the constructed nature of scientific knowledge, and the differing uses that have been made of both biophysical and societal arguments in dividing space. Such divisions of space were shown to be more than differently scaled constructions, instead revealing ontologically different standpoints. I discussed the different approaches to defining the boundaries of each entity, laying emphasis on socio-cultural meanings of space. This implied that they were not self-evident, discernible spatial units, but instead had dynamic and shifting boundaries, requiring multiple levels of identification, belonging to radically different ontologies.

Yet if space is viewed as fundamentally hybrid, is it still pertinent to attempt to divide it into distinct entities? Castree, Braun, MacMillan and Whatmore all refer in various guises to territorialisation. They do not however explicitly suggest within what spatial entity this takes place, referring to networks rather than explicitly spatialised concepts. Yet surely this approach requires a reconceptualisation of space as the site of de/re-territorialisation, while avoiding defining spatial entities using either societal or biophysical boundaries?

In the social nature perspective, equating certain privileged spaces with 'nature' is problematic. This means that no spatial entity can be wholly defined by biophysical criteria. Such a binarist approach would imply that certain delimited geographical territories existed where environmental politics could be exclusively focussed. Yet these spaces 'are clearly neither wholly natural nor merely zones where certain social actors impose their culturally specific ideas of what nature is supposed to be' (Castree & MacMillan 2001 : 221). Instead, a spatial entity is seen to result from a process of reterritorialisation that does not settle on it either side of the societal/biophysical divide. Instead, it is the result of many different associations that weave it together in a dynamic process (Castree & MacMillan 2001 : 216). This (re)territorialisation corresponds to a politics of purification which, like politics, is always geographical. In this perspective, the construction of space is therefore fundamentally hybrid, involving *both* societal and biophysical elements, human and non-human.

While it is exciting to imagine that not all actors in the process are human, I do not go as far as some authors who suggest that all actors hold equal standing in the process. Instead, I follow Castree and MacMillan in suggesting that while there is a need to remain critical of binarist thinking, these processes are not 'social' and 'natural' in equal measures. This implies 'that actors, while social, natural, and relational, vary greatly in their powers to influence others; that power, while dispersed, can be directed by some (namely, specific "social" actors) more than others; and that a politics of nature attuned to the needs and rights of both human and natural entities must ultimately be orchestrated through putatively "social" actors' (Castree & MacMillan 2001 : 222).

Hybrid boundaries

If spatial entities need revisiting, what then of the boundaries that define them? I suggested earlier (Chapter 2) that boundaries must be understood as complex spatial and social phenomena that structure territories. The limits of a territorial project are defined by a boundary which participates in structuring it (Raffestin 1980 : 148). This boundary crystallises existing power relations between groups, focussing conflicting territorialities along one line. 'The boundary does not only delimitate a territory strictu sensu but rather a spatio-temporal envelope, that is to say that it organises an operational time and space, a place within which a relational system can exist'[17] (Raffestin 1974 : 27). In the discussion, I described the ontologically distinct *fiat* and *bona fide* boundaries, suggesting that different schools of thought had developed distinct conceptual frameworks based around these different standpoints. These two ontologically distinct forms of boundaries echoed the dichotomy between natural and social sciences, in line with the scientific traditions of their proponents.

In consequence of this dichotomy, I noted that arguments for boundary definitions tended to follow these two trends. The attempts to define 'natural' or 'rational' boundaries, for example, implicitly used bona fide objects to uphold the definition of a fiat boundary. On a more practical level, I alluded to the temptation within the protected area movement to align political boundaries with 'natural' ones. In doing so, I illustrated the difficulties and dangers of translating what was essentially a form of fact-based, scientific and applied knowledge of biophysical processes into political projects. This was shown to rapidly lead back, in circles, to the fruitless debates in the 19th and early 20th century regarding determinism and natural boundaries. However, in the light of the discussion of 'social nature', this position must be further discussed. Instead of taking this distinction between societal and biophysical boundaries to be absolute, it must be recognised to be the result of a 'purification' carried out by the actors involved in the process.

Protected area managers offered binarist accounts of their actions. However, this did not indicate an unproblematic distinction. Instead, it was the result of the multiple purifications that took place in the process of constructing knowledge about space. In Chapter 5, belief in the objectivity of the scientific method granted tremendous power to decision-makers. When I asked protected area managers to describe the process of defining boundaries, indicating in particular who was involved at which stage, all stressed the scientific nature of the process. This was seen as rational, with specialists revealing what was already there. It assumed pre-existing natural boundaries: *bona fide* boundaries revealed by a rationalised process. Because the resulting zonation was 'scientifically' established, resistance by other actors was seen as irrational or untenable. This meant that a belief in the validity and legitimacy of the method of constructing mediated representations of the world conferred substantial authority. This power relationship between the expert and other actors was a key to understanding the discourse of many managers, since they considered themselves imbued with a non-negotiable authority rooted in rationality. Thus science, like law, technology and property, was 'invariably complicit in effecting such distinctions' (Whatmore 2002 : 61). Each was 'characterized by a universalising ambition that fashions the world as a *terra incognita* or *terra nullius*, to which they alone bring order by effacing or subsuming all other modes of knowledge or regulation' (Whatmore 2002 : 61) (emphasis in original). Science was inclined to efface its own practices, masquerading its fabrications as self-evident accomplishments, negating the purifications that were inherent in the process of constructing knowledge.

Defining protected areas was an act of territorialisation and institutionalisation: restricting human impact or access through an intrinsically political process. Managers desired an increased institutionalisation of protected areas that led to the emergence of territorial units that were established and clearly identified as part of the spatial structure of a society. The boundaries of such areas reflected this attempt at territorial control and came to incarnate the power exercised over prescribed space. This reified power manifested itself in territoriality, in power made visible, deflecting the attention from the power relationship to the territory, away from the controller and controlled. This attempt to provoke a geohistorical process was inevitably contested in a variety of ways, reflecting the inherently political nature of boundaries. Yet rather than seeing this as domesticating nature, the prevalent discourse and rationale among protected area managers linked it to the opposite end of the spectrum: managers saw themselves as defining areas of wilderness, setting nature free from human intervention. The creation and where applicable the internal zonation of such spatial entities implied institutionalisation. In order to cope with the uncertainties inherent in such spatially complex scenarios, managers of protected areas sought to reinforce decision-making structures and administrations and thereby their own authority over the land. Thus the purifications involved in defining boundaries were far from innocent. Instead, they directly served to perpetuate the existing power balance, further asserting the authority of protected area managers over the land. However, the authority of managers was repeatedly challenged by the specifically transboundary nature of the interactions. In the next section, I discuss the conceptual consequences of this spatialised encounter with the Other.

Suggesting the unpronounceable: hybrid neo-Medievalisms?

The discussion of transboundary situations illustrated the multifaceted nature of boundaries, indicating that transboundary protected areas were much more complex than simple encircled areas stretching across international boundaries. I used the metaphor of New Medievalism to grasp at the difficulty of ascribing one administration or institution to one space, illustrating the intrinsically contested nature of boundaries. New Medievalism was described as an overlapping of various authorities on the same territory. This reflected changes in the world that established vocabularies could not reflect. New vocabularies were only in the process of being devised and searched for in order to provide explanations. This complex rewriting of space implied that existing boundaries were not only contested but also conflicting. Thus rather than unproblematically defining one Self and one Other, these 'neo-medieval' scenarios created multiple Selves and multiple Others in a overlapping and conflicting patchwork of multi-scalar identities.

The discussion of boundaries and identity started by suggesting that Paasi's concepts of integration and distinction offered a framework for understanding processes of confrontation with the Other. The Other was on the 'other side', while any process that led to common work implied combining the language of integration ('building bridges across the boundary') with coming to terms with the language of difference, of self-definition and identity. These two aspects were seen as necessarily intertwined since boundaries not only separated groups and social communities from each other, but also mediated contacts between them. By dwelling on the socio-spatial dimensions of integration and distinction, I attempted to identify elements of reterritorialisation that emerged during the construction of transboundary entities, describing it as a contested process. The framework for understanding this lay in the identification of discourses of othering and the problematic issues of defining social and spatial integration. This implied that rather than being unproblematic 'boundary zones' of localised interaction, transboundary protected areas were contested spaces constructed through relations of power between international, national and local actors. As such, they were prime examples of the emergence of new territorial units illustrating the multi-scaled complexities of regional transformation.

The examples at the beginning of this chapter highlighted further difficulties in creating transboundary protected areas. Following on from the discussion of the boundary between nature/culture, the intrinsically 'neo-medieval' nature of these entities was further complicated. The Other was understood to be even more multiple than initially thought, further hybridising transboundary entities. What constituted 'nature' was seen to be contingent on social practices and culturally-specific definitions. This meant that the boundaries between the distinct ontological domains of 'nature' and 'culture' were defined and negotiated differently in different contexts, leading to the creation of different 'natures'. Transboundary protected areas were no longer considered objects that existed independently from the maps, tables, techniques and practices. Instead, they were constructed discursively by these different elements, by the different relations and links within heterogeneous social networks that included both human and non-human actors.

Conclusions

This section began by recounting the encounter with a chimera: a beast that was simultaneously lion, goat and dragon. This fire-breathing monster was taken to illustrate the changing hybrid geographies that stem from questioning taken-for-granted dichotomies. I suggested that the opposition between the fundamentally binarist discourses of managers and the apparently hybrid protected areas needed confronting. Drawing from the results of the study as a whole, the key was seen to reside in understanding the nature of the purifications that took place during the process of defining boundaries. Initially, by drawing on the authority conferred by belief in the self-evident accomplishments of science, protected area managers were able to reinforce their own position as legitimate custodians of the land. In order to cope with the uncertainties inherent in such spatially complex scenarios, managers of protected areas sought to reinforce decision-making structures and administrations through appropriate zonation and thereby their own authority over the land.

However, the authority of managers was repeatedly challenged by the specifically 'transboundary' nature of the interactions. The discussion of transboundary situations illustrated the multifaceted nature of boundaries, indicating that transboundary protected areas were much more complex than simple encircled areas stretching across international boundaries. Rather than unproblematically defining one Self and one Other, these 'hybrid neo-medieval' or perhaps more elegantly, these 'chimeric' scenarios created multiple Selves and multiple Others in a overlapping and conflicting patchwork of multi-scalar identities.

Notes

1 The literature on 'performative' approaches is very diverse and leads down radically different ways of engaging with the world. Further suggested reading would include Dewsbury, Harrison, Rose, Wylie 2002; Thrift 2000 and Rose 2002.

2 Personal translation from: 'Je pense que (la réserve de biosphère) c'est une façon de faire un aménagement du territoire durable, cela suppose d'abord une profonde réflexion sur les différentes zones et ce que l'on peut leur attribuer comme enjeux et comme objectifs et je trouve que c'est quelque chose qui permet d'être opérationnel. Et, en plus, si on fait une réserve de biosphère à deux, cela permet d'avoir un noyau à l'intérieur qui est beaucoup plus fort et beaucoup plus intéressant. C'est quelque chose qui correspond bien aux objectifs des parcs nationaux, c'est-à-dire d'avoir vraiment une zone très fortement protégée et par contre une réflexion sur l'aménagement des zones périphériques'.

3 Personal translation from notes taken in German and English.

4 Personal translation from: 'La nouveauté est aussi que l'on a pris les zones évidemment naturelles, et on a mis tous les monuments historiques, notamment des châteaux qui sont en pleine forêt. On a estimé qu'autour, de toute manière, la zone qui s'y trouve et le château lui-même étant porteurs de nature, on est quand même en Europe, qui est fortement modifiée par l'homme, où nature et culture sont fortement imbriquées. Si tu ne prends pas en compte la nature qui fait l'objet d'une modification ou d'aménagements, il n'y a plus beaucoup de nature que tu peux prendre en compte, contrairement aux zones qui ont plus de l'espace comme le Nord sibérien ou quelques secteurs des Etats-Unis ou même en Australie. Donc, on a rajouté les monuments historiques'.

5 Personal translation from: 'Mais c'est pas dans l'esprit français, ou latin, d'avoir des zones laissées à la nature parce que l'homme a trop envie d'intervenir. Il estime que s'il laisse la nature sauvage, c'est une perte de pouvoir et d'influence, toi tu connais sûrement ces problèmes-là, alors que nous on est plus partagés. Dans l'équipe, il y a des gens qui sont "Kultur" et puis il y a des gens qui pensent qu'on devrait faire comme les allemands, mettre la barre plus haut, et mettre en zone centrale les zones réellement ... qui éliminerait les châteaux, encore que ... (...) Moi je pense qu'ils auraient tort de changer: moi je sais reconnaître quand les Allemands font bien les choses. Non, c'est vrai. J'estime que leur zonage tient bien'

6 Personal translation from notes taken in German and English.

7 Personal translation from: 'On a rajouté les vergers traditionnels, qui sont des zones tout à fait nature-culture aussi, en fort déclin, pour lequel on a un programme d'action. On trouvait logique donc de les avoir en zones tampon, je dirais les zones un peu naturelles qui ont encore du patrimoine fort mais qui, de toutes façons, font l'objet d'une gestion par l'homme, sur lequel on doit justement expérimenter des pratiques alternatives. L'aire de transition est en revanche, elle, devenue tout petite : c'est toutes les communes, toutes les zones d'activité, les villes, enfin, les zones urbaines. (...) Voilà pour le zonage'.

8 Personal translation from: 'Et l'outil "parc" (parc naturel régional) est parti de là, quoi. (...) C'était vraiment "Kulturlandschaft", comme le disent les Allemands, mais le concept était là, l'idée était là'.

9 Personal translation from: 'Et les zones tampon donc sont des zones, mais il y a une logique quand même, attention, les Allemands ont adopté une logique, la logique, c'est l'épine dorsale du Palatinat'.

10 Personal translation from: 'C'est difficile quand il y a des sujets épineux dans le style du loup qu'on a dans les parcs, qui n'est pas un sujet très facile et donc ça entraîne pas mal de polémique. (...) En plus c'est les loups qui sont venus d'Italie donc, c'est pas facile !'

11 Comment noted during the Expert seminar in Entracque (Italy) and Menton (France) on 'Un park Européen pour le 21ème Siècle' ('Un Parco Europeo per il 21° Secolo'), 14/15 October 1999.

12 Personal translation from: '[JF] Est-ce que, dans ce genre de choses, les attitudes des populations sont comparables des deux côtés de la frontière ? [Chloé] Pour le loup non, pas du tout, puisque pour le loup, au niveau italien, ils n'ont pas du tout le même problème dans la mesure où ils ont moins de troupeaux et puis ils ont dans leur culture le culte du loup, Remus et Romulus, donc c'est la louve qui a élevé Remus et Romulus, c'est une autre culture, nous c'est le Chaperon rouge, c'est pas tout à fait pareil ! Donc, et en plus bon, les bergers ont toujours eu le loup donc ils ont gardé des petits troupeaux et ils ont toujours gardé leurs troupeaux, alors qu'ici l'aspect économique et la facilité de garder les troupeaux, ils avaient des troupeaux plus grands, donc les conditions avaient changé, évidemment ...'

13 Personal translation from: 'Sur les panneaux il y a une présentation, une carte commune des espaces, une présentation de tout ce qu'on a fait ensemble, une présentation des points d'information des deux parcs et donc les panneaux en français, en italien, mais c'est exactement le texte et on explique la réglementation des deux côtés parce que ça aussi, au niveau réglementation, c'est que, d'un côté ou de l'autre on n'avait pas tout à fait la même réglementation donc là c'est une explication au moins de ces différences pour que la personne qui se trouve à un col sache un peu mieux les différences'.

14 Themes linked to human-animal relations are not explored more in depth here. There is however a fascinating literature emerging on the topic, endeavouring to 'discern the many ways in which animals are "placed" by human societies in their local material spaces (settlements, fields, farms, factories and so on), as well as in a host of imaginary, literary, psychological and even virtual spaces' (Philo & Wilbert 2000 : 5). Suggested reading would include Anderson 1997; Philo & Wilbert 2000.

15 Taken from Theogony lines 319–324 (online: http://www.perseus.tufts.edu).

16 The fact that it was Monsieur Adam's garden seemed to add an almost biblical and mythical dimension to the tale ... A conceptual Garden of Eden?

17 Personal translation from: 'la frontière ne délimite pas seulement un territoire stricto sensu mais bien davantage une enveloppe spatio-temporelle, c'est à dire tout à la fois un aménagement du temps et de l'espace opératoire, lieu de la réalisation d'un système de relations'.

Chapter 11

Conclusions

Chimeras challenge sacred cows

This book has argued for a comprehensive conceptual and practical link between writings on boundaries and reflections on nature. By arguing that the complex 'transboundary' character of the objects studied required it, I explored how critical geopolitical approaches to boundaries and social approaches to nature could be used to shed light on each other, ultimately pointing towards fruitful future avenues for research. The concept of boundary was the bridge that I used to bring these traditions together: both a starting point and a key. A series of thematic chapters, strongly grounded in field research and drawing heavily both from literature and interviews, indicated the fecundity of such a hybrid path of research.

The last chapter of the discussion (Chapter 10) explicitly drew the two threads of literature together, suggesting that transboundary protected areas were unique examples through which to introduce the 'political' into discourses of 'nature'. This conclusion is an attempt to build on this and revisit the spatial concepts of territory, region and bioregion (Chapter 2) in the light of the hybridisation of the biophysical and the societal. This conceptual conclusion therefore avoids repeating the specific conclusions laid out in successive chapters, instead addressing the conceptual consequence(s) of the emergence of the notion of hybrid-neo-medievalism. Such a dreadfully unpronounceable notion cannot be left unexamined. Ghastly neologisms often contain grains of truth that can be reformulated subsequently. In celebration of the encounter in the botanical garden, the term chimeric territory might assist in capturing some of this complexity.

The questions that have emerged are simple to set out: firstly, if space is viewed as fundamentally hybrid, is it still pertinent and conceptually meaningful to attempt to divide it into distinct entities? In other words, if we return to Paasi's suggestion that geography has to 'establish the stories of creating a distinction between us and the Other' (Paasi 1996 : 21), what spatial framework is this likely to take place in? Secondly, if geography takes the need to develop a political theory of nature and environment seriously, what spatial concepts can this be based around? This section draws on the ideas that have emerged in this book, indicating future paths that require additional exploration.

Space or territory?

Chapter 2 dealt with the differing uses that were made of both biophysical and societal arguments in constructing space. In order to do this, key spatial concepts were chosen as illustrations of existing spatial discourses: territory, region and bioregion. The purpose of this was both to nod to some of the existing sacred cows of the discipline and to indicate the remaining tensions between what were taken to be conflicting biophysical and societal orientations. The subsequent discussion of transboundary protected areas indicated that neither of these concepts reflected the complexity of processes taking place. These spatial concepts were more than bovine shrines since their status, both within and outside the discipline, were unequal.

The rewriting of space described in this book built on the idea that boundaries were not only contested but also conflicting. Rather than defining one Self and one Other, 'neo-medieval' scenarios created multiple Selves and multiple Others in a overlapping and conflicting patchwork of multi-scalar identities. In such a context, identifying one territory or one idealised spatial entity was not pertinent. This meant that throughout the book, I repeatedly referred to 'territoriality' or 'territorialisation', all the while largely avoiding the use of the term 'territory', preferring the term 'spatial entity' to 'territorial entity'. Even if the possibility of overlapping territories was suggested, the concept of territory remained largely counterproductive, conceptually reintroducing the idea of rigid spatial entities.

I initially discussed how the limits of a spatial project or spatial ideal were defined by boundaries (Chapter 2). Boundaries were seen to crystallise existing power relations between groups, focussing conflicting territorialities along one line within the created territory. However, while defining a territory may have been conceptually interesting, it was no longer pertinent on a concrete level in a context of hybridity. Whatmore (2002) and to a certain extent Ò Tuathail (1996), used the term territoriality in different guises. The former, most tellingly, altogether avoided defining what a 'territory' might be. This use of concepts spawned by the term territory also stretched to authors referring to territorialisation (see for instance Braun 2003, Castree 2003). Again, neither of these authors suggested within what spatial entity this took place, referring, if anything, to networks rather than to explicitly spatial concepts.

The spatial entities on which territoriality and territorialisation were based appeared no longer pertinent when hybrid scenarios were considered. Reified power manifested itself in territoriality, in power made visible, yet paradoxically attention was deflected from the power relationship to the territory, away from the controller and controlled. By focussing first on the territory, not on the power relations, the view risked being obscured. Because territory was so intimately linked up to territoriality and territorialisation, it was no longer pertinent as a stand-alone spatial concept. Furthermore, considering territories to be fixed objects benefited dominant actors, reinforcing the status quo and the existing distribution of power.

The concept of territory therefore requires revisiting as it appears no longer to be the ultimate tool in the geographical toolbox – the sacred cow – but then neither are region, bioregion, landscape or even network. Forging a link between territory and network appears promising, yet is far from straightforward. In fact, the (a)spatial dimensions of networks also need revisiting in the light of hybridity. Territories are more than containers of relational networks since they construct and are constructed by them. Quoting Lévy, November suggests that 'the territory/network pair allows two terms to be put on the same level, to be compared and articulated. Unlike what has been done for a long time, this does not reduce networks to being either material frameworks or else, on the contrary, to being abstractions removed from the notion of distance'[1] (Levy in November 2000 : 273). She adds that only by putting both on the same level can situations be truly understood, although she hints that both cannot be apprehended using exactly the same tools. Lévy further suggests that this poses a challenge to the usual Euclidian approach to space: 'several networks intermingle, but also different kinds of networks: topological networks (that create discontinuity) with topographical networks (that create continuity). At a certain level, the densification of networks creates territory'[2] (Lévy in November 2000 : 274). This builds on the idea that boundaries (discontinuities) are more than the simple limits to a spatial entity, but instead actively participate in structuring the whole.

If territory is revisited as a concept, it is tempting to replace it solely with a constructed yet non-specific concept of space. However, as discussed in this book, the behaviour of protected area managers on the ground indicated that creating spatially-defined entities was far from over and done with. Instead, in this specific context, it was the fundamental way they thought about and engaged with space. Throughout the discussion, protected area managers sought to institutionalise specific areas, leading to the emergence of spatial units that were established and clearly identified. Defining protected areas was thus a textbook example of territorialisation and institutionalisation. However, it was the act of defining an area – more than the resulting area itself – that was meaningful. The boundaries of such areas reflected this attempt at territorial control and came to incarnate the power exercised over space. Thus, on a conceptual level, it seemed as though the processes and purifications were to be considered over and above the resulting spatial entity. If this were the case, what concepts and notions should be explored more fully in future?

Territoriality without territory?

In the light of this book, it initially appeared appropriate to consider territoriality without territories. A similar conclusion is being reached in several emerging bodies of literature, notably in France around people such as Jacques Lévy and Michel Lussault. They suggest using space as a starting point, clarifying the meaning of the existing term 'spatialité' (spatiality) to solve the conundrum. Lussault, in particular, suggests an interesting link between space/spatiality (espace/spatialité), different from that binding territory/territoriality. This avoids assuming a hierarchy or sequence between the two, and conceptually transcends the need for a defined spatial entity.

'Pre-existing space (*i.e.* both the shapes of spatial configurations and the ideologies and values attached to space) that predates one spatial action or a series of spatial actions carried out by an operator, constitutes a new occurrence of spatiality. Symmetrically, every action that stems from spatiality produces space (within a new arrangement). This is written within and enriches a pre-existing spatial configuration, leading to an increase in the quantity of space present within a society. In the framework of such an approach, space is neither considered a neutral container of functions, nor is it a surface on which social relations are projected, or a simple political attribute. It is a reality that is constructed by spatial action and that means something for someone, for an actor. Space stems from spatiality and constitutes it (. . .). There is an incessant dialogue between space and spatiality. By addressing such an issue, the usual habit of separating that which relates to space (too often reduced to simple material shapes) and that which relates to society is replaced by an understanding of the consubstantiality of both: spatiality constitutes the concept that allows both domains to be bridged'[3] (Lussault 2003 : 867). Thus spatiality here is not unlike 'territoriality', although it is explicitly removed from the notion of a spatial entity. Further exploration of this concept would be interesting, particularly in the context of 'chimeric' characteristics, where a reliance on an established spatial entity is counterproductive.

Territoriality and Otherness

However, discarding territory is not so easy and may in fact itself be counterproductive. On the ground, the identification of discourses of othering (distinction) and their links to social and spatial integration allowed existing processes of (re)territorialisation to be understood and located. Rather than being unproblematic boundary zones of localised interaction, transboundary protected areas were contested spaces constructed through relations of power between international, national and local actors, as well as between different ontological positions related to boundaries and space. As such, they were not only prime examples of the multi-scaled complexities of regional transformation but also sites of ontological struggle between biophysical and societal conceptions of space.

The complex territorialities that emerged during, within and through the construction of transboundary entities were intrinsically contested and dynamic. The fluid nature of such a process is expressed in the awkward yet pertinent terms (re)territorialisation or de/re-territorialisation. If one accepts that territoriality (or spatiality?) is always dynamic, then such spelling is superfluous. In effect, it is considered dynamic because there is implicitly a dialogue between elements and actors: a ritualised confrontation with an Other. The Other was not only the foreigner. Otherness, or alterity, encompassed wider features that stretched across a wide range of ontological positions, further hybridising transboundary entities. What constituted 'nature' (and thereby a dimension of otherness) was seen to be contingent on social practices and culturally-specific definitions. This meant that the boundaries between the distinct ontological domains of 'nature' and 'culture' were defined and negotiated differently in different contexts, leading to the creation of different 'natures'. The intrinsically neo-medieval institutional and interpersonal characteristics of these entities were further complicated by the fluidity of the ontological boundary between nature/culture. Spatial entities were chimeric yet spatialised, constructed dynamically as sites of encounter between Self and Other, 'nature' and 'culture': they were, in effect, chimeric territories.

These transboundary protected areas did not exist independently from the maps, tables, techniques and practices that constituted them. Instead, they were constructed discursively by these different elements, by the different relations and links within heterogeneous social networks that included both human and non-human actors. The belief in the validity and legitimacy of the method of constructing these mediated representations of the world conferred substantial authority. This power relationship between the expert and other actors was a key to understanding the discourse of many managers, since they considered themselves imbued with a non-negotiable authority rooted in rationality. Science effaced its own practices, masquerading its fabrications as self-evident accomplishments, negating the purifications inherent in the process of constructing knowledge and space by upholding a dualistic discourse.

The dynamic process of (re)territorialisation is therefore a politics of purification which, like politics, is always geographical. The construction of space is fundamentally hybrid, involving *both* societal and biophysical elements, human and non-human. Nature, rather than being separate from the societal, is always social and political, always an intrinsically discursive construction. If the complex implications of such hybridity are to be fully understood and addressed within geography, then a clear (re)formulation of spatial and political concepts is necessary. The concept of chimeric territory is a first step down this path.

Notes

1 Free translation from: 'le couple territoire/réseau permet de mettre sur le même plan, de comparer, d'articuler deux termes, sans réduire les réseaux, comme on l'a longtemps fait, a des supports matériels ou, au contraire, a des abstractions détachés de la notion de distance'.

2 Free translation from: 'plusieurs réseaux s'imbriquent, mais aussi des réseaux (topologiques, créant de la discontinuité) avec des territoires (topographiques, engendrant de la continuité). A un certain degré, la densification des réseaux produit du territoire'.

3 Free translation from: 'l'espace déjà-là (*i.e.* à la fois les formes des configurations spatiales et les idéologies et valeurs afférentes à l'espace), antérieures à une action ou une série d'actions spatiales d'un opérateur, constitue une nouvelle occurrence de la spatialité. Symétriquement, chaque action, qui procède de la spatialité est productrice d'espace (sous forme d'un agencement nouveau), qui s'inscrit et enrichit une configuration spatiale préexistante, accroît en quelque sorte la quantité d'espace présente dans une société donnée. Dans le cadre d'une telle approche, l'espace n'est pensé ni comme un contenant neutre de fonctions, ni comme une surface de

projection des rapports sociaux, ni comme un simple attribut du politique. Il s'agit d'une réalité construite dans l'action spatiale et qui signifie quelque(s) chose(s) pour quelqu'un, pour un acteur. L'espace procède de la spatialité, tout en l'autorisant (...). Entre l'espace et la spatialité existe une dialogique incessante. En abordant ainsi une telle question, on affirme, loin des habitudes à séparer ce qui est de l'ordre des espaces (trop souvent rabattus sur leurs seules formes matérielles) et ce qui ressortit aux actions sociales, la consubstantialité de ceux-la et de celles-ci : la spatialité constitue le concept qui permet la jonction entre ces deux domaines'.

Appendices

Appendix I. Five transboundary protected areas in Europe

Please note that all data refers to the periods during which the fieldwork was carried out. This means that in several cases situations and contexts may have changed substantially, particularly regarding the permeability of political borders and the administrative arrangements within sites.

1. Poland – Slovakia: the Tatra Biosphere Reserve

The national park councils in both countries started to work on a concept for a future common protected area covering areas in both countries since the 1980s. The Tatras transboundary biosphere reserve was approved by UNESCO in 1993.

Table A.1 The Tatra biosphere reserve

	Polish entity	**Slovak entity**
	Tatrzanski Park Narodowy – 'Tatra National Park' *21,164 hectares* Established in 1954	Tatranský národný park – 'Tatra National Park' *Actual size unclear* Established in 1949
Local name of biosphere reserve	Rezerwat Biosfery Tatry	
Seat of administration	Zakopane	Tatranska Lomnica
Total area	21,164 hectares	*Actual size unclear*
Percentage of total	Unclear (roughly 1/4)	Unclear (roughly 3/4)
Total transboundary area	*Unclear*	

Transboundary Biosphere Reserve and Governance:

- *No formal mechanism* for contacts between the two parks. The existing cooperation rested on individual contacts.
- Scarce financial resources, formalised transboundary *cooperation not identified a priority*. Costs of cooperation (car petrol and telecommunications etc) meant that joint meetings were few and far between; contacts sporadic between managers.

- Acrimonious institutional reform in Slovakia caused a split between 'national park' and 'biosphere reserve'.
- Increasingly different, *'bottom-up' forms of contact* between practitioners and scientists.
- Pragmatic choices meant that the Poles cooperated with their most likely partners
- Absence scientific board on Slovak side no obstacle to scientists from either side working together. Several common projects were underway, monitoring long-term change, or working on more practical issues such as bark beetle management.

Transboundary cooperation not encouraged before 1990. Border areas were strategically sensitive and out of bounds. However, before 1990, *official* contacts – between directors of national parks or between ministers for example – were more frequent. Contacts with Western countries offered more attractive possibility of international experience. Border closed within biosphere reserve borders, one seasonal passage point at the top of the Rysy mountain for pedestrians. The closest permanent border passage point was on the outer border of the park, in Lysa Polana: by road under an hour and a half.

2. Poland – Slovakia – Ukraine: the East Carpathians Biosphere Reserve

A local tale told of a man who was born in the Austro-Hungarian Empire, was christened in Czechoslovakia, married in Hungary, had his first child in the USSR, and died in Ukraine, without ever leaving his village.

Table A.2 The East Carpathians biosphere reserve

	Polish entity	Slovak entity	Ukrainian entity
	'Bieszczady National Park' *29,200 hectares* The national park was established in 1973. The original protected area was enlarged in 1989, 1991 and 1999.	'Poloniny National Park' *29,805 hectares* The national park was formerly part of the Slovak Eastern Carpathian Landscape Protected Area (Vychodné Karpaty). This was split in 1997 to make a 'landscape protected area' and a 'national park'.	'Uzhanski National Nature Park' *39,159 hectares* The biosphere reserve was established in 1999 on the basis of the existing Stuzhitsa Protected Landscape Area and the Nadsianski Regional Landscape Park.

	Cisniansko-Wetlinski Landscape Park *51,165 hectares* Dolina Sanu Landscape Park *33,480 hectares*	Poloniny NP buffer zone *10,973 hectares*	Nadsianski Regional Landscape Park *19,428 hectares*
Local name of biosphere reserve	*Rezerwat Biosfery Karpaty Wschodnie*	*Biosférická Rezervácia Východné Karpaty*	*(Cyrillic script*)*
Seat of administration	Ustrzyki Gorne	Snina	Velykyj Berezny
Total area	113,845 hectares	40,778 hectares	58,587 hectares
Percentage of total	53%	19%	28%
Total transboundary area 213,211 hectares			

*Words and names originally written in Cyrillic script are offered in the most commonly found transcriptions.

The transboundary protected area:

- Idea of transboundary protected area formally presented in 1990 UNESCO-MAB meeting in Kiev.
- The proposal was approved by the ministers of all three countries, who signed an agreement in October 1991.
- The project was inscribed in the World Network of Biosphere Reserves in two stages, with the Slovak and Polish applications being registered in *1993*, joined by the Ukrainian one in *1999*.
- Biosphere reserve model chosen as federating unit: the only internationally-recognised programme that could apply to countries in and out of European Union European

Governance:

- *No official biosphere reserve body* responsible for running common activities. Coordination done on an ad-hoc basis by concerned managers.
- Series of different attempts at governance: a transboundary *Coordinating* Council (representatives of relevant ministries and institutions involved such as protected areas, state forests, scientific institutions and local authorities), last meeting was held in 1994.
- On the Polish side, an informal *Consultative* Council imagined as a substitute but at the time of the fieldwork had never met.
- A *Foundation* (legally based in Switzerland) set up at the same time as the trilateral biosphere reserve to encourage, organise, conduct and promote activities serving to protect the overall biodiversity of the Eastern Carpathian Mountains zone. Initial assets of US$ 600,000 from the

American MacArthur Foundation and the Global Environment Trust of the GEF under the administration of the World Bank: a sum rather small to yield substantial revenue for joint operations as planned.

- The Foundation's annual meeting provided the only established and fully funded meeting between the three sides, providing intense contact in both a formal and informal setting between senior managers for a couple of days.
- The revenue from the Foundation's funds used to fund small-scale local projects in each country and not joint initiatives as originally planned. Projects were selected on an *ad hoc* basis by local managers.

Despite measures of goodwill and fledgling transboundary initiatives, political boundaries were *real obstacles* in the Eastern Carpathians. Eight hours by bus from the Slovak to the Polish administrative centre, even longer to join up the Polish and Ukrainian ones, despite the short distance on a map. Reaching the Ukrainian side from Slovakia by car was relatively quick, although *long waits at the border* were not uncommon, especially when leaving Ukraine. Some progress in accessibility between Poland and Slovakia: two seasonal pedestrian crossing points opened in July 1999 between the Bieszczady national park and the Slovak Protected Landscape Area. Some degree of control on movement between provinces within Ukraine, with armed soldiers manning road blocks, much like in the days of the USSR. Concern that crossing the Slovak/Ukrainian border would get more complicated in the future with the introduction of visas for citizens of both countries, costing in excess of 40 US$. If this were the case, then cooperation would be seriously hampered in the future. Poland to become the Eastern boundary of the European Union: border controls becoming more of an issue, with *tighter controls* linked to the fear of illegal immigration.

3. France – Germany: the Vosges du Nord – Pfälzerwald Biosphere Reserve

In the forest, old boundary markers remained, often maintained as witnesses to previous divisions. The famous fortified 'Ligne Maginot' and 'Ligne Siegfried' passed respectively through the French and German parks, with many of the remaining monuments open to the public, contrasting in epoch but not in meaning with the remaining Medieval castles.

The transboundary protected area and governance:

- Initially designated as two separate biosphere reserves based around existing protected areas: the Vosges du Nord in 1989 and the Pfälzerwald in 1992.
- Joint designation as a transboundary biosphere reserve in *1998*.
- Since 1995, institutional and legislative reform in Germany has meant that the Federal State of Rheinlandpfalz has contested the Verein's ability to manage the BR officially.

- Transboundary cooperation suffered from German institutional reform as the French managers had increasing difficulty in identifying their opposite number.
- *No formal structure for cooperation*, replaced by attempts to create transboundary 'pairs' of managers, assisted by coordinators. Management goals and practical means were very different on the two sides.
- Various use of one or two coordinators during a series of Interreg-funded cooperation programmes. Lots of debate about ideal structure: currently one coordinator, employed by German administration.

Table A.3 The Vosges du Nord – Pfälzerwald biosphere reserve

	French entity	**German entity**
	'Parc Naturel Régional des Vosges du Nord' *122,000 hectares* The park was established by Charter in 1975. It was redesignated within a Charter in 1994 and 2002.	'Naturpark Pfälzerwald' *179,800 hectares* The park was established in 1958 and classified in 1967 as a landscape protected area (Landschafts-schutzzone).
Administrative structure of existing protected area	A local public administration with no legislative power (Public law)	Own management body since 1982: a non-governmental organisation (NGO) know as the 'Verein' (Association) made up of local authorities, NGOs and foundations (Private law)
Local name of biosphere reserve	Réserve de Biosphère des Vosges du Nord/ Palatinat	Biosphärenrezervat Nordvogesen/ Pfälzerwald
Seat of administration	La Petite Pierre	Lambrecht
Total area	122,000 hectares	179,800 hectares
Percentage of total	40%	60%
Total transboundary area	301,800 hectares	

Both countries signed Schengen agreement: border crossing unrestricted. Most hiking trails connected to those in the other country, despite challenges in making the signposts agree.

4. Romania – Ukraine: the Danube Delta Biosphere Reserve

Table A.4 The Danube Delta biosphere reserve

	Romanian entity	**Ukrainian entity**
	'The Danube Delta Biosphere Reserve' Designated a biosphere reserve by the Romanian government in 1990, receiving official recognition from UNESCO – MAB in 1992. In 1991, designated both a Natural World Heritage site and a Ramsar site.	'The Danube Biosphere Reserve' Designated in 1998 on the basis of the 'Natural Reserve Dunaiski Plavni', formerly part of the Chernomorsky (Black Sea) Nature Reserve, the other part of which became the Chernomorsky Biosphere Reserve in 1982. In 1991, designated as a Ramsar site.
Local name of biosphere reserve	Reservatia Biosferei Delta Dunarii	(*Cyrillic*)
Management structure	The *Danube Delta Biosphere Reserve* (DDBR) *Authority* established in 1990, formalised through legislation passed in 1993: 'Law Regarding the Establishment of the Danube Delta Biosphere Reserve' (No. 82, December 1993). Also an independent research institute, the *Danube Delta Institute* ('Institutul National de Cercetare – Dezvoltare Delta Danarii – Tulcea').	Managed by a state administration under the responsibility of the Ministry of the Environment.
Seat of administration	Tulcea	Vilkovo
Total Area	580,000 hectares	46,492 hectares
Percentage of total	83%	17%
Total transboundary area	696,492 hectares	

The transboundary protected area and governance:

- Inscribed in the World Network of Biosphere Reserves in two stages, with the Romanian side being recognised as a BR in 1992, joined by the Ukrainian side in a transboundary entity in *1998*.
- Contacts intensified within the framework of *two national World Bank projects* both entitled 'Danube Delta Biodiversity', managed through the Global Environment Facility (GEF) project: 1994 to 1998 (Ukraine) and 1994 to 2000 (Romania). Contacts subsequently decreased.
- There was *no formal structure for coordinating contacts* between the two sides. Contacts were maintained informally or in the context of specific projects.

Substantial prior preparation was needed to cross boundary: Citizens from each country did not require a visa, but US$ 50 was required by the Ukrainian authorities as 'harbour tax': cost of any return trip at US$ 100, plus fuel. Travellers crossing the border were subject to medical checks at both ports. However, commercial boats transporting people across the boundary started again in 2001 further up the river, running once or twice a week, offering an alternative possibility for transport.

5. France – Italy: the Parc National du Mercantour – Parco Naturale Alpi Marittime

The fifth case study was the only site visited that was not a transboundary biosphere reserve, chosen as a contrast to the other sites. These two protected areas had however been working on joint projects for a number of years and had been examining the possibility of seeking international recognition: either as a World Heritage Site or as a transboundary biosphere reserve.

Table A.5 The Parc National du Mercantour – Parco Naturale Alpi Marittime

	French entity	Italian entity
	'Parc National du Mercantour' *68,500 hectares* The Parc National du Mercantour was created by decree in 1979, created on the basis of a pre-existing Réserve Nationale de Chasse (national hunting reserve) and the protected Vallée des Merveilles.	'Parco Naturale Alpi Marittime' *28,078 hectares* Established in 1995, bringing together the pre-existing Parco Naturale dell'Argentera and the Riserva del Bosco e dei Laghi di Palanfrè (forest and lake reserve of Palanfrè) which both existed since 1979.
Administrative structure	The park was divided into two zones of differing legislative status. It had formal decision-making powers over the 'central' zone, maintaining an advisory role within the 'peripheral' zone.	
Seat of administration	Nice	Valdieri
Percentage of total	71%	29%
Total transboundary area	96,578 hectares	

The transboundary protected area and governance:

- Sustained history of contact between local protected areas in the two countries, dating back to a first meeting between park directors in 1984.
- *No formal body for coordination* between the two protected areas. Differences in work patterns and bureaucratic traditions did not allow for the creation of an informal common body, although substantial contacts were ongoing between individuals in both protected areas.
- Key role played by Italian director, who provided *continuity* by remaining in her position for over twenty years.
- Formal signing of '*Twinning Charters*' (Chartes de Jumelage or Carta di Gemellaggio) in 1987 and 1998, yet collaboration relied heavily on personal contacts. Sustained attempts to formalise cooperation on legal and institutional levels to create a 'European Park': a 'comparative jurisdictional, administrative and organisational' study carried out in 1999.

- *Common projects* mainly concerned the common management of charismatic species such as ibex (managed jointly since 1987) or lammergeyer (reintroduced jointly since 1993).
- Substantial institutional resistance on the French side to creating common structures not explicitly defined in the law.

Both countries within the Schengen area of the European Union: boundary open to free passage.

Appendix II: List of interviews

An 'interview' consisted of an semi-structured discussion loosely based around a set of key questions, that lasted from roughly ten minutes to a couple of hours. Despite a framework of questions being laid out initially, interviews in fact were relatively unstructured and were adapted to the role and responsibility of the person interviewed. Individual situations, both personal and linked to location, were so different that at times it seemed impossible to seek a more formal structure. Despite having laid out a standard framework of questions, time and time again I found myself casting this aside in favour of a more *ad hoc* technique that allowed for flows of ideas to blossom unhindered. This less structured technique turned out to be the most effective way of conducting interviews, allowing for maximum flexibility. It was often when the conversation apparently wandered off that real insights were gained. Additionally, the very varied speech patterns and linguistic abilities among the respondents did not allow a specific set of questions to be used in all cases. In cases where local people did not speak English, French, Italian or German, I used informal interpreters, usually from the administrations visited or in one case from the local school. This multi-lingual fieldwork means that the interviews are transcribed in four languages. In order to keep the text readable for all, all quotes are translated into English in the text, with the original language appearing in an endnote. While this makes it more readable, it creates an additional filter. The inclusion of the original within easy reach hopefully lessens this. The translations are literal, not literary, carried out with no attempt to correct or 'improve' the flow of words, including non sequiturs. This leads to sentences that may appear heavy yet truthfully reflect the original.

The case study areas are referred to numerically in the following table, with an additional reference to three interviews carried out with UNESCO staff. Anonymity of individuals has been maintained by using 'aliases' of appropriate national origin and gender. The names assigned to individuals were, when available, taken from tables of the 'top 10' most popular names for each respective country, usually for the year 2000.

1 Danube Delta Biosphere Reserve (Romania) and Danube Biosphere Reserve (Ukraine)
2 East Carpathians Biosphere Reserve (Poland/Slovakia/Ukraine)
3 Tatry National Park (Poland) and Tatras National Park (Slovakia)

4 Vosges du Nord Natural Regional Park (France)/Pfälzerwald Nature Park (Germany)
5 Mercantour National Park (France)/Alpi Marittime Nature Park (Italy)
6 Additional interviews with UNESCO staff

Table A.6 List of interviews and format of data

Site	Alias	Administration and Country	Interviews	
			Language	Format of Data
1	Elena	Department of Labour and Social Solidarity, Tulcea, Romania	English (Translated by 'Nicolae')	
1	Tatiana	Danube Biosphere Reserve, Vilkovo, Ukraine	English (Translated by 'Svetlana')	
1	Cristian	University of Bucarest, Bucarest & Romanian MAB Committee, Romania	English	Interview with notes taken simultaneously 04.04.2000
1	Nikolaï	Danube Biosphere Reserve, Vilkovo, Ukraine	English (Translated by 'Svetlana')	Interview with notes taken simultaneously
1	Natalia	Danube Biosphere Reserve, Vilkovo, Ukraine	English	
1	Boris	Danube Biosphere Reserve, Vilkovo, Ukraine	English (Translated by 'Svetlana')	
1	Nicolae	Danube Delta Biosphere Reserve Authority, Tulcea, Romania	English	Interview with notes taken simultaneously
1	Ion	Danube Delta Biosphere Reserve Authority, Tulcea, Romania	English	Interview with notes transcribed simultaneously 05.05.2000

Site	Alias	Administration and Country	Interviews	
			Language	Format of Data
1	Grigore	Danube Delta Institute, Tulcea, Romania	English	Interview with notes taken simultaneously 05.05.2000
1	Aleksandr	Danube Biosphere reserve, Vilkovo, Ukraine	English	
1	Svetlana	Translator, Vilkovo, Ukraine	English (served as interpreter in Ukraine)	
1	Gheorghe	Danube Delta Biosphere Reserve Authority, Tulcea, Romania	English	Interview with notes transcribed simultaneously 04.05.2000
2	Lech	Regional Directorate of State Forests, Krakow, Poland	English	
2	Philipp	WWF & Foundation for Eastern Carpathian Biodiversity Conservation	English	
2	Stanislaw	Foundation for Eastern Carpathian Biodiversity Conservation, Poland	English	
2	Dominik	Poloniny National Park, Slovakia	English, (translated by 'Lukáš')	Interview with notes taken simultaneously 30.05.2000
2	Lukáš	Poloniny National Park, Slovakia	English (served as interpreter in Slovakia)	
2	Petr	Ministry of the Environment of the Slovak Republic, Slovakia	English	

Site	Alias	Administration and Country	Interviews	
			Language	Format of Data
2	Konrad	Forest Research Institute, Krosno, Poland	English	
2	Katharina	WWF Vienna, Austria	English	
2	Vladimir	Lviv University, Ukraine	English	
2	Jozef	Bieszczady National Park, Poland	English (translated by 'Andrzej')	Interview with notes taken simultaneously 29.06.2000
2	Ivan	Uzhansky Nature National Park, Ukraine	English (translated by 'Andrzej')	Interview with notes taken simultaneously
2	Zbigniew	Politician Lutowisca, Poland	English (translated by 'Andrzej')	Interview with notes taken simultaneously
2	Andrzej	Bieszczady National Park, Poland	English (served as interpreter in Poland and Ukraine)	Interview with notes taken simultaneously, Summer 1998 and 28.06.2000
3	Jan	Tatry National Park, Poland	English (translated by 'Krzysztof')	
3	Michal	Tatry National Park, Slovakia	English	Interview with notes taken simultaneously Summer 1998 and 26.06.2000
3	Krzysztof	Tatry National Park, Poland	English (served as interpreter in Poland)	Interview with notes taken simultaneously 23.05.2000
3	Kazimierz	Tatry National Park, Poland	English	
3	Vaclav	Forestry Division of Tatry National Park, Slovakia	English	
3	Staszek	Tatry National Park, Poland	English	

Site	Alias	Administration and Country	Interviews	
			Language	Format of Data
3	Tomasz	Tatry National Park, Poland	English	
3	Jurek	Tatry National Park, Poland	English (translated by 'Krzysztof')	Interview with notes taken simultaneously 24.03.2000
3	Karol	Tatry National Park, Poland	English (translated by 'Krzysztof')	
3	Tadeusz	Tatry National Park, Poland	English	Interview with notes taken simultaneously 23.05.2000
4	Niklas	Naturpark Pfälzerwald, Germany	English	Taped group interview 19.09.2000
			German	Taped individual interview 28.06.02
4	Hugo	Parc Naturel Régional des Vosges du Nord, France	French	Taped interview 22.02.2000 Taped interview 26.06.02
4	Théo	Parc Naturel Régional des Vosges du Nord, France	French	Taped interview 22.09.2000
4	Lukas	Ministry of the Environment & Naturpark Pfälzerwald, Germany	English	Taped group interview 19.09.2000
			German	Taped individual interview 28.06.02
4	Kathrin	Forschungsanstalt für Waldökologie und Forstwirtschaft Rheinland-Pfalz, Trippstadt, Germany	German	Interview with notes taken simultaneously 01.07.02

Site	Alias	Administration and Country	Interviews	
			Language	Format of Data
4	Daniel	Naturpark Pfälzerwald, Germany	English	Taped group interview 19.09.2000 Taped individual interview 27.06.02
4	Alexander	Ministry of the Environment, Neustadt, Germany	German	Taped interview 27.06.02
4	Manon	Parc Naturel Régional des Vosges du Nord, France	French	Interview with notes taken simultaneously 24.06.02
4	Maximilian	Naturpark Pfälzerwald, Germany	English/German	Taped group interview 19.09.2000
5	Matteo	Parco Naturale Alpi Marittime, Italy	Italian	
5	Lia	Department of Architecture, University of Torino, Italy	Italian	Interview with notes taken simultaneously 08.04.2002
5	Enzo	Parco Naturale Alpi Marittime, Italy		
5	Chloé	Parc National du Mercantour, France	French	Taped interview 26.03.2002
5	Thomas	Consultant to Parc National du Mercantour, France	French	Taped interview 27.03.2002
5	Aldo	Parco Naturale Alpi Marittime, Italy	Italian	
5	Aurelio	Parco Naturale Alpi Marittime, Italy	Italian	
5	Alexandre	Parc National du Mercantour, France	French	Taped interview 25.03.2002

Site	Alias	Administration and Country	Interviews	
			Language	Format of Data
5	Fabrizia	Department of Architecture, University of Torino, Italy	Italian	Taped interview 08.04.2002
5	Maddalena	Parco Naturale Alpi Marittime, Italy	Italian	
5	Chiara	Parco Naturale Alpi Marittime, Italy	Italian	
5	Camille	Parc National du Mercantour, France	French	Taped interview 28.03.2002
5	Carlo	Parco Naturale Alpi Marittime, Italy	Italian	
5	Pietro	Parco Naturale Alpi Marittime, Italy	Italian	
5	Alessia	Parco Naturale Alpi Marittime, Italy	Italian	Taped interview 09.04.2002
6	JR	UNESCO	English	Interview with notes taken simultaneously
6	MB	UNESCO	French	Taped interview, 23.02.2000
6	MJ	UNESCO	French	

Bibliography

Adams, A. B. (1962) 'How it began', in A. B. Adams (ed) *First World Conference on National Parks*. Proceedings of a conference organized by The International Union for the Conservation of Nature and Natural Resources, cosponsored by UNESCO, FAO, US National Parks Service, Natural Resource Council of America, Seattle, Washington June 30–July 7: National Park Service, United States Department of the Interior.

— (1962) 'First World Conference on National Parks', in A. B. Adams (ed) Proceedings of a conference organized by The International Union for the Conservation of Nature and Natural Resources. Cosponsored by UNESCO, FAO, US National Parks Service, Natural Resource Council of America, Seattle, Washington June 30–July 7: National Park Service, United States Department of the Interior.

Aggarwal, V. K. and Dupont, C. (1999) 'Goods, games and Institutions', *International Political Science Review* 20(4): 393–409.

Agnew, J. (1999) 'Regions on the mind does not equal regions of the mind', *Progress in Human Geography* 23(1): 91–96.

— (2000) 'Commentary 1 on Sack, R.D. 'Human territoriality: its theory and practice', Classics in Human Geography revisited', *Progress in Human Geography* 24(1): 91–93.

— (2001) 'Regions in revolt', *Progress in Human Geography* 25(1): 103–110.

— (2002) *Making Political Geography*, London: Arnold.

Albert, M. (1998) 'On boundaries, territory and postmodernity: an international relations perspective', in D. Newman (ed) *Geopolitics*, Vol. 3: Frank Cass.

Allen, J. (2003) *Lost Geographies of Power*, Oxford: Blackwell.

Allen, T. F. H. and Starr, T. B. (1982) *Hierarchy, Perpectives for Ecological Complexity*, Chicago: University of Chicago Press.

Allen, T. F. H. and Hoekstra, T. W. (1992) *Toward a Unified Ecology*, New York: Columbia University Press.

Amaral, D. (1994) 'New reflections on the theme of international boundaries', in C. H. Schofield (ed) *World Boundaries*, Vol. 1, London: Routledge.

Ancel, J. (1938) *Géographie des Frontières*, Paris: Gallimard.

Anderson, J. (1995) 'The exagerrated death of the Nation-State', in J. Anderson, C. Brook and A. Cochrane (eds) *A Global World? Re-ordering Global Space*, Oxford: Open University.

Anderson, M. (1996) *Frontiers: Territory and State Formation in the Modern World*, Cambridge: Polity Press in association with Blackwell Publishers.

Anderson, J. (1996) 'The shifting stage of politics: new medieval and postmodern territorialities?' *Environment and Planning D: Society and Space* 14: 133–153.

Anderson, K. (2001) 'The nature of 'race'' *Social Nature: Theory, Practice, and Politics*, Malden & Oxford: Blackwell.

Anon (1975) 'Les Régions Transfrontalières de l'Europe', in Association des Instituts d'Etudes Européennes (ed) *Les Régions Transfrontalières de l'Europe*, Geneva: AIEE.

Anon (1982) 'Council of Europe work on protected areas', in J. A. McNeely and K. R. Miller (eds) *World Congress on National Parks : National Parks, Conservation and Development* – The Role of Protected Areas in Sustaining Society, Bali, Indonesia: Smithsonian, Washington.

Anon (1996) Entdeckungsroute – *Landschaft über Grenzen*, La Petite Pierre & Lambrecht: Parc Naturel Régional des Vosges du Nord – Biosphärenreservat Naturpark Pfälzerwald.

Anon (1998) 'Chronique d'un mariage annoncé', Monts et Merveilles – *Le Journal du Parc National du Mercantour* (Numéro Spécial Jumelage): 7.

Anon (1998) *Montagnes Sans Frontières – Montagne Senza Frontiere*: Parc National du Mercantour – Parco Naturale Alpi Marittime.

Anon (1999) 'Transcending borders – Parks for Europe', *Europarc European Bulletin*.

Anon (2000) 'Malentendu', *Pro Natura* Magazine(1): 34.

Anon (2003) 'Transboundary Conservation: Promoting Peaceful Cooperation and Development While Protecting Biodiversity', in IUCN and ITTO (eds) *International Workshop on Increasing the Effectiveness of Transboundary Conservation Areas in Tropical Forests*, Ubon Ratchathani, Thailand, 17–21 February 2003: IUCN.

Arnold, M. (1909) *Empedocles on Etna, and Other Poems*, by A: 1852, Oxford: Oxford University Press.

American Anthropological Association (2001) 'Results of the workshop on Conservation and Community' Conservation and Community Working Group, Anthropology and Environment Section, 28th November 2001: Electronic listserv discussion list.

Axelrod, R. (1984) *The Evolution of Cooperation*, New York: Basic Books.

Bachelet, M. (1995) *L'ingérence écologique*, Paris: Frison-Roche.

Badshah, M. A. (1962) 'National parks: their principles and purposes', in A. B. Adams (ed) First World Conference on National Parks. Proceedings of a conference organized by The International Union for the Conservation of Nature and Natural Resources, cosponsored by UNESCO, FAO, US National Parks Service, Natural Resource Council of America, Seattle, Washington June 30–July 7: National Park Service, United States Department of the Interior.

Bailly, A. and Scariati, R. (1998) 'L'humanisme en géographie', in A. Bailly (ed) *Les Concepts de la Géographie*, Paris: Colin.

Balmford, A., Mace, G. M. and Ginsberg, J. R. (1998) 'The challenges to conservation in a changing world: putting processes on the map', in G. M.

Mace (ed) *Conservation in a Changing World*, Cambridge: Cambridge University Press.

Barth, F. (1995) 'Les groupes ethniques et leurs frontières', in P. Poutignat and J. Streiff-Fenart (eds) *Théories de l'Ethnicité*, Paris: Centre de Documentation Universitaire.

— (1998), first published 1969 *Ethnic Groups and Boundaries: The Social Organization of Cultural Difference*, Prospect Heights: Waveland Press.

Barzetti, V. (1993) Parks and Progress, IVth World Congress on National Parks and Protected Areas, Gland: IUCN – The World Conservation Union in collaboration with the Inter-American Development Bank.

Batisse, M. (1985) 'Action Plan for Biosphere Reserves', *Environmental Conservation* 12(1): 17–27.

— (1986) 'Developing and focusing the biosphere reserve concept', *Nature & Resources* 22(3): 2–11.

— (1993) 'Biosphere Reserves: an overview', *Nature & Resources* 29(1–4): 3.

— (1996) 'Biosphere Reserves and regional planning: a prospective vision', *Nature & Resources* 32(3).

— (1997) 'Biosphere reserves: a challenge for biodiversity conservation and regional development', *Environment* 39(5).

— (1999) 'Patrimoine mondial et réserves de biosphères: des instruments complémentaires' Un parco europeo per il XXI secolo, Entraque, Italy, 13 October 1999: unpublished.

Beck, P. J. (1994) 'Antarctica: the Antarctic Treaty system after thirty years', in C. H. Schofield (ed) *World Boundaries*, Vol. 1, London: Routledge.

Beltran, E. (1962) 'Use and conservation: two conflicting principles', in A. B. Adams (ed) *First World Conference on National Parks*. Proceedings of a conference organized by The International Union for the Conservation of Nature and Natural Resources, cosponsored by UNESCO, FAO, US National Parks Service, Natural Resource Council of America, Seattle, Washington June 30–July 7: National Park Service, United States Department of the Interior.

Beltran, J. (2000) *Indigenous and Traditional Peoples and Protected Areas: Principles, Guidelines and Case Studies*, Vol. 4, Gland and Cambridge: IUCN – The World Conservation Union.

Berque, A. (1984) 'Paysage-empreinte, paysage-matrice. Eléments pour une géographie culturelle', *Espace Géographique* 1: 33–34.

Bertrand, C. and Bertrand, G. (1992) 'La géographie et les sciences de la nature', in A. Bailly, R. Ferras and D. Pumain (eds) *Encyclopédie de Géographie*, Paris: Economica.

Bertrand, G. and Dollfus, O. (1973) 'Le paysage et son concept', *Espace Géographique* 3: 161–163.

Bibby, C. J., Collar, N. J., Crosby, M. J., Heath, M. F., Imboden, C., Johnson, T. H. *et al.* (1992) *Putting Biodiversity on the Map: Priority Areas for Global Conservation*, Cambridge: International Council for Bird Preservation.

Bierce, A. (1911), reprinted 1984 *The Enlarged Devil's Dictionary*, London: Dover Publications.

Bilderbeek, S. (1992) *Biodiversity and International Law*, Netherlands: IOS Press.

Bioret, F., Cibien, C., Génot, J.-C. and Lecomte, J. (1998) *A Guide to Biosphere Reserve Management: a Methodology Applied to French Biosphere Reserves*, Vol. 19, Paris: UNESCO.

Blake, G. (1998) 'Globalisation and the paradox of enduring national boundaries', in L. Boon-Thong and T. Shamsul Bahrin (eds) *Vanishing Borders: The New International Order of the 21st Century*, Aldershot: Ashgate.

Blunt, A. and Wills, J. (2000) *Dissident Geographies: An Introduction to Radical Ideas and Practice*, Harrow: Pearson Education.

Bock, B. C. and Soles, R. E. (1996) 'Networking using biosphere reserves', in IUCN (ed) *Biosphere Reserves – Myth or Reality, Proceedings of the Workshop on Biosphere Reserves*, World Conservation Congress, Montreal: MaB – UNESCO – IUCN.

Bodénès, S. (1990) 'Théorie Limologique et Identités Frontalières: le Cas Franco-Genevois' Thèse de Doctorat, Département de Géographie, Faculté des Sciences Economiques et Sociales, Genève: Université de Genève.

Bouwer, K. (1981) 'Landscape and environment in human geography', in S. P. d. V. Tjallingii, A.A. (ed) *Perpectives in Landscape Ecology, Contributions to Research, Planning and Management of Our Environment*, Proceedings of the International Congress organised by the Netherlands Society for Landscape Ecology, April 6–11, Veldhoven: Pudoc.

Brandt, J. (1998) 'Key concepts and interdisciplinarity in landscape ecology: a summing-up and outlook', in J. W. Dover and R. G. H. Bunce (eds) *Key Concepts in Landscape Ecology*, IALE (UK): Garstang.

Braun, B. and Wainwright, J. (2001) 'Nature, postructuralism, and politics' *Social Nature: Theory, Practice, and Politics*, Malden & Oxford: Blackwell.

Brechin, S. R., Wilshusen, P. R., Fortwangler, C. L., West, P. C. (2002) 'Beyond the square wheel: toward a more comprehensive understanding of biodiversity conservation as social and political process', *Society and Natural Resources*, 15: 41–64.

Breymeyer, A. (1999) *The East Carpathians Biosphere Reserve*, Warsaw: UNESCO-MaB Committee of Poland.

Bridgewater, P. B. and Cresswell, I. D. (1996) 'The reality of the world network of biosphere reserves: its relevance for the implementation of the convention on biological diversity', in IUCN (ed) *Biosphere Reserves – Myth or Reality*, Proceedings of the Workshop on Biosphere Reserves, World Conservation Congress, Montreal: MaB – UNESCO – IUCN.

Bridgewater, P., Phillips, A., Green, M. and Amos, B. (1996) *Biosphere Reserves and the IUCN System of Protected Area Management Categories*, Canberra.

Bridgewater, P. (2001) 'Epilogue: main results and thoughts for the future of Biosphere Reserves', *Parks* 11(1): 46–47.

Brock, L. (1991) 'Peace through parks: the environment on the peace research agenda', *Journal of Peace Research* 28(4): 407–423.

Brockman, F. (1962) 'Supplement to report of committee on problems of nomenclature', in A. B. Adams (ed) *First World Conference on National Parks*. Proceedings of a conference organized by The International Union for the Conservation of Nature and Natural Resources, cosponsored by UNESCO, FAO, US National Parks Service, Natural Resource Council of America, Seattle, Washington June 30–July 7: National Park Service, United States Department of the Interior.

Brown, V., Smith, D. I., Wiseman, R. and Handmer, J. (1995) *Risks and Opportunities: Managing Environmental Conflict and Change*, London: Earthscan.

Brunckhorst, D. J. and Rollings, N. M. (1999) 'Linking ecological and social functions of landscapes: I. Influencing resource governance', *Natural Areas Journal* 19(1): 57–64.

— (1999) 'Linking ecological and social functions of landscapes: II. Scale and modeling of spatial influence', *Natural Areas Journal* 19(1): 65–72.

Brunet, R. (1967) *Les Phénomènes de Discontinuités en Géographie*, Vol. 7, Paris: Centre National de la Recherche Scientifique.

Brunner, R. (1998) *Parke für das Leben – Unterstützung für Grenzüberschreitende Schutzgebiete*, Wien: Bundesministerium für Umwelt, Jugend und Familie.

Bull, H. (1977) *The Anarchical Society : a Study of Order in World Politics*, New York: Columbia University Press.

Burel, F. and Baudry, J. (1990) 'Hedgerow network patterns and processes in France', in I. S. F. Zonneveld, R.T.T. (ed) *Changing Landscapes: An Ecological Perspective*, New York: Springer-Verlag.

Calhoun, C. (1993) 'Nationalism and ethnicity', *Annual Review of Sociology* 19: 211–239.

Campbell, D. (1999) 'Apartheid cartography: the political anthropology and spatial effects of international diplomacy in Bosnia', *Political Geography* 18: 395–435.

Castree, N. (2001) 'Socializing nature: theory, practice, and politics' *Social Nature: Theory, Practice, and Politics*, Malden & Oxford: Blackwell, 1–19

Castree, N. (2003) 'Environmental issues: relational ontologies and hybrid politics', *Progress in Human Geography* 27(2): 203–211.

Castree, N. and Braun, B. (2001) *Social Nature: Theory, Practice, and Politics*, Malden & Oxford: Blackwell.

Castree, N. and MacMillan, T. (2001) 'Dissolving dualisms: actor-networks and the reimagination of nature' *Social Nature: Theory, Practice, and Politics*, Malden & Oxford: Blackwell.

Chivallon, C. (2000) 'D'un espace appelant forcément les sciences sociales pour le comprendre', *Logiques de l'Espace, Esprit des Lieux*, Géographies à Cerisy, Lévy, J., Lussault, M. (Eds) Belin : Paris 299–317

Cibien, C. (1998) 'Les réserves de biosphères (MAB), des stratégies originales et globales', *Naturopa* 87: 10.

Cisneros, J. A. and Naylor, V. J. (1999) 'Uniting La Frontera: the ongoing efforts to establish a transboundary park', *Environment* 41(3): 12–21.

Claval, P. (1979) 'Préface', in P. Claval (ed) *Tableau de la Géographie de la France*, facsimile edition, first edition 1903, Mayenne: Talendier.

— (1998) *An Introduction to Regional Geography*, Oxford: Blackwell.

CNPPA (1982) 'Categories, objectives and criteria for protected areas', in J. A. McNeely and K. R. Miller (eds) *Proceedings of the Third World Congress on National Parks*, International Union for the Conservation of Nature and Natural Resources in cooperation with United Nations Environment Programme, United Nations Educational, Scientific and Cultural Organization, World Wildlife Fund-US, Parks Canada, United States National Park Service, Bali, Indonesia, 11–22 October: Smithsonian Institution Press, Washington DC.

Coffman, M. S. (1997) 'Globalized Grizzlies', *The New American* (August 18): 1–8.

Collins, L. (1999) 'Geopolitical change and regional identity in Alsace', in E. Bort and R. Keat (eds) *The Boundaries of Understanding: Essay in Honour of Malcolm Anderson*, Edinburgh: University of Edinburgh, International Social Science Institute.

Conrad, M. (2002) *'Ansprache' Veranstaltung zur offiziellen Gründung des Biosphärenreservats Pfälzerwald – Nordvogesen, Schönau*, Germany: Biosphärenreservat Naturpark Pfälzerwald.

Cordell, J. (1993) 'Boundaries and bloodlines: tenure of indigenous homelands and protected areas', in E. Kemf (ed) *The Law of the Mother: Protecting Indigenous Peoples in Protected Areas*, San Francisco: Sierra Club Books.

Cordero, M. (1999) 'Cultura, patrimonio e nuove frontiere da tracciare' Un parc européen pour le 21e siècle, Menton & Tende: Parc National du Mercantour, Parco Naturale Alpi Marittime.

Cosgrove, D. (2001) 'Apollo's Eye: a Cartographic Genealogy of the Globe in the Western Imagination' Baltimore, Johns Hopkins Univ. Press

Crampton, J. W. (2001) 'Maps as social constructions: power, communication and visualization', *Progress in Human Geography* 25(2): 235–252.

Crang, M. (2002) 'Qualitative methods: the new orthodoxy?' *Progress in Human Geography* 26(5): 647–655.

Crang, M. and Thrift, N. (2000) *Thinking Space*, London: Routledge.

Craven, I. and Wardoyo, W. (1993) 'Gardens in the forest', in E. Kemf (ed) *The Law of the Mother: Protecting Indigenous Peoples in Protected Areas*, San Francisco: Sierra Club Books.

Crowe, S. (1972) 'The master plan for national parks and their regional setting', in H. Elliott (ed) *Second World Conference on National Parks*. Proceedings of a Conference sponsored and organized by National Parks Centennial Commission of the United States of America, National Park Service of the US Department of the Interior, and International Union for the Conservation of Nature and Natural Resources. Cosponsored by UNESCO, FAO, Natural Resources Council of America, Yellowstone and Grand Teton National Park, September 18–27: IUCN for National Parks Centennial Commission, Morges.

Council of Europe (1999) 'Conclusions', in Council of Europe (ed) *Nature does not have any borders: towards tranfrontier ecological networks*, Paris, Histoire National d'Histoire Naturelle, 2–3 September 1999.

Croze, H. (1982) 'Monitoring within and outside protected areas', in J. A. McNeely and K. R. Miller (eds) *Proceedings of the Third World Congress on National Parks*, International Union for the Conservation of Nature and Natural Resources in cooperation with United Nations Environment Programme, United Nations Educational, Scientific and Cultural Organization, World Wildlife Fund-US, Parks Canada, United States National Park Service, Bali, Indonesia, 11–22 October: Smithsonian Institution Press, Washington DC.

Curry-Lindahl, K. (1972) 'Projecting the future in the worldwide national park movement', in H. Elliott (ed) *Second World Conference on National Parks.* Proceedings of a Conference sponsored and organized by National Parks Centennial Commission of the United States of America, National Park Service of the US Department of the Interior, and International Union for the Conservation of Nature and Natural Resources. Cosponsored by UNESCO, FAO, Natural Resources Council of America, Yellowstone and Grand Teton National Park, September 18–27: IUCN for National Parks Centennial Commission, Morges.

Curzon, (Lord Curzon of Kedleston) (1907) 'Romanes Lecture on the subject of frontiers', Online 2001: International Boundaries Research Unit, Durham.

Dalby, S. (1992) 'Ecopolitical discourse: 'environmental security' and political geography', *Progress in Human Geography* 16(4): 503–522.

Davey, A. G. (1998) *National System Planning for Protected Areas*, Vol. 1, Gland and Cambridge: IUCN – The World Conservation Union.

Debarbieux, B. (1999) 'Le territoire: histoires en deux langues – a bilingual (his-) story of territory', *Discours Scientifiques et Contextes Culturels : Géographies Françaises et Britanniques à l'Epreuve Postmoderne*, Bordeau : MSHA 33–46.

de Blij, H. J. (1980) (first edition 1973) *Systematic Political Geography*, Wiley : New York

de Blij, H. J. and Capone, D. L. (1969) 'Wildlife conservation areas in East Africa: an application of field theory in political geography', *Southeastern Geographer* 9(2): 94–107.

de l'Harpe, A. (2002) 'Cultures, Territoires et Tourisme: Les Enjeux du Projet Transfrontalier "Espace Mont-Blanc" ', Thèse de Doctorat, Department of Geography, Geneva: University of Geneva.

Delcourt, P. A. and Delcourt, H. R. (1992) 'Ecotone dynamics in space and time', in A. J. D. C. Hansen, F. (ed) *Landscape Boundaries, Consequences for Biotic Diversity and Ecological Flows*, Vol. 92, New York: Springer-Verlag.

Deleuze, G. and Guattari, F. (1992) *A Thousand Plateaus: Capitalism and Schizophrenia*, London: Athlone Press.

Demeritt, D. (2001) 'Being constructive about nature' *Social Nature: Theory, Practice, and Politics*, Malden & Oxford: Blackwell.

Demeritt, D. (2002) 'What is the 'social construction of nature'? A typology and sympathetic critique', *Progress in Human Geography*, 26(6): 767–760.

Dewsbury, J. D., Harrison, P., Rose, M. and Wylie, J. (2002) 'Enacting geographies: Introduction', *Geoforum* 33: 437–440.

Di Castri, F. and Hansen, A. J. (1992) 'The environment and development crises as determinants of landscape dynamics', in A. J. D. C. Hansen, F. (ed) *Landscape Boundaries, Consequences for Biotic Diversity and Ecological Flows*, Vol. 92, New York: Springer-Verlag.

Diamond, J. M. (1975) 'The island dilemma: lessons of modern biogeographic studies for the design of natural reserves', *Biological Conservation* 7: 129–146.

Dodds, K. (1998) 'Political geography I: the globalization of world politics', *Progress in Human Geography* 23(4): 595–606.

— (2000) 'Political geography II: some thoughts on banality, new wars and the geopolitical tradition', *Progress in Human Geography* 24(1): 119–129.

Dudley, N., Hockings, M. and Stolton, S. (1999) 'Measuring the effectiveness of protected areas management', in S. Stolton and N. Dudley (eds) *Partnerships for Protection: New Strategies for Planning and Management for Protected Areas*, London: Earthscan.

Duffy, R., (2002) 'Peace Parks: The Paradox of Globalisation' *Geopolitics* 6(2) 1–26.

Duncan, J. and Ley, D. (1993) *Place/Culture/Representation*, London & New York: Routledge.

Duppré, H.-J. (2002) *'Eröffnung und Begrüssung' Veranstaltung zur offiziellen Gründung des Biosphärenreservats Pfälzerwald – Nordvogesen, Schönau*, Germany: Biosphärenreservat Naturpark Pfälzerwald.

Eidsvik, H. K. (1983) 'Evolving a new approach to biosphere reserves', *Conservation, Science and Society*, Contributions to the First International Biosphere Reserve Congress, Minsk, Byelorussia, 26 Sept.– 2 Oct, UNESCO and UNEP (ed) in collaboration with FAO and IUCN at the invitation of the USSR Conservation.

Elazar, D. J. (1999) 'Political science, geography, and the spatial dimension of politics', *Political Geography* 18: 875–886.

Elliott, H. (1972) 'Introduction', in H. Elliott (ed) *Second World Conference on National Parks*. Proceedings of a Conference sponsored and organized by National Parks Centennial Commission of the United States of America, National Park Service of the US Department of the Interior, and International Union for the Conservation of Nature and Natural Resources. Cosponsored by UNESCO, FAO, Natural Resources Council of America, Yellowstone and Grand Teton National Park, September 18–27: IUCN for National Parks Centennial Commission, Morges.

Entrikin, J. N. (1994) 'Moral geographies: the planner in place', *Geography Research Forum* 14: 113–119.

Erz, W. (1972) 'The broad aspect of planning and management for the future, with emphasis on physical and living resources', in H. Elliott (ed) *Second World Conference on National Parks*. Proceedings of a Conference sponsored and organized by National Parks Centennial Commission of the United States of America, National Park Service of the US Department of the Interior, and International Union for the Conservation of Nature and

Natural Resources. Cosponsored by UNESCO, FAO, Natural Resources Council of America, Yellowstone and Grand Teton National Park, September 18–27: IUCN for National Parks Centennial Commission, Morges.

Europarc (2000) 'Proceedings of the EUROPARC Federation 1999 General Assembly and Conference', in H. Fürst and R. Gray (eds) *Transcending Borders – Parks for Europe*, Zakopane, Poland, 15th–19th September 1999: Ministry of Environment, the Board of Polish National Parks and the Tatra National Park (Poland).

Eva, F. (1998) 'International boundaries, geopolitics and the (post)modern territorial discourse: the functional fiction', in D. Newman (ed) *Geopolitics*, Vol. 3: Frank Cass.

Ezcurra, E. (1983) 'Planning a system of biosphere reserves', *Conservation, Science and Society*, Contributions to the First International Biosphere Reserve Congress, Minsk, Byelorussia, 26 Sept.–2 Oct, UNESCO and UNEP (ed) in collaboration with FAO and IUCN at the invitation of the USSR Conservation.

Fabos, J. G. and Hendrix, W. (1981) 'Regional ecosystem assessment: an aid for ecologically compatible land use planning', in S. P. d. V. Tjallingii, A.A. (ed) *Perpectives in Landscape Ecology*, Contributions to Research, Planning and Management of Our Environment, Proceedings of the International Congress organised by the Netherlands Society for Landscape Ecology, April 6–11, Veldhoven: Pudoc.

Fall, J. J. (1997) 'Mostar: une Territorialité Contrainte' Département de Géographie, Université de Genève, Mémoire de Licence (Degree thesis).

Fall, J. J. (1998) 'Beyond Boundaries : Transboundary Cooperation in the East Carpathians Biosphere Reserve', Department of Geography, University of Oxford, MSc Dissertation.

Fall, J. J. (1999) 'Transboundary biosphere reserves: a new framework for cooperation', *Environmental Conservation* 26(4): 1–3.

Fall, J. J. (1999) 'Transboundary biosphere reserves: applying landscape ecological arguments to protected area planning', in P. Agger and R. Bitsch (eds) *Management of Biodiversity in a Landscape Ecological Perspective*, Vol. 14, Sominestationen, 5–10 September: Center for Landscape Research, Roskilde University.

Fall, J. J. (2003) 'Planning protected areas across boundaries: new paradigms and old ghosts', in U. Manage Goodale, M. Stern, C. Margoluis, A. Lanfer and M. Fladeland (eds) *Transboundary Protected Areas: The Viability of Regional Conservation Stategies*, New York: Food Products Press.

Fall, J. J. (2003) 'Drawing the Line: Boundaries, Identity and Cooperation in Transboundary Protected Areas', Département de Géographie, Université de Genève, PhD Thesis.

Fall, J. J. (2004) 'Divide and rule: constructing human boundaries in 'boundless nature'', *GeoJournal* 58:243–251

Fall, J. J. and Jardin, M. (2003) *Five Transboundary Biosphere Reserves in Europe*, Paris: UNESCO.

Farina, A. (1998) *Principles and Methods in Landscape Ecology*, Cambridge: Chapman and Hall.

Ferras, R. and Hussy, C. (1991) 'Les concepts de la cartographie: leur rôle dans la recherche géographique', in A. Bailly (ed) *Les Concepts de la Géographie*, Paris: Masson.

Finke, L. (1986) *Landschaftsökologie*, Braunschweig: Westerman.

Fjeldsa, J. and Rahbek, C. (1998) 'Continent-wide conservation priorities and diversification processes', in G. M. Mace (ed) *Conservation in a Changing World*, Cambridge: Cambridge University Press.

Forman, R. T. T. (1981) 'Interactions among landscape elements: a core of landscape ecology', in S. P. d. V. Tjallingii, A.A. (ed) *Perpectives in Landscape Ecology*, Contributions to Research, Planning and Management of Our Environment, Proceedings of the International Congress organised by the Netherlands Society for Landscape Ecology, April 6–11, Veldhoven: Pudoc.

— (1990) 'Ecologically sustainable landscapes: the role of spatial configuration', in I. S. F. Zonneveld, R.T.T. (ed) *Changing Landscapes: An Ecological Perspective*, New York: Springer-Verlag.

Forman, R. T. T. and Moore, P. N. (1992) 'Theoretical foundations for understanding boundaries in landscape mosaics', in A. J. D. C. Hansen, F. (ed) *Landscape Boundaries, Consequences for Biotic Diversity and Ecological Flows*, Vol. 92, New York: Springer-Verlag.

Foucher, M. (1991) *Fronts et Frontières: un Tour du Monde Géopolitique*, Paris: Fayard.

Frame, R. (1995) *The Political Development of the British Isles 1100-1400*, Oxford: Clarendon Press.

Frankel, O. H. and Soulé, M. E. (1981) *Conservation and Evolution*, Cambridge: Cambridge University Press.

Frei, N. (1998) 'Lebensraum Entlebuch' Etnologischen Institut, Bern: University of Bern.

Freyfogle, E. T. (1998) 'Bounded people, boundless land', in R. L. Knight and P. B. Landres (eds) *Stewardship Across Boundaries*, Washington DC: Island Press.

Frost, R. (1955) 'Mending Wall' in North of Boston, reprinted in The Weekend Book, London: Nonesuch Press.

Gadgil, M. (1996) 'Managing biodiversity', in K. J. Gaston (ed) *Biodiversity: a Biology of Numbers and Difference*, Oxford: Blackwell Science.

Galbadon, M. (1992) 'Corridors, transition zones and buffers: tools for enhancing the effectiveness of protected areas', in J. A. McNeely (ed) *Parks for Life: Report of the Fourth Congress on National Parks and Protected Areas*, Caracas, Venezuela: IUCN – The World Conservation Union in collaboration with WWF – World Wildlife Fund.

Game, M. and Peterken, G. F. (1984) 'Nature reserve selection strategies in the woodlands of central Lincolnshire, England', *Biological Conservation* 29: 157–181.

Ganser, P. (2001) *Cooperation, Environment and Sustainability in Border Regions*, San Diego: Institute for Regional Studies of the Californias.

Garratt, K. (1982) 'The relationship between adjacent lands and protected areas: issues of concern for the protected area manager', in J. A. McNeely and K. R. Miller (eds) *World Congress on National Parks*: National Parks, Conservation and Development – The Role of Protected Areas in Sustaining Society, Bali, Indonesia: Smithsonian, Washington.

Gaskell, G. (2000) 'Individual and group interviewing', in M. W. Bauer and G. Gaskell (eds) *Qualitative Researching with Text, Image and Sound: a Practical Handbook*, London: Sage.

Gaston, K. J. (1996) 'What is biodiversity?' in K. J. Gaston (ed) *Biodiversity: a Biology of Numbers and Difference*, Oxford: Blackwell Science.

Gay, J.-C. (1995) *Les Discontinuités Spatiales*, Paris: Economica.

Gay, P. (1983) 'Proposals for improved mechanisms to develop the biosphere reserve system as an integrated network', *Conservation, Science and Society*, Contributions to the First International Biosphere Reserve Congress, Minsk, Byelorussia, 26 Sept.–2 Oct, UNESCO and UNEP (ed) in collaboration with FAO and IUCN at the invitation of the USSR Conservation.

George, P. (1974) *Dictionnaire de la Géographie*, Paris: Presses Universitaires de France.

Ginésy, C. (1998) 'Commentaire', Monts et Merveilles – *Le Journal du Parc National du Mercantour* (Numéro 7 Spécial Jumelage): 2.

Given, D. R. (1994) *Principles and Practice of Plant Conservation*, London: Chapman and Hall.

Glassner, M. I. and De Blij, H. J. (1980) *Systematic Political Geography*, Third Edition Edition, New York: John Wiley & Sons.

Glowka, L., Burhenne-Guilmin, F., Synge, H., McNeely, J. and Gündling, L. (1996) *A Guide to the Convention on Biological Diversity*, Gland: IUCN.

Goetel, W. (1962) 'Parks between countries', in A. B. Adams (ed) *First World Conference on National Parks*. Proceedings of a conference organized by The International Union for the Conservation of Nature and Natural Resources, cosponsored by UNESCO, FAO, US National Parks Service, Natural Resource Council of America, Seattle, Washington June 30–July 7: National Park Service, United States Department of the Interior.

Goodwin, P. P. (1999) 'The end of consensus? The impact of participatory initiatives on conceptions of conservation and the countryside in the United Kingdom', Environment and Planning D: *Society and Space* 17: 383–401.

Gosz, J. R. (1992) 'Ecological functions in a biome transition zone: translating local responses to broad-scale dynamics', in A. J. D. C. Hansen, F. (ed) *Landscape Boundaries, Consequences for Biotic Diversity and Ecological Flows*, Vol. 92, New York: Springer-Verlag.

Gottman, J. (1973) *The Significance of Territory*, USA: University Press of Virginia.

Gottmann, J. (1951) 'Geography and international relations', *World Politics* II(2): 153–173.

Gregory, D. (2001) '(Post)colonialism and the production of nature' *Social Nature: Theory, Practice, and Politics*, Malden & Oxford: Blackwells.

Groom, A. J. R. and Taylor, P. (1990) *Frameworks for International Cooperation*, London: Pinter Publishers.
Grumbine, R. E. (1994) 'What is ecosystem management?' *Conservation Biology* 8(1): 27–38.
Guibal, J.-C. (1999) 'L'espace transfrontalier: 'une terre d'hommes'' Un parc européen pour le 21e siècle, Menton & Tende: Parc National du Mercantour, Parco Naturale Alpi Marittime.
Gurusamy, L. D. (1999) 'The Convention on Biological Diversity: exposing the flawed foundations', *Environmental Conservation* 26(2): 79–82.
Guziovà, Z. (1996) 'Across the frontiers: biosphere reserves in bioregional management of shared ecosystems in Central Europe', MAB – UNESCO – IUCN (ed) *World Conservation Congress – Proceedings of the Workshop on Biosphere Reserves*, Montreal: IUCN.
Haber, W. (1990) 'Using landscape ecology in planning and management', in I. S. F. Zonneveld, R.T.T. (ed) *Changing Landscapes: An Ecological Perspective*, New York: Springer-Verlag.
Hacking, I. (2001) *The Social Construction of What?*, Cambridge & London: Harvard University Press.
Haggett, P. (1990) *The Geographer's Art*, Oxford: Blackwell.
Haïla, Y. and Kouki, J. (1994) 'The phenomenon of biodiversity in conservation biology', Ann. Zoo. Fennici 31: 5–18.
Hajer, M. A. (1995) *The Politics of Environmental Discourse: Ecological Modernization and the Policy Process*, Oxford: Clarendon Press.
Hales, D. F. (1982) 'The World Heritage Convention: status and directions', in J. A. McNeely and K. R. Miller (eds) *World Congress on National Parks* : National Parks, Conservation and Development – The Role of Protected Areas in Sustaining Society, Bali, Indonesia: Smithsonian, Washington.
Hales, D. (1989) 'Changing concepts of national parks', in D. Western and M. Pearl (eds) *Conservation for the Twenty-First Century*, Oxford: Oxford University Press.
Hall, S. (1997) *Representation: Cultural Representations and Signifying Practices*, London: Sage.
Hambler, C. and Speight, M. R. (1995) 'Biodiversity conservation in Britain: science replacing tradition', *British Wildlife* 6(3): 137–147.
Hamilton, L. S. (1991) 'Introduction', in L. S. Hamilton, D. P. Bauer and H. F. Takeuchi (eds) *Parks, Peaks and People – a Collection of Papers Arising from an International Consultation on Protected Areas in Mountain Environments*, Hawaii Volcanoes National Park, 26 Oct–2 Nov 1991: East-West Centre Programme on Environment with assistance from the Woodlands Mountain Institute, US National Parks Service, and the IUCN Commission on National Parks and Protected Areas (The World Conservation Union).
Hamilton, L. S., Mackay, J. C., Worboys, G. L., Jones, R. A. and Manson, G. B. (1996) *Transborder protected area cooperation*, Canberra: AALC and IUCN.
Hamilton, L. (1997) 'Maintaining ecoregions in mountain conservation corridors', *Wild Earth* Fall: 63–66.

Hamilton, L. (1997) 'Guidelines for effective transboundary cooperation: philosophy and best practice (Draft Proceedings)' *Parks for Peace: International Conference on Transboundary Protected Areas as a Vehicle for International Cooperation*, 1998 draft edition Edition, 16–18 September 1997, Somerset West, near Cape Town, South Africa: IUCN – The World Conservation Union.

Hamilton, L. S., Fall, J. J., Rosabal, P., Rossi, P., Cisneros, J. and Salas, A. (1998) 'Transboundary cooperation in protected areas: good practice guidelines' *International Symposium on Parks for Peace*, Bormio, Stelvio National Park, Italy, 18–21 May 1998: IUCN – The World Conservation UnionIUCN – The World Conservation Union. First draft.

Hamilton, L. (1998) 'Enlarging and linking Andean protected areas into corridors' *Symposium on Sustainable Mountain Development: Understanding Interfaces of Andean Cultural Landscapes for Management*, Quito, Ecuador, December 10–17 1998.

Hanks, J. (2001) 'Preface', in A. Philips (ed) *Transboundary Protected Areas for Peace and Co-operation*, Best Practice Protected Area Guidelines Series No. 7, Gland & Cambridge: IUCN.

Hansen, A. J., Risser, P. G. and Di Castri, F. (1992) 'Epilogue: biodiversity and ecological flows across ecotones', in Di Castri, F., Hansen, A.J. (ed) Landscape Boundaries, Consequences for Biotic Diversity and Ecological Flows, Vol. 92, New York: Springer-Verlag.

Haraway, D. (1985) *Modest Witness @ Second Millennium FemaleMan Meets Onco Mouse: Feminism and Technoscience*, London: Routledge.

Harley, J. B. (1989) 'Deconstructing the map', *Cartographica* 26: 1–20.

— (1997) 'Deconstructing the map', in T. Barnes and D. Gregory (eds) *Reading Human Geography*, London: Arnold.

— (2001) *The New Nature of Maps: Essays in the History of Cartography*, Laxton, Paul Edition, Baltimore & London: Johns Hopkins University Press.

Harrison, J., Miller, K. and McNeely, J. (1982) 'The world coverage of protected areas: development goals and environmental needs', in J. A. McNeely and K. R. Miller (eds) *Proceedings of the Third World Congress on National Parks*, International Union for the Conservation of Nature and Natural Resources in cooperation with United Nations Environment Programme, United Nations Educational, Scientific and Cultural Organization, World Wildlife Fund-US, Parks Canada, United States National Park Service, Bali, Indonesia, 11–22 October: Smithsonian Institution Press, Washington DC.

Harroy, J.-P. (1972) 'A century in the growth of the national park concept throughout the world', in H. Elliott (ed) *Second World Conference on National Parks*. Proceedings of a Conference sponsored and organized by National Parks Centennial Commission of the United States of America, National Park Service of the US Department of the Interior, and International Union for the Conservation of Nature and Natural Resources. Cosponsored by UNESCO, FAO, Natural Resources Council of America,

Yellowstone and Grand Teton National Park, September 18–27: IUCN for National Parks Centennial Commission, Morges.
Hartog, F. (1996) *Mémoire d'Ulysse: Récits sur la Frontière en Grèce Ancienne*, Paris: Gallimard.
Haushofer, K. (1986) *De la Géopolitique*, Paris: Fayard.
Heylighen, F. (1992) 'Selfish Memes and the Evolution of Cooperation', *Journal of Ideas* 2(4): 77–84.
Hilary, E. (1993) 'Foreword', in E. Kemf (ed) *The Law of the Mother: Protecting Indigenous Peoples in Protected Areas*, San Francisco: Sierra Club Books.
Hocknell, P. R. (2001) *Boundaries of Cooperation: Cyprus, de facto Partition and the Delimitation of Transboundary Resource Management*, Vol. 5, The Hague: Kluwer Law International.
Hodges, J. (1992) 'The Tatra National Park – A Report to the Director of the Tatra National Park and the Pembrokeshire Coast National Park Officer': Pembrokeshire Coast National Park, Great Britain.
Holdgate, M. (1999) *The Green Web*, London: Earthscan.
Hooper, M. D. (1981) 'Planning for the management of national parks', in S. P. d. V. Tjallingii, A.A. (ed) *Perpectives in Landscape Ecology*, Contributions to Research, Planning and Management of Our Environment, Proceedings of the International Congress organised by the Netherlands Society for Landscape Ecology, April 6–11, Veldhoven: Pudoc.
Hotham, P. and Stein, R. (2001) *Europarc Expertise Exchange Working Group*, Grafenau: Europarc.
House, J. W. (1957) 'The Franco-Italian boundary in the Alpes Maritimes', unknown: 107–131.
Hubert, J.-P. (1993) *La Discontinuité Critique: Essai sur les Principes a priori de la Géographie Humaine*, Paris: Publications de la Sorbonne.
Hudson, A. (1998) 'Beyond the borders: globalisation, sovereignty and extra-territoriality', in D. Newman (ed) *Geopolitics*, Vol. 3: Frank Cass.
Huntington, S. (2001) *Le Choc des Civilisations*, Paris: Odile Jacob.
Hussy, C. (1988) 'Introduction', in C. Hussy (ed) *Politische Geographie*, Paris: Economica.
Ingram, G. B. (1999) 'Conservation of biological diversity as landscape architecture', in P. Agger and R. Bitsch (eds) *Management of Biodiversity in a Landscape Ecological Perspective*, Vol. 14, Sominestationen, 5–10 September: Center for Landscape Research, Roskilde University.
IUCN, MAB and UNESCO (1979) *The Biosphere Reserve and its Relationship to Other Protected Areas*, Gland: IUCN.
IUCN (1980) *The World Conservation Strategy: Living Resource Conservation for Sustainable Development*, Gland: IUCN – The World Conservation Union/UNEP – United Nations Environment Programme/WWF – World Wildlife Fund.
— (1994) *Guidelines for Protected Area Management Categories*, Gland: IUCN, The World Conservation Union.
— (1998) 'Parks for Peace' *International Conference on Transboundary Protected Areas as a Vehicle for International Cooperation*, 1998 draft

edition Edition, Somerset West, near Cape Town, South Africa: IUCN – The World Conservation Union.

Jancowski, J. and Niewiadomski, Z. (2000) 'Proposed priority joint actions in the "East Carpathians" International Biosphere Reserve', Ustrizyki Gorne: Foundation for the Eastern Carpathians Biodiversity Conservation.

Johnston, C. A., Pastor, J. and Pinay, G. (1992) 'Quantitative methods for studying landscape boundaries', in F. di Castri, A. J. Hansen (ed) *Landscape Boundaries, Consequences for Biotic Diversity and Ecological Flows*, Vol. 92, New York: Springer-Verlag.

Jones, S. B. (1945) *Boundary Making*, Vol. 8, Washington: Carnegie Endowment for International Peace.

Jones, E. and Eyles, J. (1977) *An Introduction to Social Geography*, Oxford: Oxford University Press.

Joukov, B. (2000) 'Les 'écorégions', ou comment concilier nature et économie sereine', *Courrier International* 484: 42–43.

Jönsson, C., Tägil, S. and Törnqvist, G. (2000) *Organizing European Space*, London: Sage.

Jumelage, C. d. (1998) *Charte de Jumelage – Carta di Gemellaggio*, Nice: Parc National du Mercantour – Parco Naturale Alpi Marittime.

Kaiser, W. (1998) 'Penser la frontière – notions et approches', *Histoire des Alpes Mobilité Spatiale et Frontières*, Räumliche Mobilität und Grenzen(3): 63–74.

Karpowicz, Z. (1992) 'Transboundary Protected Areas', in J. A. McNeely (ed) *Parks for Life: Report of the Fourth Congress on National Parks and Protected Areas*, Caracas, Venezuela: IUCN – The World Conservation Union in collaboration with WWF – World Wildlife Fund.

Kaus, A. (1993) 'Environmental perceptions and social relations in the Mapimí Biosphere Reserve', *Conservation Biology* 7(2): 398–406.

Kay, A. (1997) 'A walk on the wild side: a critical geography of domestication', *Progress in Human Geography* 21(4): 463–485.

Kelleher, G. (1999) *Guidelines for Marine Protected Areas*, Gland and Cambridge: IUCN – The World Conservation Union.

Keohane, R. O. (1991) 'Cooperation and international regimes', in R. Little and M. Smith (eds) *Perspectives on World Politics*, Second Edition, London and New York: Routledge.

Kim, K.-G. (2000) 'East Asian Biosphere Reserve Network Experience on Transboundary Biosphere Reserves', Seoul: UNESCO-MAB National Committee.

Kimball, R. (1992) *The Complete Lyrics of Cole Porter*, Cambridge & New York: Da Capo Press.

Kirby, K. M. (1996) *Indifferent Boundaries: Spatial Concepts of Human Subjectivity*, New York & London: The Guildford Press.

Kliot, N. (2001) 'Transborder Peace Parks: the political geography of cooperation (and conflict) in borderlands', in C. Schofield, D. Newman, A. Drysdale and J. A. Brown (eds) *The Razor's Edge: International Boundaries and Political Geography*, London: Kluwer Law International.

Knight, D. B. (1982) 'Identity and territory: geographical perspectives on nationalism and regionalism', *Annals of the Association of American Geographers* 72(4): 514–531.

Knight, R. L. and Landres, P. B. (1998) *Stewarship Across Boundaries*, Washington D.C.: Island Press.

Knippenberg, H. and Markusse, J. (1999) '19th and 20th Century borders and border regions in Europe: some reflections', in H. Knippenberg and J. Markusse (eds) *Nationalising and Denationalising European Border Regions: Views from Geography and History*, Vol. 53, Dordrecht: Kluwer Academic Publishers.

Knobel, R. (1962) 'Scientific and popular use: a conflict', in A. B. Adams (ed) *First World Conference on National Parks*. Proceedings of a conference organized by The International Union for the Conservation of Nature and Natural Resources, cosponsored by UNESCO, FAO, US National Parks Service, Natural Resource Council of America, Seattle, Washington June 30–July 7: National Park Service, United States Department of the Interior.

Korinman, M. (1990) *Quand l'Allemagne Pensait le Monde: Grandeur et Décadence d'une Géopolitique*, Paris: Fayard.

Krasinski, Z. (1990) 'The border where the bison roam', *Natural History* 6: 62–63.

Kratochwil, F. (1986) 'Of systems, boundaries and territoriality: an inquiry into the formation of the state system', *World Politics* 39(1): 27–52.

Krinitskii, V. V. (1972) 'Protected areas in the world's industrially advanced regions: importance, progress, and problems', in H. Elliott (ed) *Second World Conference on National Parks*. Proceedings of a Conference sponsored and organized by National Parks Centennial Commission of the United States of America, National Park Service of the US Department of the Interior, and International Union for the Conservation of Nature and Natural Resources. Cosponsored by UNESCO, FAO, Natural Resources Council of America, Yellowstone and Grand Teton National Park, September 18–27: IUCN for National Parks Centennial Commission, Morges.

Krzan, Z. (1998) 'A long-term protection plan for the Tatra National Park (Poland)', in M. Mugica and C. Munoz (eds) *Proceedings of the EUROPARC Federation*.

— (1999) 'Tatry National Park and Biosphere Reserve (Poland)', in I. Voloschuk (ed) *The National Parks and Biosphere Reserves in Carpathians*, Poprad: WWF, IUCN, UNESCO, Carpathi.

— (2000) 'Polish-Slovak cooperation in the protection of the Tatra mountains', in H. Fürst and R. Gray (eds) *Proceedings of the EUROPARC Federation 1999 General Assembly and Conference*, Transcending Borders – Parks for Europe, Zakopane, Poland, 15th–19th September 1999: Ministry of Environment, the Board of Polish National Parks and the Tatra National Park (Poland).

Krzan, Z., Skawinski, P. and Kot, M. (1994) 'Problems of natural diversity protection in the Tatra National Park and Biosphere Reserve', in A. Breymeyer and R. Noble (eds) *International Workshop on Biodiversity*

Conservation in Transboundary Protected Areas, Bieszczady and Tatra National Parks, Poland, May 15–25, 1994: Natinonal Academy Press, Washington DC.

Lachapelle, P. R., McCool, S. F. and Patterson, M. E. (2000) 'Barriers to effective natural resource planning in a "messy" world' *Eighth International Symposium on Society and Resource Management*, Bellingham, Washington 17–22 June 2000.

Lacoste, Y. (1985) *La Géographie, Ça Sert d'abord à Faire la Guerre*, Paris: La Découverte.

Landres, P. B., Knight, R. L., Pickett, S. T. A. and Cadenasso, M. L. (1998) 'Ecological effects of administrative boundaries', in R. L. Knight and P. B. Landres (eds) *Stewarship Across Boundaries*, Washington D.C.: Island Press.

Latour, B. (1999) *Pandora's Hope*, Cambridge: Harvard Press.

Leach, M., Mearns, R. and Scoones, I. (1997) 'Challenges to community-based sustainable development', IDS Bulletin 28(4): (unknown).

Lees, A. (1993) 'Melanasia's sacred inheritance', in E. Kemf (ed) *The Law of the Mother: Protecting Indigenous Peoples in Protected Areas*, San Francisco: Sierra Club Books.

Lewis, M. W. (1991) 'Elusive societies: a regional-cartographic approach to the study of human relatedness', *Annals of the Association of American Geographers* 81(4): 605–626.

Lewis, M. W. and Wigen, K. (1997) *The Myth of Continents: A Critique of Metageography*, Berkeley, Los Angeles, London: University of California Press.

Little, R. and Smith, M. (1991) *Perspectives on World Politics*, Second Edition Edition, London and New York: Routledge.

Llwyd, L. (Pseudonym of Keith Buchanan) (1999) (first published 1968) 'A preliminary contribution to the geographical analysis of a Pooh-scape', *Progress in Human Geography* 23(2): 257–266.

Loenen, M. (1981) 'Planning and research for new nature-areas', in S. P. d. V. Tjallingii, A.A. (ed) *Perpectives in Landscape Ecology, Contributions to Research, Planning and Management of Our Environment*, Proceedings of the International Congress organised by the Netherlands Society for Landscape Ecology, April 6–11, Veldhoven: Pudoc.

Loomba, A. (1998) *Colonialism/Postcolonialism*, London & New York: Routledge.

Lorot, P. (1995) *Histoire de la Géopolitique*, Paris: Economica.

Lucas, P. H. C. (1990) 'A glasnost-era park is born', *Natural History* 6: 61.

— (1992) *Protected Landscapes: a Guide for Policy-makers and Planners*, Chapman & Hall.

Lussault, M. (1997) 'Espace, société, nature', in R. Knafou (ed) *L'État de la Géographie*, Paris: Belin.

Lussault, M. (2000) *'Actions !' Logiques de l'Espace, Esprit des Lieux, Géographies à Cerisy*, Lévy, J., Lussault, M. (Eds) Belin : Paris 11–36.

Lutz, E. and Caldecott, J. (1996) *Decentralization and Biodiversity Conservation*, Washington: World Bank.

Lynch, O. J. and Alcorn, J. B. (1994) 'Tenurial Rights and Community-based Conservation', in R. M. Wright and S. C. Strum (eds) *Natural Connections: Perspectives in Community-based Conservation*, Washington: Island Press.

MacArthur, R. H. (1972) *Geographical Ecology: Patterns in the Distribution of Species*, New York: Harper & Row.

MacLeod, G. and Jones, M. (2001) 'Renewing the geography of regions', Environment and Planning D: *Society and Space* 19: 669–695.

Machlis, G. E. (2000) 'The two cultures: scientists and natural resource managers in contemporary America' *Eighth International Symposium on Society and Resource Management*, Western Washington University, Bellingham, Washington.

Malausa, J.-C. (1999) 'La richesse biologique et paysagère de l'espace transfrontalier' Un parc européen pour le 21e siècle, Menton & Tende: Parc National du Mercantour, Parco Naturale Alpi Marittime.

Malik, A. (1982) 'Opening address: protected areas and political reality', in J. A. McNeely and K. R. Miller (eds) *Proceedings of the Third World Congress on National Parks*, International Union for the Conservation of Nature and Natural Resources in cooperation with United Nations Environment Programme, United Nations Educational, Scientific and Cultural Organization, World Wildlife Fund-US, Parks Canada, United States National Park Service, Bali, Indonesia, 11–22 October: Smithsonian Institution Press, Washington DC.

Maltby, E. (1999) 'What is ecosystem management?' in R. Crofts, E. Maltby, R. Smith and L. Maclean (eds) *Integrated Planning: International Perspectives*, Battleby, Scotland 7–9 April 1999: IUCN & Scottish Natural Heritage.

Marchal, G. P. (1996) Grenzen und Raumvorstellungen – Frontières et Conceptions de l'Espace, Luzern: Chronos.

Martin, C. (1993) 'Introduction', in E. Kemf (ed) *The Law of the Mother: Protecting Indigenous Peoples in Protected Areas*, San Francisco: Sierra Club Books.

Martinez, O. J. (1994) 'The dynamics of border interaction', in C. H. Schofield (ed) *World Boundaries*, Vol. 1, London: Routledge.

Maser, C. (1992) *Global Imperative: Harmonizing Culture and Nature*, Walpole: Stillpoint.

Massey, D. (2001) 'Geography on the agenda', *Progress in Human Geography* 25(1): 5–17.

Maux, I. (2002) 'Comment est née la conception française des parcs nationaux?' *Revue de Géographie Alpine* 2, Numéro Spécial Espaces Protégés et Espaces de Recherche: le cas de l'Arc Alpin(90): 33–44.

McCloskey, M. (2000) 'Is this the course you want to be on?' *Eighth International Symposium on Society and Resource Management*, Bellingham, Washington, 17–22 June 2000.

McDowell, D. (1998) 'Inaugural speech' *Parks for Peace International Conference on Transboundary Protected Areas as a Vehicle for International Cooperation*, 1998 draft edition Edition, Somerset West, near Cape Town, South Africa: IUCN – The World Conservation Union.

McGinnis, M. V. (2001) 'The bioregional quest for community', *Landscape Journal* 1: 84–88.

McNeely, J. A. (1982) 'The World Heritage Convention: protecting natural and cultural wonders of global importance – a slide presentation', in J. A. McNeely and K. R. Miller (eds) *World Congress on National Parks : National Parks, Conservation and Development – The Role of Protected Areas in Sustaining Society*, Bali, Indonesia: Smithsonian, Washington.

— (1982) 'National Parks, Conservation, and Development: the Role of Protected Areas in Sustaining Society', in J. A. McNeely and K. R. Miller (eds) *Proceedings of the Third World Congress on National Parks*, International Union for the Conservation of Nature and Natural Resources in cooperation with United Nations Environment Programme, United Nations Educational, Scientific and Cultural Organization, World Wildlife Fund-US, Parks Canada, United States National Park Service, Bali, Indonesia, 11–22 October: Smithsonian Institution Press, Washington DC.

— (1982) 'Introduction: protected areas are adapting to new realities', in J. A. McNeely and K. R. Miller (eds) *Proceedings of the Third World Congress on National Parks*, International Union for the Conservation of Nature and Natural Resources in cooperation with United Nations Environment Programme, United Nations Educational, Scientific and Cultural Organization, World Wildlife Fund-US, Parks Canada, United States National Park Service, Bali, Indonesia, 11–22 October: Smithsonian Institution Press, Washington DC.

— (1992) 'Onward from Caracas', in J. A. McNeely (ed) *Parks for Life: Report of the Fourth Congress on National Parks and Protected Areas*, Caracas, Venezuela: IUCN – The World Conservation Union in collaboration with WWF – World Wildlife Fund.

— (1999) 'Bioregional planning and ecosystem based management: commonalities, contrasts, constraints and convergences', in R. Crofts, E. Maltby, R. Smith and L. Maclean (eds) *Integrated Planning: International Perspectives*, Battleby, Scotland 7–9 April 1999: IUCN & Scottish Natural Heritage.

Merriam, G. (1990) 'Ecological processes in the time and space of farmland mosaics', in I. S. F. Zonneveld, R.T.T. (ed) *Changing Landscapes: An Ecological Perspective*, New York: Springer-Verlag.

Merriam, G. and Wegner, J. (1992) 'Local extinctions, habitat fragmentation, and ecotones', in A. J. D. C. Hansen, F. (ed) *Landscape Boundaries, Consequences for Biotic Diversity and Ecological Flows*, Vol. 92, New York: Springer-Verlag.

Milich, L. and Varady, R. G. (1998) 'Managing transboundary resources', *Environment* 40(8): 10–41.

Miller, K. R. (1982) 'The Bali Action Plan: a framework for the future of protected areas', in J. A. McNeely and K. R. Miller (eds) *World Congress on National Parks : National Parks, Conservation and Development – The Role of Protected Areas in Sustaining Society*, Bali, Indonesia: Smithsonian, Washington.

— (1983) 'Biosphere reserves and the global network of protected areas', in U. a. U. i. c. w. F. a. I. a. t. i. o. t. USSR (ed) *Conservation, Science and Society,*

Contributions to the First International Biosphere Reserve Congress, Minsk, Byelorussia, 26 Sept.–2 Oct: UNESCO.

— (1999a) 'Bioregional planning and biodiversity conservation', in S. Stolton and N. Dudley (eds) *Partnerships for Protection: New Strategies for Planning and Management for Protected Areas*, London: Earthscan.

— (1999b) 'What is bioregional planning?' in R. Crofts, E. Maltby, R. Smith and L. Maclean (eds) *Integrated Planning: International Perspectives*, Battleby, Scotland 7–9 April 1999: IUCN & Scottish Natural Heritage.

Monod, T. (1962) 'The strict nature reserve and its role', in A. B. Adams (ed) *First World Conference on National Parks*. Proceedings of a conference organized by The International Union for the Conservation of Nature and Natural Resources, cosponsored by UNESCO, FAO, US National Parks Service, Natural Resource Council of America, Seattle, Washington June 30–July 7: National Park Service, United States Department of the Interior.

Morehouse, B. (1995) 'A functional approach to boundaries in the context of environmental issues', *Journal of Borderlands Studies* X(2): 53–73.

Mougenot, P. (1968) *Atlas Historique*, Paris: Stock.

Mucciarelli, R. (1998) 'Editorial', *Monts et Merveilles – Le Journal du Parc National du Mercantour* (Numéro 7 Spécial Jumelage): 2.

Mugica, M. and Munoz, C. (1998) 'Proceedings of the EUROPARC Federation 1998 General Assembly and Conference', in Europarc (ed) *Parks For People: Challenges For Protected Landscapes at the Turn of the Century*, Mallorca, 7th–11th October 1998.

Muhr, T. (1997) *Atlas.ti Short User's Guide*, Berlin: Scientific Software Development.

Murphy, A. B. (1999) '"Living together separately" Thoughts on the relationship between political science and political geography', *Political Geography* 18: 887–894.

Muscarà, L. (1998) 'Les mots justes de Jean Gottmann', Cybergeo, *European Journal of Geography* www.cybergeo.presse.fr (online).

Myers, N. (1982) 'Eternal values for the Parks movement and the Monday Morning World', in J. A. McNeely and K. R. Miller (eds) *World Congress on National Parks : National Parks, Conservation and Development – The Role of Protected Areas in Sustaining Society*, Bali, Indonesia: Smithsonian, Washington.

— (1986) 'Threatened biotas: 'hotspots' in tropical forests', *The Environmentalist* 8(3): 1–20.

Myers, N., Mittermeier, R. A., Mittermeier, C. G., da Fonseca, G. A. B. and Kent, J. (2000) 'Biodiversity hotspots for conservation priorities', *Nature* 403: 853–858.

National Parks Service (2001) 'Yellowstone National Park', 2001: National Park Service, www.nps.gov.

Natter, W. and Schultz, H.-D. (2003) 'Imagining Mitteleuropa: conceptualisations of 'its' space in and outside German geography', unpublished.

Nauber, J. and Pokorny, D. (1993) 'Establishment of biosphere reserves in Germany: a case study of the Rhoen Biosphere Reserve', *Nature & Resources* 29(1–4): 29–34.

Naveh, Z. and Lieberman, A. (1994) *Landscape Ecology: Theory and Application*, Second Edition, New York: Springer-Verlag.

Neef, E. (1981) 'Stages in the development of landscape ecology', in S. P. d. V. Tjallingii, A.A. (ed) *Perpectives in Landscape Ecology, Contributions to Research, Planning and Management of Our Environment*, Proceedings of the International Congress organised by the Netherlands Society for Landscape Ecology, April 6–11, Veldhoven: Pudoc.

Newman, D. and Paasi, A. (1998) 'Fences and neighbours in the postmodern world: boundary narratives in political geography', *Progress in Human Geography* 22(2): 186–207.

Nicholson, E. M. (1972) 'What is wrong with the national park movement?' in H. Elliott (ed) *Second World Conference on National Parks*. Proceedings of a Conference sponsored and organized by National Parks Centennial Commission of the United States of America, National Park Service of the US Department of the Interior, and International Union for the Conservation of Nature and Natural Resources. Cosponsored by UNESCO, FAO, Natural Resources Council of America, Yellowstone and Grand Teton National Park, September 18–27: IUCN for National Parks Centennial Commission, Morges.

Nicol, J. I. (1972) 'International parks and cooperation for more effective management', in H. Elliott (ed) *Second World Conference on National Parks*. Proceedings of a Conference sponsored and organized by National Parks Centennial Commission of the United States of America, National Park Service of the US Department of the Interior, and International Union for the Conservation of Nature and Natural Resources. Cosponsored by UNESCO, FAO, Natural Resources Council of America, Yellowstone and Grand Teton National Park, September 18–27: IUCN for National Parks Centennial Commission, Morges.

Niewiadomski, Z. (1996) 'Bieszczady National Park', Ustriki Dorne: Bieszczady National Park authorities.

— (1999) 'The First European Trilateral MaB Biosphere Reserve in the Eastern Carpathians (Poland/Slovakia and Ukraine) and the Polish Experience', in I. Eisto, T. J. Hokkanen, M. Ohman and A. Repola (eds) *Local Involvement and Economic Dimensions in Biosphere Reserve Activities* – Proceedings of the 3rd EuroMab Biosphere Reserve Coordinators' Meeting, Vol. 7, Helsinki: Academy of Finland.

November, V. (2002) 'Les Territoires du Risque: Le Risque Comme Objet de Réflexion Géographique', Berne : Peter Lang

O'Loughlin, J. and Heske, H. (1991) 'From 'geopolitik' to 'géopolitique': converting a discipline for war to a discipline for peace', in N. Kliot and S. Waterman (eds) *The Political Geography of Conflict and Peace*, London: Belhaven Press.

Ohmane, K. (1995) *The End of the Nation-State*, New York: Free Press.

Olsen, J. (2001) 'The perils of rootedness: on bioregionalism and right wing ecology in Germany', *Landscape Journal* 1: 73–83.

Olson, S. F. (1962) 'A philosophical concept', in A. B. Adams (ed) *First World Conference on National Parks*. Proceedings of a conference organized by The

International Union for the Conservation of Nature and Natural Resources, cosponsored by UNESCO, FAO, US National Parks Service, Natural Resource Council of America, Seattle, Washington June 30–July 7: National Park Service, United States Department of the Interior.

Oneka, M. (1996) On Park Design: Looking Beyond the Wars, Doctoral Thesis, Wageningen: Wageningen Agricultural University.

Ovington, J. D. (1982) 'Ecological processes and national park management', in J. A. McNeely and K. R. Miller (eds) *World Congress on National Parks* : National Parks, Conservation and Development – The Role of Protected Areas in Sustaining Society, Bali, Indonesia: Smithsonian, Washington.

Paasi, A. (1991) 'Deconstructing regions: notes on the scales of spatial life', *Environment and Planning* 23: 239–256.

— (1996) *Territories, Boundaries and Consciousness: the Changing Geographies of the Finnish-Russian Border*, London: Wiley.

— (1998) 'Boundaries as social processes: territoriality in the world of flows', in D. Newman (ed) Geopolitics, Vol. 3: Frank Cass.

— (2000) 'Commentary 2 on Sack, R.D. 'Human territoriality: its theory and practice', Classics in Human Geography revisited', *Progress in Human Geography* 24(1): 93–95.

Parker, G. (1991) 'The geopolitics of dominance and international cooperation', in N. Kliot and S. Waterman (eds) *The Political Geography of Conflict and Peace*, London: Belhaven Press.

Passet, S. (1994) 'S'unir pour mieux protéger: parcs nationaux de la vanoise et du grand paradis', *Parchi* 3 (special issue): 50–61.

Pearce, D. (1993) *Blueprint* 3, London: Earthscan.

Peck, S. (1998) *Planning for Biodiversity*, Washington: Island Press.

Penrose, J. (2002) 'Nations, states and homelands: territory and territoriality in nationalist thought', *Nations and Nationalisms* 8(3): 277–297.

Perkins, C. (2003) 'Cartography: mapping theory', *Progress in Human Geography* 27(3): 341–351.

Peuquet, D., Smith, B. and Brogaard, B. (1998) 'The ontology of fields' Report of a Specialist Meeting Held under the Auspices of the Varenius Project Panel on Computational Implementations of Geographic Concepts, Vol. http://www.ncgia.ucsb.edu, Talks and Discussions at the Specialist Meeting, Bar Harbor, Maine, 11–13 June 1998: Online.

Phillips, A. (1996) 'Biosphere Reserves and protected areas: what is the difference?' in MAB – UNESCO – IUCN (eds) *World Conservation Congress – Proceedings of the Workshop on Biosphere Reserves*, Montreal: IUCN.

Phillips, A. (1996) 'Foreword', in MAB – UNESCO – IUCN (eds) *World Conservation Congress – Proceedings of the Workshop on Biosphere Reserves*, Montreal: IUCN.

Philo, C. (1991) 'Foucault's Geography', *Environment and Planning D: Society and Space* 10: 137–162.

Philo, C., Wilbert, C. (2000) 'Animal spaces, beastly places: an introduction', *Animal Spaces, Beastly* Places, 1–34.

Pickett, S. T. A. (1978) 'Patch dynamics and the design of nature reserves', *Biological Conservation* 13: 27–37.

Poore, D. (1992) Guidelines for Mountain Protected Areas, Gland: IUCN – The World Conservation Union.
Pratt, G. (1999) 'Geographies of identity and difference: marking boundaries', in D. Massey (ed) *Human Geography Today*, Cambridge: Polity Press.
Pratt, M. and Brown, J. A. (2000) 'Borderlands Under Stress', in M. Pratt and J. A. Brown (eds) *5th International Conference of the International Boundary Research Unit*, University of Durham, 15–17th July 1998: Kluwer Law International.
Prescott, J. R. V. (1978) *Boundaries and Frontiers*, London: Croom Helm.
— (1987) *Political Frontiers and Boundaries*, London: Allen & Unwin.
Price, M. F. (1996) 'People in biosphere reserves: an evolving concept', *Society & Natural Resources* 9: 645–654.
Price, M. F. (2000) 'Proceedings of the First Joint Meeting of EuroMAB National Committees and Biosphere Reserve Coordinators', in M. F. Price (ed) *EuroMAB 2000*, 10–14 April, Cambridge, UK: UNESCO.
Primack, R. B. (1993) *Essentials of Conservation Biology*, Sunderland: Sinauer Associates.
Proctor, J. and Pincetl, S. (1996) 'Nature and the reproduction of endangered species: the spotted owl in the Pacific Northwest and southern California', *Environment and Planning D: Society and Space* 14: 683–708.
Racine, J.-B., Raffestin, C. and Ruffy, V. (1980) 'Echelle et action, contributions à une interprétation du méchanisme de l'échelle dans la pratique de la géographie', *Geographica Helvetica* 35(5 (numéro spécial)): 87–94.
Raffestin, C. (1974) 'Espace, temps et frontière', *Cahiers de Géographie du Québec* 18(43): 23–34.
— (1980) *Pour une Géographie du Pouvoir*, Paris: Litec.
— (1986) 'Territorialité: concept ou paradigme de la géographie sociale?' *Geographica Helvetica* 2: 91–96.
— (1988) 'Le rôle de la carte dans une société moderne', *Vermessung, Photogrammetrie, Kulturtechnik* 4.
— (1991) *Géopolitique et Histoire*, Lausanne: Payot.
— (1997) 'Le rôle des sciences et des techniques dans les processus de territorialisation', *Revue Européenne des Sciences Sociales* XXXV(108): 93–106.
Ratzel, F. (1897) (translated 1988) *Politische Geographie*, Paris: Economica.
Reed, N. P. (1972) 'How well has the United States managed its national park system? The application of ecological principles to park management', in H. Elliott (ed) *Second World Conference on National Parks*. Proceedings of a Conference sponsored and organized by National Parks Centennial Commission of the United States of America, National Park Service of the US Department of the Interior, and International Union for the Conservation of Nature and Natural Resources. Cosponsored by UNESCO, FAO, Natural Resources Council of America, Yellowstone and Grand Teton National Park, September 18–27: IUCN for National Parks Centennial Commission, Morges.

Reid, P. J. and Church, A. (1998) 'Transfrontier cooperation and the borders in the European Union', in L. Boon-Thong and T. Shamsul Bahrin (eds) *Vanishing Borders: The New International Order of the 21st Century*, Aldershot: Ashgate.

Reinert, M. (1990) 'Alceste – une méthodologie d'analyse des données textuelles et une application: Aurélia de Gérard de Nerval', *Bulletin de Méthodologie Sociologique* 26: 24–54.

Reynaud, A. (1992) 'Centre et périphérie', in A. Bailly, R. Ferras and D. Pumain (eds) *Encyclopédie de Géographie*, Paris: Economica.

Ricq, C. (1998) *Handbook on transfrontier cooperation for local and regional authorities in Europe*, Vol. 4, 3rd Edition, Strasbourg: Council of Europe Publishing.

Robertson Vernhes, J. (1997) 'Biosphere reserves: old and new', in IUCN (ed) *Protected Areas in the 21st Century; from islands to networks*, Albany, Australia, 24–29 November 1997: IUCN – The World Conservation Union.

Rose, G. (1997) 'Situating knowledges: positionality, reflexivities and other tactics', *Progress in Human Geography* 21(3): 305–320.

Rossi, P. (1998) 'Histoire et évolution d'une coopération', *Monts et Merveilles – Le Journal du Parc National du Mercantour* (Numéro 7 Spécial Jumelage).

Roy, A. (1997) *The God of Small Things*, London: Flamingo.

Ruegg, J. (1999) *Cours d'introduction à l'aménagement du territoire* (unpublished), Genève: Université de Genève.

Rumley, D. and Minghi, J. V. (1991) 'Introduction: the border landscape concept', in D. Rumley and J. V. Minghi (eds) *The Geography of Border Landscapes*, London & New York: Routledge.

Ruzicka, M. and Miklos, L. (1990) 'Basic premises and methods in landscape ecological planning and optimization', in I. S. F. Zonneveld, R.T.T. (ed) *Changing Landscapes: An Ecological Perspective*, New York: Springer-Verlag.

Ryszkowski, L. (1992) 'Energy and material flows across boundaries in agricultural landscapes', in A. J. D. C. Hansen, F. (ed) *Landscape Boundaries, Consequences for Biotic Diversity and Ecological Flows*, Vol. 92, New York: Springer-Verlag.

Sack, R. D. (1980) *Conceptions of Space in Social Thought: A Geographic Perspective*, London and Basingstoke: Macmillan.

— (1997) *Homo Geographicus*, London: Johns Hopkins Press.

— (2000) 'Commentary 2 on Sack, R.D. 'Human territoriality: its theory and practice', Classics in Human Geography revisited', *Progress in Human Geography* 24(1): 96–99.

Said, E. (1995) (first edition 1978) *Orientalism: Western Conceptions of the Orient*, London: Penguin Modern Classics.

Sandwith, T., Shine, C., Hamilton, L. and Sheppard, D. (2001) *Transboundary Protected Areas for Peace and Co-operation*, Gland & Cambridge: IUCN.

Schama, S. (1995) *Landscape and Memory*, London: HarperCollins.

Schreiber, K.-F. (1990) 'The history of landscape ecology in Europe', in I. S. F. Zonneveld, R.T.T. (ed) *Changing Landscapes: An Ecological Perspective*, New York: Springer-Verlag.

Scott, J., Sweedler, A., Ganster, P. and Eberwein, W.-D. (1997) 'Dynamics of transboundary interaction in comparative perspective', in J. Scott, A. Sweedler, P. Ganster and W.-D. Eberwein (eds) *Borders and Border Regions in Europe and North America*, San Diego: San Diego University Press.

Scott, J. W. (2000) 'Transboundary cooperation on Germany's borders: strategic regionalism through multilevel governance', *Journal of Borderlands Studies* XV(1): 143–167.

Servat, J. (1999) 'L'espace transfrontalier: 'une terre de patrimoine'' Un parc européen pour le 21e siècle, Menton & Tende: Parc National du Mercantour, Parco Naturale Alpi Marittime.

Seymour, F. J. (1994) 'Are successful community-based conservation projects designed or discovered?' in R. M. Wright and S. C. Strum (eds) *Natural Connections: Perspectives in Community-based Conservation*, Washington: Island Press.

Shafer, C. L. (1990) *Nature Reserves: Island Theory and Conservation Practice*, Washington and London: Smithsonian Institution Press.

— (1999) 'US National Park buffer zones: historical, scientific, social and legal aspects', *Environmental Management* 23(1): 49–73.

Sheail, J., Treweek, J. R. and Mountford, J. O. (1997) 'The UK transition from nature preservation to 'creative conservation'', *Environmental Conservation* 24(3): 224–235.

Sheppard, D. (2000) 'Conservation without frontiers – the global view', in H. Fürst and R. Gray (eds) *Proceedings of the EUROPARC Federation 1999 General Assembly and Conference*, Transcending Borders – Parks for Europe, Zakopane, Poland, 15th–19th September 1999: Ministry of Environment, the Board of Polish National Parks and the Tatra National Park, Poland.

Sherpa, L. N., Peniston, B., Lama, W. and Richard, C. (2003) *Hands Around Everest: Transboundary Cooperation for Conservation and Sustainable Livelihoods*, Kathmandu: The Mountain Institute & ICIMOD.

Sidaway, J. D. (2001) 'Rebuilding bridges: a critical geopolitics of Iberian cooperation in a European context', *Environment and Planning D: Society and Space* 19: 743–778.

Sjoberg, J. (1998) '"Europe of the Regions": what do sub-state boundaries imply in different European countries?' in E. Bort (ed) *Borders and Borderlands in Europe*, Edinburgh: The University of Edinburgh, International Social Sciences Institute.

Skawinski, P. (1993) 'Human impact on nature in the Tatra mountains', in W. Cichoki (ed) *The Endangered Nature of the Polish Tatra Mountains*, Zakopane: Tatrzanski Park Narodowy.

Slayter, R. (1982) 'The World Heritage Convention: introductory comments', in J. A. McNeely and K. R. Miller (eds) *World Congress on National Parks : National Parks, Conservation and Development – The Role of Protected Areas in Sustaining Society*, Bali, Indonesia: Smithsonian, Washington.

Sletto, B. (2002) 'Boundary making and regional identities in a globalized environment: rebordering the Nariva Swamp, Trinidad', *Environment and Planning D: Society and Space* 20: 183–208.

Smith, B. (1995) 'On drawing lines on a map', in A. U. Frank, W. Kuhn and D. M. Mark (eds) *Spatial Information Theory, Proceedings of COSIT '95*, Berlin/Heidelberg/Vienna/New York/london/Tokyo: Springer Verlag.

— (1997) 'The cognitive geometry of war', in P. Koller and K. Puhl (eds) *Current Issues in Political Philosophy*, Vienna: Hölder-Pichler-Tempsky.

— (2000) 'Fiat and bona fide boundaries', *Philosophy and Phenomenological Research* 60(2): 401–420.

Snow, C. P. (1993) *The Two Cultures* (from a lecture given in 1959). Cambridge, Cambridge University Press.

Soles, R. E. (1998) '"Open letter to concerned citizens"', 2000: US MAB (online).

Spellerberg, I. F. (1992) *Evaluation and Assessment for Conservation: Ecological Guidelines for Determining Priorities for Nature Conservation*, London: Chapman and Hall.

Spencer Sochaczewski, P. (1999) 'Across a divide', *International Wildlife* July/August: 34–41.

Stea, D. (1996) 'Romancing the line: edges and seams in Western and Indigenous mindscapes with special reference to Bedouin', in Y. Gradus and H. Lithwick (eds) *Frontiers in Regional Development*, Lanham: Rowman & Littlefields.

Stoiko, S. (1996) 'Problems in transboundary protected areas in Ukraine', in A. Breymeyer and R. Noble (eds) *Biodiversity Conservation in Transboundary Protected Areas*, May 1994, Poland: National Research Council and the National Academy Press Washington.

Stolton, S. and Dudley, N. (1999) *Partnerships for Protection: New Strategies for Planning and Management for Protected Areas*, London: Earthscan.

Storey, D. (2001) *Territory: the claiming of space*, Harlow: Prentice Hall.

Strauss, A. and Corbin, J. (1990) 'Techniques for enhancing theoretical sensitivity' *Basics of Qualitative Research*, Thousand Oaks & London: Sage.

Sumartojo, S. (1997) 'The pursuit of economic advantage: a realist interpretation of cross-border co-operation', in M. Anderson and E. Bort (eds) *Schengen and EU Enlargement: Security and Cooperation at the Eastern Frontier of the European Union*, Edinburgh: The University of Edinburgh, International Social Sciences Institute.

Swanson, F. J., Wondzell, S. M. and Grant, G. E. (1992) 'Landforms, disturbance and ecotones', in A. J. D. C. Hansen, F. (ed) *Landscape Boundaries, Consequences for Biotic Diversity and Ecological Flows*, Vol. 92, New York: Springer-Verlag.

Talbot, L. (1982) 'The role of protected areas in the implementation of the World Conservation Strategy', in J. A. McNeely and K. R. Miller (eds) *Proceedings of the Third World Congress on National Parks*, International Union for the Conservation of Nature and Natural Resources in cooperation with United Nations Environment Programme, United Nations Educational, Scientific and Cultural Organization, World Wildlife Fund-US, Parks Canada, United States National Park Service, Bali, Indonesia, 11–22 October: Smithsonian Institution Press, Washington DC.

Tansley, A. G. (1935) 'The use and abuse of vegetational concepts and terms', *Ecology* 16: 284–307.

Taylor, B. (2001) 'Bioregionalism: an ethics of loyalty to place', *Landscape Journal* 1: 50–72.

Taylor, P. G (1990) 'Frameworks for International Cooperation', Groom A. J. R.; Taylor (eds), P. G., London: Pinter.

Taylor, P. J. (1985) *Political Geography: World Economy, Nation-state and Locality*, New York: Longman.

Terborgh, J. and Winter, B. (1983) 'A method for siting parks and reserves with special reference to Colombia and Ecuador', *Biological Conservation* 27: 45–58.

Thompson, I. (1998) 'Translator's Preface', in P. Claval (ed) *An Introduction to Regional Geography*, Oxford: Blackwell.

Thorsell, J. (1990) 'Through hot and cold wars, parks endure', *Natural History* 6: 54–58.

Thorsell, J. and Harrison, J. (1991) 'National parks and nature reserves in the mountain regions of the world', in L. S. Hamilton, D. P. Bauer and H. F. Takeuchi (eds) *Parks, Peaks and People – a Collection of Papers Arising from an International Consultation on Protected Areas in Mountain Environments*, Hawaii Volcanoes National Park, 26 Oct–2 Nov 1991: East-West Centre Programme on Environment with assistance from the Woodlands Mountain Institute, US National Parks Service, and the IUCN Commission on National Parks and Protected Areas (The World Conservation Union).

Trautman, W. (1983) 'Prerequisites for a representative network of biosphere reserves for Europe', in *Conservation, Science and Society*, Contributions to the First International Biosphere Reserve Congress, Minsk, Byelorussia, 26 Sept.–2 Oct, UNESCO and UNEP (ed) in collaboration with FAO and IUCN at the invitation of the USSR Conservation.

Udall, S. L. (1962) 'Nature islands for the world', in A. B. Adams (ed) *First World Conference on National Parks*. Proceedings of a conference organized by The International Union for the Conservation of Nature and Natural Resources, cosponsored by UNESCO, FAO, US National Parks Service, Natural Resource Council of America, Seattle, Washington June 30–July 7: National Park Service, United States Department of the Interior.

Udvardy, M. D. F. (1982) 'A biogeographical classification system for terrestrial environments', in J. A. McNeely and K. R. Miller (eds) *Proceedings of the Third World Congress on National Parks*, International Union for the Conservation of Nature and Natural Resources in cooperation with United Nations Environment Programme, United Nations Educational, Scientific and Cultural Organization, World Wildlife Fund-US, Parks Canada, United States National Park Service, Bali, Indonesia, 11–22 October: Smithsonian Institution Press, Washington DC.

UNESCO (1971) 'Rapport final', in UNESCO (ed) *Conseil international de coordination du Programme sur l'homme et la biosphère* (MAB), Vol. 1, Paris 9–19 November 1971: UNESCO.

— (1974) 'Rapport final', UNESCO and UNEP (ed) *Groupe de concertation: les critères et les lignes directives du choix et de la constitution de réserves de la biosphère*, Vol. 22, Paris 9–19 November 1971: UNESCO.
— (1986) 'Report of the Scientific Advisory Panel On Biosphere Reserves', in UNESCO (ed) *Final Report of the International Coordinating Council of the Programme Man and the Biosphere*, Vol. 60, Paris: UNESCO.
— (1996) *Biosphere Reserves: The Seville Strategy and the Statutory Framework of the World Network*, Paris: UNESCO.
— (1996) 'La Stratégie de Séville pour les réserves de biosphère', *Nature & Ressources* 31(2): 2–15.
— (1999) 'Frequently Asked Questions on Biosphere Reserves', accessed 2000, http://www.unesco.org/mab.
— (1999) 'Planification Instruments in Biosphere Reserves', in UNESCO (ed) *International Workshop*, Sierra de la Nieves, Malaga, Spain, 16–19 June 1999: Centro Internacional de Desarrollo Sostenible.
— (2003) 'List of biosphere reserves', accessed 2003, http://www.unesco.org/mab.
— (2000b) 'Seville + 5 International Meeting of Experts, Proceedings' Seville + 5, Pamplona, Spain, 23–27 October 2000: UNESCO.
— (2000c) *Solving the Puzzle: The Ecosystem Approach and Biosphere Reserves*, Paris: UNESCO.
— 2002 *Biosphere Reserves: Special Places for People and Nature*, Paris: UNESCO.
Unwin, P. (1992) *The Place of Geography*, Harlow: Longman.
Van der Linde, H., Oglethorpe, J., Sandwith, T., Snelson, D. and Tessema, Y. (2001) *Beyond Boundaries: Transboundary Natural Resource Management in Sub-Saharan Africa*, Washington DC: Biodiversity Support Program.
Van der Maarel, E. (1981) 'Biogeographical and landscape-ecological planning of nature reserves', in S. P. d. V. Tjallingii, A.A. (ed) *Perpectives in Landscape Ecology*, Contributions to Research, Planning and Management of Our Environment, Proceedings of the International Congress organised by the Netherlands Society for Landscape Ecology, April 6–11, Veldhoven: Pudoc.
Van Djik, H. (1999) 'State borders in geography and history', in H. Knippenberg and J. Markusse (eds) *Nationalising and Denationalising European Border Regions: Views from Geography and History*, Vol. 53, Dordrecht: Kluwer Academic Publishers.
Van Houtum, H. (2000) 'An overview of European geographical research on borders and border regions', *Journal of Borderlands Studies* XV(1): 57–83.
Velasco-Graciet, H. (1998) 'La Frontière, le Territoire et le Lieu' Thèse de Doctorat, UFR de Lettres: Université de Pau et des Pays de l'Adour.
Verschuren, J. (1962) 'Science and nature reserves', in A. B. Adams (ed) *First World Conference on National Parks*. Proceedings of a conference organized by The International Union for the Conservation of Nature and Natural Resources, cosponsored by UNESCO, FAO, US National Parks Service, Natural Resource Council of America, Seattle, Washington June 30–July 7: National Park Service, United States Department of the Interior.

Vidal de la Blache, P. (1903) (facsimile edition published in 1979) *Tableau de la Géographie de la France*, Mayenne: Talendier.
Vila, P. (2000) *Crossing Bords, Reinforcing Borders: Social Categories, Metaphors, and Narrative Identities on the U.S. – Mexico Frontier*, Austin: University of Texas Press.
Vink, A. P. A. (1981) 'Anthropocentric landscape ecology in rural areas', in S. P. d. V. Tjallingii, A.A. (ed) *Perpectives in Landscape Ecology*, Contributions to Research, Planning and Management of Our Environment, Proceedings of the International Congress organised by the Netherlands Society for Landscape Ecology, April 6–11, Veldhoven: Pudoc.
Voller, J. and Harrison, S. (1998) *Conservation Biology Principles for Forested Landscapes*, Vancouver: UCB Press.
Voloschuk, I. (1999) 'Tatry/Tatras National Park and Biosphere Reserve (Slovak Republic)', in I. Voloschuk (ed) *The National Parks and Biosphere Reserves in Carpathians*, Poprad: WWF, IUCN, UNESCO, Carpathi.
Von Droste, B. (1982) 'How UNESCO's Man and the Biosphere Programme is contributing to human welfare', in J. A. McNeely and K. R. Miller (eds) *World Congress on National Parks : National Parks, Conservation and Development – The Role of Protected Areas in Sustaining Society*, Bali, Indonesia: Smithsonian, Washington.
Wachtel, P. S. and McNeely, J. A. (1991) *Eco-bluff Your Way to Greenism*, Chicago: Bonus Books.
Waitt, G. and Head, L. (2002) 'Postcards and frontier mythologies: sustaining views of the Kimberley as timeless', *Environment and Planning D: Society and Space* 20(3): 319–344.
Weber, J. (1996) 'Conservation, développement et coordination: peut-on gérer biologiquement le social?' Gestion communautaire des ressources naturelles renouvelables et développement durable, Harare, Zimbabwe, 24–27 juin 1996: unpublished.
Welch, R. V. (1996) 'Redefining the frontier: regional development in the postwelfare era', in Y. Gradus and H. Lithwick (eds) *Frontiers in Regional Development*, Lanham: Rowman & Littlefields.
Wendl, T. and Rösler, M. (1999) 'Frontiers and borderlands: the rise and relevance of an anthropological research genre', in M. Rösler and T. Wendl (eds) *Frontiers and Borderlands: Anthropological Perspectives*, Münich: Peter Lang.
Westing, A. (1989) 'Comprehensive human security and ecological realities', *Environmental Conservation* 16(4).
— (1993) 'Biodiversity and the challenge of national borders', *Environmental Conservation* 20(1).
— (1998) 'International conference on transboundary protected areas as a vehicle for international cooperation', *Environmental Conservation* 25(1): 78–79.
— (1998) 'A transfrontier reserve for peace and nature on the Korean peninsula', International *Environmental Affairs* 10(1): 8–177.

— (1998) 'Establishment and management of transfrontier reserves for conflict prevention and confidence building', *Environmental Conservation* 25(2): 91–94.

Westphal, J. (2002) 'Schlussworte' Veranstaltung zur offiziellen Gründung des Biosphärenreservats Pfälzerwald – Nordvogesen, Schönau, Germany: Biosphärenreservat Naturpark Pfälzerwald.

Whatmore, S. (2002) *Hybrid Geographies: Nature Cultures Spaces*, London, Thousand Oaks, New Dehli: Sage Publications.

Wiens, J. A. (1992) 'Ecological flows across landscape boundaries: a conceptual overview', in A. J. D. C. Hansen, F. (ed) *Landscape Boundaries, Consequences for Biotic Diversity and Ecological Flows*, Vol. 92, New York: Springer-Verlag.

Wilshusen, P. R., Brechin, S. R., Fortwangler, C.L., West, P.C. (2002) 'Reinventing the square wheel: critique of a resurgent protection paradigm in international biodiversity conservation', *Society and Natural Resources*, 15: 17–40.

Wirth, C. L. (1962) 'National Parks', in A. B. Adams (ed) *First World Conference on National Parks*. Proceedings of a conference organized by The International Union for the Conservation of Nature and Natural Resources, cosponsored by UNESCO, FAO, US National Parks Service, Natural Resource Council of America, Seattle, Washington June 30–July 7: National Park Service, United States Department of the Interior.

Wolmer, W. (2003) 'Transboundary Conservation: the Politics of Ecological Integrity in the Great Limpopo Transfrontier Park', *Journal of Southern African Studies* 29(1): 261–278.

Wolmer, W. (2003) 'Transboundary protected area governance: tensions and paradoxes' Paper prepared for the workshop on Transboundary Protected Areas in the Governance Stream of the 5th World Parks Congress, Durban, South Africa, 12–13 September 2003.

World Commission on Environment and Development. (1987) *Our Common Future*, Oxford: Oxford University Press.

World Resources Institute. (2000) 'What is a Bioregion?' accessed 2001, http://www.wri.org.

Woodward, R. (2004) *Military Geographies*, RGS-IBG Book Series, Nick Henry, Jon Sadler (eds), Oxford: Blackwell.

Young, O. (1994) *International Governance: Protecting the Environment in a Stateless Society*, Ithaca and London: Cornell University Press.

Young, T. (2001) 'Belonging not containing: the vision of bioregionalism', *Landscape Journal* 1: 46–49.

Zaccaï, E. (1999) 'Sustainable development: characteristics and interpretations', *Geographica Helvetica*(2): 73–80.

Zanini, P. (1997) *Significati del Confine: I Limiti Naturali, Storici, Mentali*, Milan: Mondadori.

Zbicz, D. C. (1999a) 'The "nature" of transboundary cooperation', *Environment* 41(3): 15–16.

— (1999b) 'Transboundary cooperation between internationally adjoining protected areas', in D. Harmon (ed) On the Frontiers of Conservation: Proceedings of the Tenth Conference on Research and Resource Management in Parks and on Public Land, The George Wright Society Biennial Conference, March 22–26, 1999 Ashville, North Carolina: The George Wright Society.

— (1999c) 'Transfrontier ecosystems and internationally adjoining protected areas': World Conservation Monitoring Centre – UNEP.

— (1999d) 'Transboundary Cooperation in Conservation: A Global Survey of Factors Influencing Cooperation between Internationally Adjoining Protected Areas' Nicolas School of the Environment, Durham: Duke University.

Zbicz, D. C. and Green, M. J. B. (1997) 'Status of the world's transfrontier protected areas', Parks – *Protected Areas Programme Journal* 7(3): 5–10.

Zinke, A. (1994) 'Ecological bricks for Europe: integration of conservation and sustainable development along the former East-West border', in M. Munasinghe and J. McNeely (eds) *Protected Area Economics and Policy: Linking Conservation and Sustainable Development*, Washington DC: World Bank and IUCN The World Conservation Union.

Zonneveld, I. S. (1981) 'The ecological backdrop of regional geography', in S. P. d. V. Tjallingii, A.A. (ed) *Perpectives in Landscape Ecology*, Contributions to Research, Planning and Management of Our Environment, Proceedings of the International Congress organised by the Netherlands Society for Landscape Ecology, April 6–11, Veldhoven: Pudoc.

Zonneveld, I. S. (1990) 'Scope and concepts of landscape ecology as an emerging science', in I. S. F. Zonneveld, R.T.T. (ed) *Changing Landscapes: An Ecological Perspective*, New York: Springer-Verlag.

Ó Tuathail, G. (1995) 'Political geography I: theorizing history, gender and world order amidst crises of global governance', *Progress in Human Geography* 19(2): 260–272.

— (1996a) 'Political geography II: (counter) revolutionary times', *Progress in Human Geography* 20(3): 404–412.

— (1996b) Critical Geopolitics: The Politics of Writing *Global Space*, Vol. 6, Minneapolis: University of Minnesota Press.

— (1998) 'Political geography III: dealing with deterritorialization', *Progress in Human Geography* 22(1): 81–93.

Index